Springer Tracts in Advanced Robotics

Volume 74

Jorge Solis and Kia Ng (Eds.)

Musical Robots and Interactive Multimodal Systems

Professor Bruno Siciliano, Dipartimento di Informatica e Sistemistica, Università di Napoli Federico II, Via Claudio 21, 80125 Napoli, Italy, E-mail: siciliano@unina.it

Professor Oussama Khatib, Artificial Intelligence Laboratory, Department of Computer Science, Stanford University, Stanford, CA 94305-9010, USA, E-mail: khatib@cs.stanford.edu

Professor Frans Groen, Department of Computer Science, Universiteit van Amsterdam, Kruislaan 403, 1098 SJ Amsterdam, The Netherlands, E-mail: groen@science.uva.nl

Editors

Dr. Jorge Solis
Humanoid Robotics Institute
Faculty of Science and Engineering
Waseda University
3-4-1 Ookubo, Shinjuku-ku
Tokyo 168-8555
Japan
E-mail: solis@ieee.org

Dr. Kia Ng
Interdisciplinary Centre for Scientific Research in Music (ICSRiM)
University of Leeds
School of Computing & School of Music
Leeds LS2 9JT
UK
E-mail: k.c.ng@leeds.ac.uk, kia@kcng.org

ISBN 978-3-642-22290-0 e-ISBN 978-3-642-22291-7

DOI 10.1007/978-3-642-22291-7

Springer Tracts in Advanced Robotics ISSN 1610-7438

Library of Congress Control Number: 2011932322

Typeset & Cover Design: Scientific Publishing Services Pvt. Ltd., Chennai, India.

Printed on acid-free paper

5 4 3 2 1 0

springer.com

STAR (Springer Tracts in Advanced Robotics) has been promoted under the auspices of EURON (European Robotics Research Network)

Foreword

Robotics is undergoing a major transformation in scope and dimension. From a largely dominant industrial focus, robotics is rapidly expanding into human environments and vigorously engaged in its new challenges. Interacting with, assisting, serving, and exploring with humans, the emerging robots will increasingly touch people and their lives.

Beyond its impact on physical robots, the body of knowledge robotics has produced is revealing a much wider range of applications reaching across diverse research areas and scientific disciplines, such as: biomechanics, haptics, neurosciences, virtual simulation, animation, surgery, and sensor networks among others. In return, the challenges of the new emerging areas are proving an abundant source of stimulation and insights for the field of robotics. It is indeed at the intersection of disciplines that the most striking advances happen.

The *Springer Tracts in Advanced Robotics (STAR)* is devoted to bringing to the research community the latest advances in the robotics field on the basis of their significance and quality. Through a wide and timely dissemination of critical research developments in robotics, our objective with this series is to promote more exchanges and collaborations among the researchers in the community and contribute to further advancements in this rapidly growing field.

The volume edited by Jorge Solis and Kia Ng presents a collection of contributions in the new emerging field of social human-robot interaction. The coverage is on musical robots and interactive multimodal systems, which are expected to become tools to preserve cultural heritage. The seven chapters in the first section focus on the analysis, modeling and understanding of musical performance and the development of novel interfaces for musical expression, while the other seven chapters in the second section concentrate on the design of automated instruments and anthropomorphic robots aimed at facilitating human-robot interaction from a musical viewpoint and proposing novel ways of musical expression.

STAR is proud to welcome this first volume in the series dedicated to musical robotics!

Naples, Italy
May 2011

Bruno Siciliano
STAR Editor

Preface

The relation between humans and music has a long history dating from the antiquity, during which poetry, dance and music were inseparable and constituted an important mean of communication of everyday life. During the golden era of automata, music also served as a tool for understanding the human motor control while performing highly skilful tasks. The subject area of this book is inherently inter- and trans-disciplinary. Recent advances in a wide range of subject areas that contributed to the developments and possibilities as presented in this book include computer science, multimodal interfaces and processing, artificial intelligence, electronics, robotics, mechatronics and beyond.

Over recent decades, Computer Science research on musical performance issues has been very much active and intense. For example, computer-based expressive performance systems that are capable of transforming a symbolic musical score into an expressive musical performance considering time, sound and timbre deviations. At the same time, recent technological advances in robot technology, music content processing, machine learning, and others are enabling robots to emulate the physical dynamics and motor dexterity of musicians while playing musical instruments and exhibit cognitive capabilities for musical collaboration with human players. Nowadays, the research on musical robots opens many opportunities to study different aspects of humans. These include understanding human motor control, how humans create expressive music performances, finding effective ways of musical interaction, and their applications to education and entertainment.

For several decades, researchers have been developing more natural interfaces for musical analysis and composition and robots for imitating musical performance. Robotics has long been a fascinating subject area, encompassing the dreams of science fiction and industry alike. Recent progress is shifting the focus of robotics. Once it was confined to highly specialized industrial applications and now it infiltrates our everyday lives and living spaces.

This book consists of a collection of scientific papers to highlight cutting edge research related to this interdisciplinary field, exploring musical activities, interactive multimodal systems and their interactions with robots to further enhance musical understanding, interpretation, performance, education and enjoyment. It covers some of the most important ongoing interactive multimodal systems and robotics research topics. This book contains 14 carefully selected and reviewed contributions. The chapters are organized in two sections:

- The first section focuses on the development of multimodal systems to provide intuitive and effective human-computer interaction. From this, more advanced methods for the analysis, modeling and understanding of musical performance and novel interfaces for musical expression can be conceived.
- The second section concentrates on the development of automated instruments and anthropomorphic robots designed to study the human motor control from an engineering point of view, to better understand how to facilitate the human-robot interaction from a musical point of view and to propose novel ways of musical expression.

The idea for this book has been formed over several meetings during related conferences between the two editors, observing and involving the developments from a wider range of related research topics as highlighted here. We can still remember the initial discussions during the i-Maestro workshops and the International Computer Music Conferences. We would like to take this opportunity to thank all authors and reviewers for their invaluable contributions and thanks to Dr. Thomas Ditzinger from the Springer for his kind support and invaluable insights over the development of this book. We are grateful to many people including our families for their support and understanding, institutions and funding bodies (acknowledged in separate Chapters) without whom this book would not be possible. We hope you will enjoy this book and find it useful and exciting.

Tokyo, Leeds
August 2010

Jorge Solis
Kia Ng

Contents

About the Editors

Jorge Solis received his B.S. degree in Electronics Systems from the Monterrey Institute of Technology (Mexico) in 1998 and the Ph.D. degree from the Scuola Superiore Sant'Anna, Perceptual Robotics Laboratory (Italy) in 2004. He is an Assistant Professor of Robotics at Waseda University and Visiting Researcher at the Humanoid Robotics Institute, Waseda University. Prof. Solis is the author and co-author of over 130 technical papers for International Journals and Conferences. He is currently a co-chair of the IEEE-RAS TC on Biorobotics and Roboethics. He received the Finalist Awards on Entertainment Robots and Systems at the IROS2007 and IROS2009 (sponsored by the New Technology Foundation). His research involves the Humanoid Robots; Medical Robots and Systems; Education Robots; Rehabilitation Robots, Haptics Interfaces, Sound/ Gesture/Vision Recognition Systems and Tele-operation Systems.

Kia Ng received his B.Sc. degree in Computational Science and Music, and Ph.D. in Computer Science from the University of Leeds (UK). Kia is director of the Interdisciplinary Centre for Scientific Research in Music (ICSRiM) and Senior Lecturer at the University of Leeds. His research interests include interactive multimedia, gesture analysis, computer vision and computer music. He has contributed to a range of large-scale research projects, including i-Maestro on technology-enhanced learning, CASPAR, and AXMEDIS on cross-media, with over 20M Euro funding. Kia has more than 150 publications and presented talks in over 30 countries including keynotes and invited lectures in Canada, China, France, Germany, Japan, UK, and USA. He has appeared on television and radio interviews such as BBC, Sky TV, and has been featured in the New York Times, Guardian, Financial Times, and others. Kia has also organised over 15 international events including conferences, exhibitions and a convention. Kia is a chartered scientist, a fellow of the Royal Society of Arts and a fellow of the Institute of Directors. Web: www.kcng.org

List of Contributors

Rolf Inge Godøy
Department of Musicology,
University of Oslo, P.B. 1017 Blindern,
N-0315 Oslo,
Norway
r.i.godoy@imv.uio.no

Fabrizio Argenti
University of Florence,
Department of Electronics and
Telecommunications, Via S. Marta 3,
Florence, 50139,
Italy
fabrizio.argenti@unifi.it

Paolo Nesi
University of Florence,
Department of Systems and Informatics,
Via S. Marta 3, Florence, 50139,
Italy
nesi@dsi.unifi.it

Gianni Pantaleo
University of Florence,
Department of Systems and Informatics,
Via S. Marta 3,
Florence, 50139,
Italy
pantaleo@dsi.unifi.it

Antonio Camurri
Casa Paganini – InfoMus, DIST,
University of Genova, Viale Causa 13,
16145, Genova,
Italy
antonio.camurri@unige.it

Gualtiero Volpe
Casa Paganini – InfoMus, DIST,
University of Genova, Viale Causa 13,
16145 Genova,
Italy
gualtiero.volpe@unige.it

Diana S. Young
Wyss Institute for Biologically Inspired
Engineering
3 Blackfan Circle
5th floor, Center for Life Sciences Building
Harvard University
Boston, MA, 02115, USA
diana.young@wyss.harvard.edu

Joseph Malloch
Input Devices and Music Interaction
Laboratory (IDMIL),
Centre for Interdisciplinary Research in
Media Music and Technology
(CIRMMT), McGill University, 555,
Sherbrooke Street West,
H3A 1E3 Montreal, Qc, Canada
joseph.malloch@mail.mcgill.ca

Stephen Sinclair
Input Devices and Music Interaction
Laboratory (IDMIL), Centre for
Interdisciplinary Research in Media
Music and Technology (CIRMMT),
McGill University, 555,
Sherbrooke Street West,
H3A 1E3 Montreal, Qc, Canada
stephen.sinclair@mail.
mcgill.ca

Avrum Hollinger
Input Devices and Music Interaction Laboratory (IDMIL),
Centre for Interdisciplinary Research in Media Music and Technology (CIRMMT), McGill University, 555,
Sherbrooke Street West,
H3A 1E3 Montreal, Qc
Canada
avrum@music.mcgill.ca

Marcelo M. Wanderley
Input Devices and Music Interaction Laboratory (IDMIL), Centre for Interdisciplinary Research in Media Music and Technology (CIRMMT),
McGill University, 555,
Sherbrooke Street West,
H3A 1E3 Montreal, Qc
Canada
marcelo.wanderley@mcgill.ca

Kia Ng
ICSRiM – University of Leeds,
School of Computing & School of Music,
Leeds LS2 9JT,
UK
k.c.ng@leeds.ac.uk,
kia@kcng.org

Frédéric Bevilacqua
IRCAM - Real Time Musical Interactions Team, STMS IRCAM-CNRS-UPMC,
Place Igor Stravinsky, Paris, 75004
France
frederic.bevilacqua@ircam.fr

Norbert Schnell
IRCAM - Real Time Musical Interactions Team, STMS IRCAM-CNRS-UPMC,
Place Igor Stravinsky, Paris, 75004,
France
norbert.schnell@ircam.fr

Nicolas Rasamimanana
IRCAM - Real Time Musical Interactions Team, STMS IRCAM-CNRS-UPMC,
Place Igor Stravinsky, Paris, 75004
France
nicolas.rasamimanana@ircam.fr

Bruno Zamborlin
IRCAM - Real Time Musical Interactions Team, STMS IRCAM-CNRS-UPMC,
Place Igor Stravinsky, Paris, 75004,
France
bruno.zamborlin@ircam.fr

Fabrice Guédy
IRCAM - Real Time Musical Interactions Team, STMS IRCAM-CNRS-UPMC,
Place Igor Stravinsky, Paris, 75004
France
fabrice.guedy@ircam.fr

Eiji Hayashi
Kyushu Institute of Technology, 680-4 Kawazu, Iisuka-city, Fukuoka, 820-8502,
Japan
haya@mse.kyutech.ac.jp

Roger B. Dannenberg
Carnegie Mellon University, Computer Science Department, 5000 Forbes Avenue,
Pittsburgh, PA 15213,
USA
rbd@cs.cmu.edu

H. Ben Brown
Carnegie Mellon University,
Robotics Institute, 5000 Forbes Avenue,
Pittsburgh, PA 15213,
USA
hbb@cs.cmu.edu

Ron Lupish
SMS Siemag, 100 Sandusky St.,
Pittsburgh, PA 15212,
USA
ron.lupish@sms-siemag.us

Koji Shibuya
Ryukoku University, Dept. of Mechanical and Systems Engineering, 1-5 Yokotani Seta-Oe, Otsu, Shiga, 520-2194
Japan
koji@rins.ryukoku.ac.jp

Jorge Solis
Faculty of Science and Engineering, Waseda University/Humanoid Robotics Institute, Waseda University, 3-4-1 Ookubo, Shinjuku-ku, Tokyo, 168-8555
Japan
solis@ieee.org

Klaus Petersen
Graduate School of Advanced Engineering and Science, Waseda University, 3-4-1 Ookubo, Shinjuku-ku, Tokyo, 168-8555, Japan
klaus@moegi.waseda.jp

Atsuo Takanishi
Faculty of Science and Engineering, Waseda University/Humanoid Robotics Institute, Waseda University, 3-4-1 Ookubo, Shinjuku-ku, Tokyo, 168-8555, Japan
contact@takanishi.mech.waseda.ac.jp

Ajay Kapur
California Institute of the Arts, Valencia, California, USA;
New Zealand School of Music, Victoria University of Wellington, New Zealand, 24700 McBean Parkway, Valencia CA 91355, USA
ajay@karmetik.com

Guy Hoffman
Interdisciplinary Center Herzliya, P.O. Box 167, Herzliya 46150, Israel
hoffman@idc.ac.il

Gil Weinberg
Georgia Institute of Technology, 840 McMillan Street, Atlanta, GA, USA
gil.weinberg@coa.gatech.edu

Chapter 1
Musical Robots and Interactive Multimodal Systems: An Introduction

Jorge Solis and Kia Ng

1.1 Context

As early as 17th and 18th centuries different researchers have been interested in studying how humans are capable of playing different kinds of musical instruments as an approach to provide a credible imitation of life. As an early example, in 1738, Jacques de Vaucanson (1709-1782) developed the "Flute Player" automaton that mechanically reproduces the organs required for playing the flute and proposed its development as one way for understanding the human breathing mechanism.

More recently, during the early 1960's, industrial robots and autonomous mobile robots became a reality. However the greatest engineering challenge of developing a robot with human-like shape still required further technological advances. Thanks to the progress in many related subject areas including robot technology, artificial intelligence, computation power, and others, the first full-scale anthropomorphic robot, the Waseda Robot No.1 (WABOT-1), was developed by the late Prof. Ichiro Kato during the early 1980's. Following this success, the first attempt at developing an anthropomorphic musical robot was carried out at the Waseda University in 1984. The Waseda Robot No. 2 (WABOT-2) was able of playing a concert organ. One year later, in 1985, the Waseda Sumimoto Robot (WASUBOT) built also by Waseda University, could read a musical score and play a repertoire of 16 tunes on a keyboard instrument. Prof. Kato argued that the artistic activity such as playing a keyboard instrument would require human-like intelligence and dexterity.

The performance of any musical instrument is not well defined and far from a straightforward challenge due to the many different perspectives and subject areas. State-of-the-art development of interactive multimodal systems provides advancements which enable enhanced human-machine interaction and novel possibilities for embodied robotic platforms. An idealized musical robot requires many different complex systems to work together; integrating musical representation, techniques, expressions, detailed analysis and control, for both playing and listening. It also needs sensitive multimodal interactions within the context of a

J. Solis and K. Ng (Eds.): Musical Robots and Interactive Multimodal Systems, STAR 74, pp. 1–12.
springerlink.com

piece, interpretation and performance considerations, including: tradition, individualistic and stylistic issues, as well as interactions between performers, and the list grows.

Due to the inherent interdisciplinary nature of the topic, this book is a collection of scientific papers intended to highlight cutting edge research related to these interdisciplinary fields, exploring musical activities interactive multimedia and multimodal systems and their interactions with robots, to further enhance musical understanding, interpretation, performance, education and enjoyment.

This book consists of 14 chapters with different key ideas, developments and innovations. It is dichotomized into two sections:

I) Understanding elements of musical performance and expression
II) Musical robots and automated instruments

Additional digital resources of this book can be found online via: `https://www.springer.com/book/978-3-642-22290-0`

1.2 Section I: Understanding Elements of Musical Performance and Expression

The first section starts with the fundamentals of gesture, key requirements and interactive multimedia systems to understand elements of musical performance. These concepts and systems contribute to the basis of playing gesture and the understanding of musical instrument playing. Different systems and interfaces have been developed to measure, model and analyse musical performance. Building on these advancements, further approaches for modeling, understanding and simulation of musical performance as well as novel interfaces for musical expression can be conceived. These Chapters also present a range of application scenarios including technology-enhanced learning.

1.2.1 Chapter 2: Sound-Action Chunks in Music

Most people will probably agree that there are strong links between the experience of music and sensations of body movement: we can readily observe musicians' body movements in performance, i.e. see so-called *sound-producing movements*. Furthermore: we can see people move to music in dancing, in walking, at concerts and in various everyday listening situations, making so-called *sound-accompanying movements*. Such common observations and more systematic research now converge in suggesting that sensations of body movement are indeed integral to musical experience as such. This is the topic of Chapter 2 which includes an overview of current research within the field as well as an overview of various aspects of music-related body movement.

This Chapter proposes that sound-movement relationships are manifest at the timescale of the *chunk*, meaning in excerpts in the approximately 0.5 to 5 seconds range, forming what is called *sound-action chunks*. Focusing on sound-action chunks is useful because at this timescale we find many salient musical features: various rhythmical and textural patterns (e.g. dance rhythms and grooves) as well as melodic, modal, harmonic and timbral (or 'sound') patterns are found at the chunk level, as are various expressive and emotive features (such as *calm, agitated, slow, fast*, etc.). All these chunk-level musical features can be correlated with body movement features by carefully capturing and processing sound-producing and sound-accompanying movements as well as by extracting perceptually salient features from the sound. Needless to say, there are many technological and conceptual challenges in documenting such sound-action links, requiring good interdisciplinary teamwork with contributions from specialists in musicology, music perception, movement science, signal processing, machine learning and robotics.

1.2.2 Chapter 3: Automatic Music Transcription: From Monophonic to Polyphonic

In order to provide natural and effective musical interaction, listening skill is paramount. This is investigated in Chapter 3 on audio analysis and transcription. Automatic music transcription is the process of analyzing a musical recorded signal, or a musical performance, and converting it into a symbolic notation or any equivalent representation concerning parameters such as pitch, onset time, duration and intensity. It is one of the most challenging tasks in the field of Music Information Retrieval, and it is a problem of great interest for many fields and applications, from interactive music education to audio track recognition, music search on the Internet and via mobiles. This Chapter aims to analyze the evolution of music understanding algorithms and models from monophonic to polyphonic, showing and comparing the solutions.

Music transcription systems are typically based on two main tasks: the *pitch estimation* and *note tracking* (associated to the retrieval of temporal information like onset times, note durations...). Many different techniques have been proposed to cope with these problems. For pitch estimation, the most recent approaches are often based on a joint analysis of the signal in the time-frequency domain, since simple spectral amplitude has revealed to be not sufficient to achieve satisfactory transcription accuracies. Many other models have been developed: auditory model based front ends, grouped in the *Computational Auditory Scene Analysis*, have been largely studied and applied in the 90s; however, the interest toward this approach has decreased. The most used techniques in recent literature are: Nonnegative Matrix Factorization, Hidden Markov Models, Bayesian models, generative harmonic models and the use of jointed frequency and time information. Regarding temporal parameter information, the detection of note onsets and offsets is often devolved upon detecting rapid spectral energy over time. Techniques such as the phase-vocoder based functions, applied to audio spectrogram, seem to be more robust with respect to peak-picking algorithms performed upon the signal envelope. A strongly relevant and critical aspect is represented by the evaluation models and methods of the performance of music transcription systems. A remarkable effort to unify the different existing approaches and metrics has been done in these last years by the MIREX (Music Information Retrieval Evaluation eXchange) community.

1.2.3 Chapter 4: Multimodal Analysis of Expressive Gesture in Music Performance

Research on expressive gesture became particularly significant in recent years and is especially relevant for music performance. It joins, in a cross-disciplinary perspective, theoretical and experimental findings from several disciplines, from psychology to biomechanics, computer science, social science, and the performing arts.

This Chapter presents a historical survey of research on multimodal analysis of expressive gesture and of how such a research has been applied to music performance. It introduces models, techniques, and interfaces developed in several research projects involving works carried out in the framework of the EyesWeb project, and provides an overview of topics and challenges for future research.

Key results described in this Chapter include automatic systems that can classify gestures according to basic emotion categories (e.g., the basic emotions) and simple dimensional approaches (e.g., the valence-arousal space).

The chapter also discusses current research trends involving the social dimension of expressive gesture which is particularly important for group playing. Interesting topics include interaction between performers, between performers and conductor, between performers and audience.

1.2.4 Chapter 5: Input Devices and Music Interaction

In order to sense and understand the playing, input devices are one of the key requirements. This chapter discusses the conceptualization and design of digital musical instruments (DMIs). While certain guiding principles may exist and be applied globally in the field of digital instrument design, the chapter seeks to demonstrate that design choice for DMIs depends on particular goals and constraints present in the problem domain. Approaches to instrument design in 3 different contexts are presented: application to new music performance; use within specialized medical imaging environments; and interaction with virtual musical instruments.

Chapter 5 begins with a short discussion on the idea of tangibility in instrument design and how this aspect of human interfacing has been handled in the computer interaction community vs. music-specific interface development. It then builds on Rasmussen's typology of human information processing, a framework that divides human control into several categories of behaviour, and discusses how these categories can be applied to various types of musical interaction.

This Chapter presents three use-cases corresponding to different development areas. First is a description of the motivations for the design of the T-Sticks, a family of cylindrical, hand-held digital musical instruments intended for live performance. Choices of sensing, mapping, sound synthesis and performance techniques are discussed.

The next example focuses on the design of the Ballagumi, a flexible interface created to be compatible with functional magnetic resonance imaging machines. This guided the choice and integration of sensors, as well as the design of the instrument body. The Ballagumi provides a sound-controlling tool subjects can interact with inside a scanner to help neuroscientists learn about the brain during musical creation.

Finally the idea of virtual DMIs and their interaction through haptic force-feedback devices is described. A software tool for construction of virtual instruments based on rigid body simulation is introduced, which can connect to audio programming environments for use with real-time sound synthesis.

1.2.5 Chapter 6: Capturing Bowing Gesture: Interpreting Individual Technique

The myriad of ways in which individual players control bowed string instruments to produce their characteristic expressive range and nuance of sound have long been a subject of great interest by teachers, aspiring students, and researchers. The physics governing the interaction between the bow and the string are such that the sound output alone does not uniquely determine the physical input used to produce it. Therefore, a sound recording alone may be viewed as an incomplete representation of the performance of a violin, viola, cello, or double bass. Furthermore, despite our detailed understanding of the physics of the bowed string family, until recently, the physical constraints of these instruments and the performance technique they require have prevented detailed study of the intricacies of live bowing technique. Today, advancements in sensor technology now offer the ability to capture the richness of bowing gesture under realistic, unimpeded playing conditions.

This Chapter reviews the significance of the primary bowing parameters of bow force, bow speed, and bow-bridge distance (position along the length of the string) and presents a measurement system for violin to accurately capture these parameters during realistic playing conditions. This system uses inertial, force and position sensors for capturing right hand technique, and is optimized to be small, lightweight, portable and unobtrusive in realistic violin performances. Early investigations using this method elucidate the salient differences between standard bowing techniques, as well as reveal the diversity of individual players themselves. In addition to exploring how such studies may contribute to greater understanding of physical performance, a discussion of implications for gesture classification, virtual instrument development, performance archiving and bowed string acoustics is included.

1.2.6 Chapter 7: Interactive Multimedia for Technology-Enhanced Learning with Multimodal Feedback

Chapter 7 discusses an interactive multimedia system for music education. It touches on one of the key requirements for an idealized musical robot to serve as a teacher or a classmate to support the learning.

In order to understand the gesture of a player and to offer appropriate feedback or interactions, such a system would need to measure and analyze the movement of the instrumental playing. There is a wide range of motion tracking technologies including sensor, video tracking and 3D motion capture (mocap) systems but this is not straightforward with traditional instruments such as the violin and cello. Currently, the majority of musical interfaces are mainly designed as tools for multimedia performance and laboratory analysis of musical gesture. However, exciting explorations in pedagogical applications have started to appear.

This Chapter focuses on the i-Maestro 3D Augmented Mirror (AMIR) which utilizes 3D motion capture and sensor technologies to offer online and offline feedback for technology-enhanced learning for strings. It provides a survey on related pedagogical applications and describes the use of a mirror metaphor to provide a 3D visualization interface design including motion trails to visualise shapes of bowing movement. Sonification is also applied to provide another modality of feedback.

Learning to play an instrument is a physical activity. If a student develops a bad posture early on this can be potentially damaging later in his/her musical career. The technologies discussed here may be used to develop and enhance awareness of body gesture and posture and to avoid these problems. This technology can be used to capture a performance in greater detail than to a video recording and has the potential to assist both teachers and students in numerous ways.

A musical robot that can provide technology-enhanced learning with multimodal analysis and feedback such as those discussed in this chapter would be able to contribute to musical education. Example works in this context such as the systems as described in Section II have proved beneficial. It will not only motivate interests and inspire learning for learner but may also provide critical analysis for professional performers.

1.2.7 Chapter 8: Online Gesture Analysis and Control of Audio Processing

This Chapter presents a framework for gesture-controlled audio processing. While this subject has been researched since the beginning of electronic and computer music, nowadays the wide availability of cost-effective and miniature sensors creates unprecedented new opportunities for such applications. Nevertheless the current software tools available to handle complex gesture-sounds interactions remain limited. The framework we present aims to complement standard practices in gesture-sound mapping, emphasizing particularly the role of time morphology, which seems too often neglected.

Chapter 8 describes an approach based on a general principle where gestures are assumed to be temporal processes, characterized by multi-dimensional temporal profiles. Our gesture analysis is divided into two stages to clearly separate low-level processing that is specific to the sensor interface and high-level processing performed on temporal profiles. This high-level processing is based on a tool that we specifically developed for the analysis of temporal data in real-time, called the *gesture follower*. It is based on machine learning techniques, comparing the incoming dataflow with stored templates.

This Chapter introduces the notion of *temporal mapping*, as opposed to *spatial mapping*, to insist on the temporal aspects of the relationship between gesture, sound and musical structures.

The general framework can be applied to numerous data types, from movement sensing systems, sensors or sounds descriptors. The Chapter discusses a typical scenario experimented in music and dance performances, and installations. The authors believe that the methodology proposed can be applied with many other different paradigms and open a large field of experimentation which is currently being pursed.

1.3 Section II: Musical Robots and Automated Instruments

The second section consists of a collection of chapters that are focused on the development of automated instruments and anthropomorphic robots designed as benchmarks to study the human motor control from an engineering point of view, to better understand how to facilitate the human-robot interaction from a musical point of view and to propose novel ways of musical expression. It addresses the fundamental concepts for mimicking the performance and interactions of musicians.

1.3.1 Chapter 9: Automated Piano: Techniques for Accurate Expression of Piano Playing

This Chapter introduces the research on the development of a superior automatic piano designed to accurately produce the soft tones of a desired performance. In particular, the evaluation of the technical and performance aspects of the proposed automatic instrument are stressed. Undoubtedly, the performance evaluation is not an easy task. In fact, it is rather difficult to evaluate the degree of excellence (i.e. whether it is a "good" performance or not), because such kinds of judgments are subjective, and individual preferences are developed by cumulative knowledge and experiences. Moreover, music is an expression of sound in time. Although a sequence of musical notes is arranged in a piece of music by time series, the implicit relationship between sound and time depends on the performer. Of course, if sound and time are not appropriately correlated, the musical expressivity cannot be displayed. So, pianists continue to train to play "ideal" performances. Such a performance is created based on the particular "touch" of a pianist. Analysis of "touch" involves a variety of symbolic descriptions of what is performed, as well as literal descriptions such as "lightly", "heavily", etc. that may be beyond the capacity of engineering to grasp. Therefore, the "touch" of a softened tone, which is difficult even for pianists, is discussed in this chapter. The development of an automatic piano and the analysis of piano's action mechanism to produce soft tones are presented and discussed.

1.3.2 Chapter 10: McBlare: A Robotic Bagpipe Player

Chapter 3 presents McBlare, a robotic bagpipe player for the venerable Great Highland Bagpipe. There are relatively few woodwind robots, and experience has shown that controlling the air is always a critical and difficult problem. The bagpipe literature is unclear about the air regulation requirements for successful playing, so this work offers some new insights. McBlare shows that bagpipes are playable over the entire range with constant pressure, although the range of acceptable pressure is narrow and depends upon the reed. The finger mechanism of McBlare is based on electro-mechanical relay coils, which are very fast, compact enough to mount adjacent to tone holes, and inexpensive. This shows that the mechanics for woodwinds need not always involve complex linkages. One motivation for building robotic instruments is to explore new control methods. McBlare can be controlled by machine-generated MIDI files and by real-time gestural control. This opens new possibilities for composers and performers, and leads to new music that could not be created by human player. The chapter describes several modes of real-time gestural control that have been implemented. McBlare has been presented publicly at international festivals and conferences, playing both traditional bagpipe music and new compositions created especially for McBlare.

1.3.3 Chapter 11: Violin Playing Robot and Kansei

Chapter 4 focuses on the development of a violin-playing robot and the introduction of *kansei*. In order to build a robot that can produce good sounds and perform expressively, it is important to realize a tight human-robot interaction. Therefore, one of the purposes of this chapter is to develop an anthropomorphic violin playing robot that can perform expressive musical sounds. In this chapter, an anthropomorphic human sized manipulator for bowing is introduced. Also, interesting mechanisms of the left hand for fingering with three fingers are introduced. Although the robot is still under construction, both the right arm and the left hand will be connected and produce expressive sounds in the near future. The other purpose of this chapter is introduction and analysis of *kansei* in violin-playing. *Kansei* is a Japanese word similar in meaning to "sensibility," "feeling," "mood," and so on. Recently many Japanese researchers in various fields such as robotics, human-machine interface, psychology, sociology, and so on, are focusing on *kansei*. However, there is no research on musical robots considering *kansei* at the moment. To develop a robot that can understand and express human *kansei* is also very important for smooth human-robot communication. For this purpose, *kansei* is defined and an information flow from musical notes to musical sounds including *kansei* is proposed. Based on the flow, some analyses of human violin-playing were carried out and one of those results is discussed.

1.3.4 Chapter 12: Wind Instrument Playing Humanoid Robots

Chapter 5 presents the development of humanoid robots capable of playing wind instruments as an approach to study the human motor control from an engineering point of view; In particular, the mechanism design principles are discussed, as are the control strategies implemented for enabling an anthropomorphic flute robot and saxophone robot to play such instruments.

In the first part of this Chapter, an overview of the development of flute-playing robots is briefly introduced and the details of the development of an anthropomorphic flutist robot are given. This research is focused on enabling the robot to play a flute by accurately controlling the air beam parameters (width, angle, velocity and length) by mechanically reproducing the following organs: lungs, lips, vocal cord, tongue, arms, fingers, neck, and eyes. All the mechanically simulated organs are controlled by means of an auditory feedback controller. In order to quantitatively analyze the improvements of the robot's performance a sound quality evaluation function has been proposed based on the analysis of the harmonic structure of the flute's sound. As a result, the developed flutist robot is capable of playing the flute to the level of proficiency comparable to that of an intermediate flute player.

In the later part, an overview of the development of saxophone-playing robots is also introduced and the details on the development of an anthropomorphic saxophonist robot are given. This research is focused on enabling the robot to play an alto saxophone by accurately controlling the air pressure and vibration of the single reed. For this purpose, the following organs were mechanically reproduced: lungs, lips, tongue, arms and fingers. All the mechanically simulated organs are controlled by means of a pressure-pitch feed-forward controller. As a result, the developed saxophonist robot is capable of playing the saxophone to the level of proficiency comparable to that of a beginner saxophone player.

1.3.5 Chapter 13: Multimodal Techniques for Human-Robot Interaction

Chapter 6 makes contributions in the areas of musical gesture extraction, musical robotics and machine musicianship. However, one of the main novelties was completing the loop and fusing all three of these areas together. Using multimodal systems for machine perception of human interaction and training the machine how to use this data to "intelligently" generate a mechanical response is an essential aspect of human machine relationships in the future. The work in this chapter presents research on how to build such a system in the specific genre of musical applications. The body of work described in this chapter is truly an artistic venture calling on knowledge from a variety of engineering disciplines, musical traditions, and philosophical practices. Much of the research in the area of computer music has primarily been based on Western music theory. This chapter fully delves into applying the algorithms developed in the context of North Indian classical music. Most of the key contributions of this research are based on exploring the blending of both these worlds. The goal of the work is to preserve and extend North Indian musical performance using state of the art technology including multimodal sensor systems, machine learning and robotics. The process of achieving our goal involved strong laboratory practice with regimented experiments with large data sets, as well as a series of concert performances showing how the technology can be used on stage to make new music, extending the tradition of Hindustani music.

1.3.6 Chapter 14: Interactive Improvisation with a Robotic Marimba Player

Chapter 7 describes a novel interactive robotic improvisation system for a marimba-playing robot named "Shimon". Shimon represents a major step forward in both robotic musicianship and interactive improvisation. The author's previous robot, Haile, was a rhythm-only, robotic drummer engaging in a turn-based interaction with human musicians. In contrast, Shimon plays a melodic instrument, has four, instead of two percussion actuators, and is able to present a much larger gestural and musical range than previous robots. Shimon employs a unique motor control system, taking into account the special requirements of a performing robotic musician: dynamic range, expressive movements, speed, and safety. To achieve these aims, the system uses physical simulation, cartoon animation techniques, and empirical modeling of actuator movements. The robot is also *interactive*, meaning that it listens to a human musician play a live show, and improvises in real-time jointly with the human counterpart. In order to solve this seeming paradox—being both *responsive* and *real-time*, Shimon uses a novel anticipatory approach and uses a *gesture-based* method to music viewing visual performance (in the form of movement) and music generation as parts of the same core process. A traditional interactive music system abstracts the musical information away from its physical source and then translates it back to movements. By taking an embodied approach the movement and the music of the robot are one, making the stage performance a complete experience for both other musicians and the audience. Embodied cognition is gaining popularity both in the field of cognitive psychology and in that of artificial intelligence. However, this is the first time that such an approach has been used for robotic musicianship. The authors evaluate their system in a number of human-subject studies, testing how robotic presence affects synchronization with musicians, as well as the audience's appreciation of the duo. The findings show a promising path to the better understanding of the role of the physical robot's "body" in the field of computer-generated interactive musicianship.

1.3.7 Chapter 15: Interactive Musical System for Multimodal Musician-Humanoid Interaction

Finally, the last chapter introduces the concept and implementation of an interactive musical system for multimodal musician-humanoid interaction. Up to now, several researchers from different fields of Human-Robot Interaction, Musical Information Retrieval, etc. have been proposing algorithms for the development of interactive systems. However, Humanoid Robots are mainly equipped with sensors that allow them to acquire information about their environment. Based on the anthropomorphic design of humanoid robots, it is therefore important to emulate two of the human's most important perceptual organs: the eyes and the ears. For this purpose, the humanoid robot integrates vision sensors in its head and aural sensors attached to the sides for stereo-acoustic perception. In the case of a musical interaction, a major part of the typical performance is based on improvisation. In these parts musicians take turns in playing solos based on the harmonies and rhythmical structure of the piece. Upon finishing his solo section, one musician will give a visual signal, a motion of the body or his instrument, to designate the next soloist. Another situation of the musical interaction between musicians is basically where the higher skilled musician has to adjust his/her own performance to the less skilled one. After both musicians get used to each other, they may musically interact. In this chapter, toward enabling the multimodal interaction between the musician and musical robot, the Musical-based Interaction System (MbIS) is introduced and described. The MbIS has been conceived for enabling the interaction between the musical robots (or/and musicians). The proposed MbIS is composed by two levels of interaction that enables partners with different musical skill levels to interact with the robot. In order to verify the capabilities of using the MbIS, a set of experiments were carried out to verify the interactive capabilities of an anthropomorphic flute robot (introduced in Section II, Chapter 5).

1.4 Future Perspectives

Imagine an idealized integration of all these concepts and systems onto an anthropometric robot. Will this produce a robotic musician capable of playing a musical instrument with expression, interacting with co-performers or teaching how to play the instrument? Obviously, there are many more layers and complex interactions between the systems and many more challenging research avenues, including interpretations, imitations, expressions, interactions and beyond.

At the time of writing, there remain a range of scientific research and technical issues (e.g. human motion analysis and synthesis, music perception, sensing and actuation technology, etc.) to be further studied to enable musical robots to analyze and synthesize musical sounds as musicians do, to understand and reason about music, and to adapt musical appreciation, interactivities and other related behaviors accordingly. Even though most of the existing musical robots are preprogrammed or too complex to drive a "human-like" musical interaction, it is envisaged that future generation musical robots will be capable of mastering the playing of several different instruments (i.e. a flute and also a saxophone). There are many more qualities and features to be explored. For example, it would be an important feature for future musical robots to be able to improve their own musical performance by analyzing the sound produced by its own and by listening and comparing with co-performers (human or other robots) with some kind of automated musical learning strategy.

Currently, several research projects are focusing on producing robust computational models of music communication behavior (e.g. performing, composing, and listening to music). Such behavioral models, embedded in robotic platforms, will open new paradigms at the frontier between music content processing and robotics (see for example, the EURASIP Journal on Audio, Speech, and Music Processing, special issue on *Music Content Processing by and for Robots*, 2011).

Musical appreciation, expression and development are inherently interdisciplinary involving many different aspects of experiences and interactions. To further advance musicianship for robots, there are a wide range of relevant and interrelated subject areas including emotional research, music appreciation quantification, integration of music and dance, and many others. Research on emotion/expression recognition/generation through musical sounds with musical robots is still in its infancy (see Section II, Chapter 3) and there is much exciting and necessary research to be carried out. Developments in entertainment robotics have been increasing for the past few years and fundamental issues for human-robot musical interaction are starting to be addressed. It would be exciting to explore novel methods for the quantification of music appreciation to assure the effectiveness of the interaction. The combination of music with other performing arts (i.e. dance) may open new research paradigms on music behavior and planning (i.e. Human-robot mapping of motion capture data) to enhance the robot's body movement and gesture to convey meaning and expressions.

The list of related topics mentioned above is far from complete and the list continues to grow over time. Some of these issues are currently being explored by several different conferences, workshops and technical publications related to both Musical Robots and Multimodal Interactive Systems including IEEE/RSJ International Conference on Intelligent Robots and Systems (IROS), IEEE International Symposium on Robot and Human Interaction Communication (RO-MAN), International Conference on New Interfaces for Musical Expression (NIME), International Computer Music Conference (ICMC), Workshop on Robots and Musical Expressions (IWRME) and others.

This is an exciting time! We have seen many scientific and technological developments that have transformed our life on many different levels. With the continuing advancements such as those discussed in this book, we look forward to continuing research and development to realize a musical robot who is capable of different musical skills and musicianship to play and teach music that can be applied in many different application scenarios to enrich our life with the expression of music.

Section I: Understanding Elements of Musical Performance and Expression

Chapter 2
Sound-Action Chunks in Music

Rolf Inge Godøy

Abstract. One core issue in music production and perception is the relationship between sound features and action features. From various recent research, it seems reasonable to claim that most people, regardless of levels of musical expertise, have fairly good knowledge of the relationship between sound and sound production, as e.g. manifest in various cases of 'air instrument' performance and other spontaneous body movements to musical sound. The challenge now is to explore these sound-action links further, in particular at the micro-levels of musical sound such as in timbre and texture, and at the meso-levels of various rhythmical and contoural features. As suggested by the seminal work of Pierre Schaeffer and co-workers on so-called *sonic objects* several decades ago, perceptually salient features can be found on the chunk-level in music, meaning in fragments of sound-action in the approximately 0.5 to 5 seconds range. In this chapter, research on the emergence of sound-action chunks and their features will be presented together with some ideas for practical applications.

2.1 Introduction

Traditionally, musical sound has been produced by human action, such as by hitting, stroking, shaking, bowing, blowing, or singing, and it is really only with the advent of electronic instruments and playback technologies that we can dispense with most, or all, human actions in the production of musical sound. This may be an obvious observation, yet it is far from trivial when we consider the consequences of this dissociation of sound and action for the design and use of electronic musical instruments: loosing the link between action and sound also makes us loose one of the most important mental schemas for how we conceive of, and perceive, musical sound.

Given that the sound source of electronic instruments is electronic circuitry requiring some kind of data input to produce and shape the sound, a major

Rolf Inge Godøy
Department of Musicology, University of Oslo, P.B. 1017 Blindern, N-0315 Oslo, Norway
e-mail: r.i.godoy@imv.uio.no

J. Solis and K. Ng (Eds.): Musical Robots and Interactive Multimodal Systems, STAR 74, pp. 13–26.
springerlink.com

challenge in the design and use of new musical instruments is to establish mental schemas for thinking and controlling musical sound features through what is rather abstract input data. One possible answer to this challenge is to try to correlate the input data of whatever synthesis model is used (e.g. additive, frequency modulation, subtractive, granular, etc.) with perceptually salient features of the output sound signal such as its overall dynamical envelopes and its stationary and transient spectral features, e.g. with labels such as 'inharmonic', 'harmonic', 'white noise', 'pink noise' etc. [30]. Such labels on the output sound signal may be extended further into the metaphorical realm with attributes such as 'wet', 'dry', 'smooth', 'rough', 'dark', 'bright', etc. [29].

However, another and complimentary answer to this challenge is to extrapolate mental schemas of sound-action relationships from our past and presumably massive experiences of non-electronic musical instruments as well as everyday sonic environments, to new electronic instruments. It could in particular be interesting to see how what we perceive as somehow meaningful sound-action units, what could be called *sound-action chunks*, emerge in our experiences of music, and how general principles for sound-action chunk formation may be applied to novel means for producing musical sound.

Actually, relating what we hear to mental images of how we assume what we hear is produced, is a basic phenomenon of auditory perception and cognition in general. In the field of ecological acoustics, it has been documented that listeners usually have quite accurate notions of how everyday and musical sounds are produced [9, 31], including both the types of actions (e.g. hitting, scrubbing, stroking) and the materials (glass, wood, metal) as well as object shapes (plate, tube, bottle) involved. As to the perception of the actions involved in sound production, the claim of the so-called *motor theory* (and various variants of this theory) has for several decades been that auditory perception, primarily in speech but also in other areas, is accompanied by mental simulations of the assumed sound-producing actions: when listening to language, there is a (mostly) covert simulation of the phonological gestures assumed to be at the cause of the sounds [6, 26]. And it has been documented that when people listen to music, there are similar activations of the motor-related areas of the brain, in particular in the case of expert musicians [22], but also in the case of novice musicians after a quite short period of training [1].

Dependent on level of expertise, these motor images of sound production may vary in acuity: a native speaker of a language, e.g. Chinese, will usually have much finer motor images of the various sounds of the Chinese language than a foreigner, yet the foreigner will be able to perceive, albeit coarsely, the difference between the phonological gestures of Chinese and another unfamiliar language, e.g. Russian. Likewise in music, we may see significant differences in the acuity of motor images of sound production between novices and experts, yet there are still correspondences in the motor images at overall or coarse levels of acuity [14]. According to the motor theory of perception, motor images are fairly robust yet also flexible because they are based on very general motor schemas, e.g. in our case, the differences between *hitting* and *stroking*, corresponding to the difference between the whole class of *impulsive* sounds, i.e. sounds with a rather abrupt beginning followed by a shorter or longer decay as in many percussion instruments, and the whole class of *sustained* sounds, i.e. sounds that have a more gradual beginning followed by a relatively stable and enduring presence as in bowed and blown instruments. These main classes are quite distinct, yet applicable to very many variant instances (see section 2.2 below).

In the so-called *embodied* view of perception and cognition [7, 8], motor schemas are seen as basic for all cognition, not only auditory perception. This means that all perception and reasoning, even rather abstract thinking, is understood as related to images of action. In our case, projecting images of sound-action relationships from past musical and environmental sonic experiences onto new musical instruments could be seen as a case of anthropomorphic, know-to-unknown, projection, and as a matter of basic functioning of our mental apparatus, what we see as a *motormimetic* element in music perception [10]. This includes both kinematic and effort-related images, raising some intriguing questions of how electronic instrument sounds, i.e. sounds produced without human effort, still can evoke subjective sensations of effort in our minds. For this reason, we shall now first have a look at some main types of music-related actions, then consider features of musical sound at different timescales, before we turn to principles of chunking and various sonic features within chunks, and at the end of the chapter present some ideas on how concepts of sound-action chunks can be put to use in practical contexts.

2.2 Music-Related Actions

We can see music-related actions both in the production of musical sound and in peoples' behavior when listening to music at concerts, in dance, and innumerable everyday situations. As suggested in [23] it could be useful then to initially distinguish between *sound-producing* and *sound-accompanying* actions.

Sound-producing actions include both *excitatory actions* such as hitting, stroking, blowing, and *modulatory* or *sound-modifying* actions such as changing the pitch with the left hand on a string instrument or the mute position on a brass instrument, as well as in some cases *selection* actions such as pulling the stops on an organ. Also in this category of sound-producing actions we find various *sound-facilitating* or *ancillary* actions, actions that help avoid strain injury or help shape the rhythmical and/or articulatory expression of the music such as by making larger movement with wrists, elbows, shoulders, and even head and torso in piano performance. Related to this, we find various *communicative* actions that performers use variously to communicate within an ensemble or to communicate expressive intent to the audience, such as swaying the body, nodding the head, or making exaggerated or theatrical hand movements on the guitar before a downbeat [3, 38].

Sound-accompanying actions include all kinds of body movements that listeners may make to music, such as moving the whole body or parts of the body to the beat of the music, gesticulate to some feature of the music, or imitate sound-producing actions of the music. This latter category can be seen in various instances of so-called *air-instrument* performance such as air-drums, air-guitar, and air-piano. Imitations of sound-producing actions are interesting in our context because they in many cases attest to quite extensive knowledge of sound production even by untrained listeners. Such cases of air instrument performances by not formally trained listeners can be understood as a result of imitative behavior and learning, essential components of the abovementioned embodied view of perception and cognition. Another important element is the abovementioned variable acuity involved here: people who are not capable of playing percussion or guitar 'for real' still seem to have coarse, yet quite well informed, notions of how to hit and move between the drums and other instruments of an imaginary drum set, or of how to pluck the strings and move the left hand on the neck of an imaginary guitar. But also on a more general level, listeners seem to readily catch on to the kinematics of sound-production, e.g. the pitch-space of the keyboard, the amplitude of required movement, as well as the overall effort required to produce the sound in question [14]. Such general kinematic and dynamic correspondences between sound and movement can also be observed in dancers' spontaneous movements to musical sound, i.e. dancers tend to readily catch on to the overall mood and sense of motion afforded by the music, and in our observations we could hardly ever find any clear discrepancies between the overall features of the sound and the movements of the dancers [18].

The kinematic and effort-related action images afforded by the sounds can then be correlated with some basic sound-producing actions. Following the pioneering work of Pierre Schaeffer and associates [11, 33, 34], we have singled out these basic sound-action categories:

- *Impulsive sound-producing actions*: besides resulting in sounds with the abovementioned envelope of an abrupt attack followed by a shorter or longer decay (dependent on the physics of the instrument in question), impulsive actions are distinguished by having a short peak of effort, typically as a fast movement towards an impact point, followed by relaxation. Sometimes also referred to as *ballistic movement*, impulsive sound-producing actions are typically based on so-called *open loop* motor control, meaning that the entire movement trajectory is preplanned and executed without feedback because it is so fast that continuous control and adjustment in the course of the trajectory would usually not be feasible, however this has been much debated in the motor control literature [5].
- *Sustained sound-producing actions*: typically resulting in the abovementioned more stable and enduring sound, this category requires continuous energy transfer, e.g. continuous motion of the bow on string instruments or continuous blowing on woodwind and brass instruments. As opposed to impulsive sound-producing actions, sustained sound-producing actions are based on so-called *closed loop* motor control, meaning having continuous feedback control allowing for adjustment during the course of the sound. However, although there is this possibility of continuous control in sustained sound-producing actions, there is still the unanswered question as to how often such adjustments may occur for the simple reason that the musician has to listen to a certain segment of sound output in order to make a decision as to adjustment, hence that there in all likelihood is a discontinuous element here as well.
- *Iterative sound-producing actions* denote fast and repeated actions such as in a tremolo, a trill, or a drum roll. Iterative sound-producing actions typically require movement and effort distinct from the two other categories because of

speed, meaning that the amplitude of the movement is small and the source of the movement may differ from that of slower movement e.g. as a fast trill on the keyboard driven by whole arm rotation (rather than by finger movement) or a fast tremolo on a string instrument driven by a wrist shaking (rather than elbow movement).

Iterative sound-producing actions may be bordering onto sustained sound-producing actions in cases where there is an overall continuous movement of an effector on a rough surface as in the case of the washboard or the maracas, and we may also observe categorical transitions between these and the other sound-producing actions as well: changing the duration and/or density of the sound-producing actions may result in so-called *phase-transitions* [19], e.g. if we shorten a sustained sound beyond a certain threshold it may switch to become an impulsive sound (i.e. go from a 'bow-like' sound to a 'percussive' sound), and conversely, if we lengthen an impulsive sound, it may switch to become a sustained sound. We may observe similar phase-transitions between impulsive and iterative sounds, and between sustained and iterative sounds, something that significantly affects the perceived internal textures of the sound.

The point here is to realize that what is often referred to as sonic features are actually just as much action features, and that this also extends to the phenomenon of *coarticulation*. Coarticulation means the subsumption of otherwise distinct actions and sounds into more superordinate actions and sounds, entailing a contextual smearing of otherwise distinct actions and sounds, e.g. rapid playing of scales and arpeggios on the piano will necessitate finger movements included in superordinate action trajectories of the wrists, elbows, shoulders, and even whole torso, as well as entail a contextual smearing of the singular tones into superordinate contours of the scales or arpeggios. Coarticulation is a much-studied phenomenon in linguistics [21], but unfortunately not so well studied in music (see [16] for an overview). One essential element of coarticulation is that it concerns both the production and the perception of sound, hence that it clearly unites sound and action into units, into what we prefer to call *sound-action chunks in music*.

Lastly, these different sound-producing actions may be combined in more complex musical textures, typically in the case of having a sustained background sound (e.g. as an accompaniment) combined with an iterative embellishment of the sustained sounds (e.g. as a tremolo texture added to the sustained tones), punctuated by impulsive sounds in the foreground of the texture. A listener may perceive such composite musical textures holistically, or may choose to intentionally focus on some aspect of the texture, e.g. the foreground melody or the drum pattern, and hence in dance and other movements to music we may see divergences in the gestural rendering of the music, and for this reason we may speak of rich gestural affordances of musical sound [13].

2.3 Features and Timescales

The different action categories are distinguished by event duration and event density, as are also most other perceptually salient features of musical sound. It could be useful then to have a look at some different sonic features and their corresponding timescales, in order to establish some more general principles for sound-action correspondences in music.

Firstly, there is the well-known timescale for perceiving pitch and stationary timbral elements, ranging from approximately 20 to approximately 20000 events per second. This is the timescale for determining features such as pitched vs. non-pitched, harmonic vs. inharmonic, noise and/or various degrees of noise, spectral peaks, formants, etc., i.e. features that are perceived as basically stationary in the course of sound fragments. Then there is the timescale of patterns at the less than 20 events per second found in various kinds of textural fluctuations of the sound, e.g. as tremolo, vibrato, and other fluctuations in intensity, pitch, and timbre. At an even longer timescale, approximately in the 0.5 to 5 seconds range, we find perceptually significant features of the sound such as its overall dynamic envelope and its overall pitch and/or spectral evolution. This is what we call the *chunk* timescale, and this is also the timescale where we find significant style-determining features of musical sound such as rhythmical and textural patterns, e.g. waltz, tango, disco, funk, as well as the associated sense of body motion and even mood and emotion, e.g. fast, slow, calm, agitated, etc. At yet longer timescales, i.e. at timescales significantly above the 5 seconds range, we typically find more formal features such as the build of whole tunes, sections, and movements. Considering the different timescale involved here, it could be convenient to suggest the following three categories in our context of sound-action chunks:

- *Sub-chunk timescale*, typically the less than approximately 0.5 seconds timescale where features such as pitch and stationary timbral elements are perceived. At this timescale we also perceive various fast fluctuations in the sound such as tremolo, trill, and other pitch-related and timbre-related fluctuations, fluctuations that may be collectively referred to as 'grain' (see section 2.5 below).
- *Chunk timescale*, the approximately 0.5 to 5 seconds timescale where we perceive the overall dynamic, pitch-related, and timbre-related envelopes of the sound, and various rhythmical and textural patterns, as well as melodic motives. This is in our context of sound-actions the most significant timescale, and probably in very many other contexts as well (see next section).
- *Supra-chunk timescale*, ranging in duration from significantly above 5 seconds to several minutes and even hours. The perceptual and cognitive workings of large-scale forms seems to be a not well researched topic [4], however for our purposes, this timescale can be regarded as consisting of concatenations of several chunks in succession, resulting in extended sound-action scripts which may be perceived as having narrative or dramaturgical features.

2.4 Chunking

In the course of their work with a new and more universal music theory in the wake of the *musique concrète* in the 1950s and 1960s, Pierre Schaeffer and his co-workers arrived at the conclusion that fragments of musical sound in approximately the 0.5 to 5 seconds range, or what we here call the *chunk timescale*, ought to be the focus of music theory. Initially a technical necessity in the production of electroacoustic music in the days before the advent of the tape recorder, the focus on the sonic object allowed exploring subjectively experienced perceptually salient features of sound and developing criteria for selecting, processing, and combining sounds. By repeated listening to fragments of sound, it was found that several features that previously had no name in Western music theory could be distinguished and even be ordered into a conceptual taxonomic apparatus.

The first part of this apparatus consists of a *typology of sonic objects*, a rough sorting of sounds according to their overall envelopes, i.e. the abovementioned categories of *impulsive*, *sustained*, and *iterative* sounds, as well as the three attribute categories of *pitched*, *non-pitched*, and *variable*, constituting a 3 x 3 matrix. A further differentiation of the internal features of the sonic objects was made with the so-called *morphology of sonic objects* where there is a progressive top-down differentiation of features based on harmonic content, pitch-related content, and various fast and slower fluctuations within the sonic objects (more on this below in section 2.5). The essential point is that such differentiations are only possible if we consider musical sound at the chunk-level: if we go to the supra-chunk level, we loose the possibility of capturing chunk-level features [2, 11, 33, 34].

As to the criteria for selecting chunks and chunk boundaries, Schaeffer and co-workers stated that it is of course possible to make arbitrary cuts in any sound recording, but also possible to make more substance-based selections of chunks. Schaeffer's own suggestion was to use qualitative discontinuities in the sound to make 'natural' chunk boundaries as reflected in the abovementioned typology of sonic objects, i.e. that a sonic object is determined by its overall envelopes, as well as some additional criteria of suitability such as duration. The duration criterion point in the direction of generally accepted limits for attention spans, typically in the three seconds range, sometimes longer, sometimes shorter [28]. As pointed out by Pöppel, this approximately three seconds range also fits quite well with average durations of everyday actions as documented by the large-scale surveys of Schleidt and Kien [35], suggesting that there may be a mutual attuning of action and attention at this timescale.

There are a number of other arguments from cognitive psychology and human movement science in favor of the timescale of the chunk as particularly significant (see [12] for an overview), however we can also from a purely music analysis point of view make a listing of the features that are manifest on the chunk-level, i.e. features that are not on the sub-chunk level, and features that are present, but not in focus, on the supra-chunk level:

- Overall dynamic shape, meaning the perceived intensity envelope of the sound
- Overall timbral shape, including the quasi-stationary elements, various fluctuations, and the slower evolutions in the course of the sonic object
- Overall pitch-related, modal, and tonal features, meaning pitched, non-pitched, fixed, variable, tonal center(s), modal patterns, harmonic patterns, etc. in short all features concerned with pitch and constellations of pitches, simultaneously and sequentially

- Overall rhythmical patterns, meaning various metrical patterns of different kinds of music such as tango, waltz, etc., or other non-dance patterns, both cyclical and non-cyclical
- Overall textural patterns, with elements such as sustained, impulsive, iterative, and variously coarticulatory fused sounds, foreground, background, and other groupings in the sonic textures
- Overall stylistic features, as various combinations of abovementioned chunk-level features and as various clichés, i.e. typical figures and ornaments, within a style
- Overall expressive, emotive features such as sense of movement, sense of effort, and sense of acceleration or deceleration, etc.

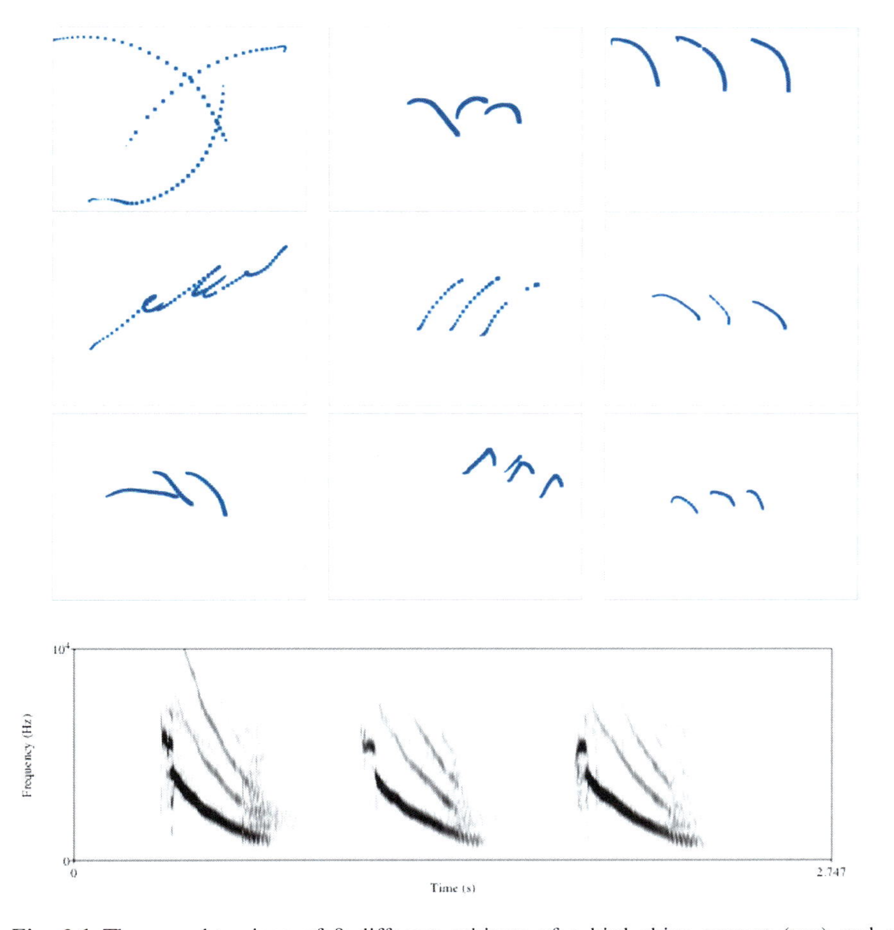

Fig. 2.1 The sound-tracings of 9 different subjects of a bird chirp excerpt (top) and a spectrogram of the excerpt (bottom). The excerpt is from [34], CD3, track 13, and for details of these sound-tracings see [15, 18].

In our sound-action context it is important to note that all these features may be conceptualized as action trajectories in multidimensional spaces. As an example, consider the 9 different drawings of a bird chirp sound in Figure 2.1. These drawings were collected from a *sound-tracing* project were we asked 9 listeners with different levels of musical training to spontaneously draw on a digital tablet the movement shapes they associated with various sounds that were played back (see [15, 18] for details). Although there were differences in the subjects' drawings, there was a fair amount of consistency in the overall, chunk-level features for sounds with few concurrent features, but greater diversity for sounds with greater number of concurrent features. Needless to say, the two-dimensional surface of the digital tablet was a restriction for rendering the multidimensionality of the sounds, so we are now developing various multidimensional hand-held controllers as well as using an infrared motion capture system allowing more freedom of movement and hence more dimensions for rendering perceptually salient features as finger/hand/arm movements.

2.5 Sub-chunk Features

When we zoom into the sub-chunk timescale, we find audio-frequency range features such as stationary pitch and stationary timbral features. It may be argued that also these features are related to mental images of actions by *effector position* and *effector shape*, e.g. as vocal folds tension and vocal tract shapes correlated to various timbral features (different formantic peaks, different degrees of white noise, etc.), or the imagined mute position on a brass instrument correlated to a particular timbre, the bow position and pressure on a string to a particular timbre on an imaginary violin, and even shapes and position of hands on an imaginary keyboard as correlated with the sounds of chords.

But in addition to these more static postures, we also have a number of significant motion-dependent features at this sub-chunk level, features that could be summarized as *textural* features, and that apply to sounds from several different sources (instrumental, vocal, environmental, electronic) as long as they fit with general motor schemas. From Schaeffer's morphology of sonic objects, we shall just mention the following two main categories here:

- *Grain*, meaning all kinds of perceived fast fluctuations in the sound, typically in pitch, intensity, or timbre. These fluctuations may be further differentiated with respect to their speed, amplitude, and relative position in the audible frequency range, as well as with respect to regularity/irregularity, non-evolving/evolving, etc. In terms of action, grain is related to very fast movement like in a rapid tremolo, *flatterzunge*, vibrato, trill, or drum roll, or to movement made on a surface such as stroking over a washboard or another rough surface resulting in a dense series of sound onsets or sound modulations.
- *Allure* (sometimes also rendered in English as *gait* or as *motion*), meaning slow kinds of fluctuation in the sound such as in a slow vibrato, a slower fluctuation in intensity or in timbral content as in the opening and closing of a *wah-wah* mute, or the undulating movement of a waltz accompaniment figure for that matter.

These sub-chunk classes of sound-actions are generic in the sense that they apply across different sounds and different sources, but where the unifying element is the general motor schemas that they are based on. Referring to the abovementioned general typological categories of *impulsive*, *sustained*, and *iterative* sounds, we see that for instance the morphological category of *grain* can be applied to both sustained friction sounds, e.g. to bowed string sounds and to surface scraping sounds, as well as to sustained blown sounds such as in the *flatterzunge* on a flute, applications that make sense both in sonic and action-related respects. Likewise, an *allure* fluctuation in a sound may just as well be produced by slow left hand movements on a string instrument as by back-and-forth movements on a DJ scratch turntable. Included in these morphological categories are also effects of coarticulation in the form of transitions between spectral shapes, as is extensively practiced by musicians on wind, brass, string instruments and in vocal performance, and as can also be simulated with so-called *diphone synthesis* [32].

2.6 Sound-Action Scripts

From the overview of chunk level and sub-chunk level features of musical sound, it could be claimed that most of these features may be correlated with actions. This means that we can map out various action types, starting with general chunk-level typological categories and principles of coarticulatory inclusion, and differentiate downwards into the action correlates of the most minute sub-chunk-level features, thus think of music as not a matter of 'pure sound' but more as a matter of *sound-action scripts*. This means that music is also a choreography of sound-producing actions, and that this choreography could be the mental schemas at the basis for a general music theory, applicable equally well to making music with traditional instruments and new electronic instruments.

This has consequences both for the input devices used and for the subsequent mapping of input data to whatever synthesis and/or sound-processing model that we might use. Ideally, input devices should somehow be designed so as to allow input actions that correspond more or less with the abovementioned sound-action couplings, e.g. a sustained sound should be the result of a sustained action, an impulsive sound the result of an impulsive action, a grain textural feature the result of small and rapid actions, etc., and also that different formantic timbral shapes should somehow be the result of different effector shapes. The significant amount of research and development concerned with new input devices that we see now testifies to this need for producing better and more 'intuitively' experienced input devices [27, 39].

But also in the *mapping* of input data to the synthesis and processing [37], the basic sound-action links outlined above could be useful in providing a more direct and intuitive link between actions and sonic features. In principle, any input data may be mapped to any control data, and this can be done according to the often-mentioned schemes of *one-to-one*, *one-to-many*, *many-to-many*, and *many-to-one* types of mappings. However, it is possible to make mappings more in line with previous non-electronic experiences, e.g. by mapping a control input for loudness to both the amplitude and the spectrum of a sound so that an increase in loudness control is not just an increase in amplitude, but also a shift in the spectrum of the sound towards more high-frequency components (e.g. by controlling the filter or the modulation index), resulting in a sound that is both louder and more brilliant.

But the topic of mapping could also be seen as that of mapping the actions of one effector to that of another effector. This is known as *motor equivalence* and means that we may perform a task with another effector than we usually do [24], and in the case of musical sounds, e.g. imitate non-vocal sounds with our vocal apparatus, or with some suitable input device, try to imitate vocal sounds by a hand held device. In this perspective, one effector may be exchanged for another effector, provided we follow some general motor scheme and provided we have suitable input devices to transmit such actions. Taking the example of sustained friction sounds, this is a rather broad class of sound-producing actions including bowing (on strings as well as on plates and other objects, e.g. the musical saw) and scraping on any surface (cymbals, tamtams, or any surface, smooth or rough), where it may be possible to use a general purpose input device such as the *Reactable* to produce a whole class of different friction sounds [20]. Likewise, a hand held device generating accelerometer data may generate input to a generic impulsive sound generator for controlling a variety of percussion-like sounds. And lastly, marker position data of hand and arm movement in open space from an infrared motion capture system may be used as multidimensional control data for controlling timbral-formantic features, e.g. for imitating vocal sounds with hand movements.

The main point is to try to enhance the relationship between what we do and the sonic results of what we do. But there are a number of impediments here, both of technical nature such as in the latency and in the poor resolution of some systems, as well as of more conceptual nature in the violation of traditional energy-transfer schemas and in the lack of haptic feedback. Also, there is the element of learning that has to be taken into account, and we need to know more about what happens with long-term practice on new musical instruments, long-term practice approaching the amount of time spent by musicians on mastering more traditional musical instruments. But all through this, having a motor-theory based understanding of sound-action chunks in music could be a strategy for improving the interfaces of new electronic instruments.

2.7 Conclusion

The relationship between music and movement has been a recurrent topic in Western writings on music since antiquity (see [36] for an overview and for some particularly interesting work on music-related body movement in 20th century). However, we have in the past decade seen a surge in research suggesting that music is an embodied phenomenon, and that music is actually a combination of sound and action [10, 25] (see in particular various chapters in [17] for information on current research directions within the field, and visit http://fourms.wiki.ifi.uio.no/Related_projects for an overview of relevant web-sites as well as http://www.fourms.uio.no/ for information on our own present research). Also, it is now generally accepted that musical sound is perceptually multidimensional and works at multiple levels of resolution, and we believe there is much evidence for that what we here call the chunk-level and the sub-chunk level, are perceptually and strategically (with respect to both analytic concepts and control possibilities) the most significant. Combining the idea of chunk-level and sub-chunk-level with the embodied view of music perception and cognition, we end up with the idea of sound-action chunks in music, and from this idea we see a number of possible and fruitful developments in both our basic understanding of music and in various practical applications in music making.

References

1. Bangert, M., Altenmüller, E.O.: Mapping perception to action in piano practice: a longitudinal DC-EEG study. BMC Neuroscience 4, 26 (2003)
2. Chion, M.: Guide des objets sonores. INA/GRM Buchet/Chastel, Paris (1983)
3. Davidson, J.W.: Visual Perception of Performance Manner in the Movements of Solo Musicians. Psychology of Music 21, 103–113 (1993)
4. Eitan, Z., Granot, R.Y.: Growing oranges on Mozart's apple tree: "Inner form" and aesthetic judgment. Music Perception 25(5), 397–417 (2008)
5. Elliott, D., Helsen, W., Chua, R.: A Century Later: Woodworth's (1899) Two-Component Model of Goal-Directed Aiming. Psychological Bulletin 127(3), 342–357 (2001)
6. Galantucci, B., Fowler, C.A., Turvey, M.T.: The motor theory of speech perception reviewed. Psychonomic Bulletin & Review 13(3), 361–377 (2006)
7. Gallese, V., Metzinger, T.: Motor ontology: The Representational Reality Of Goals, Actions And Selves. Philosophical Psychology 16(3), 365–388 (2003)
8. Gallese, V., Lakoff, G.: The Brain's Concepts: The Role of the Sensory-Motor System in Conceptual Knowledge. Cognitive Neuropsychology 22(3/4), 455–479 (2005)
9. Gaver, W.W.: What in the world do we hear? An ecological approach to auditory event perception. Ecological Psychology 5(1), 1–29 (1993)
10. Godøy, R.I.: Motor-mimetic Music Cognition. Leonardo 36(4), 317–319 (2003)
11. Godøy, R.I.: Gestural-Sonorous Objects: embodied extensions of Schaeffer's conceptual apparatus. Organised Sound 11(2), 149–157 (2006)
12. Godøy, R.I.: Reflections on chunking in music. In: Schneider, A. (ed.) Systematic and Comparative Musicology: Concepts, Methods, Findings, Hamburger Jahrbuch für Musikwissenschaft, Band, vol. 24, pp. 117–132. Peter Lang, Vienna (2008)
13. Godøy, R.I.: Gestural affordances of musical sound. In: Godøy, R.I., Leman, M. (eds.) Musical Gestures: Sound, Movement, and Meaning, pp. 103–125. Routledge, New York (2010)
14. Godøy, R.I., Haga, E., Jensenius, A.R.: Playing "Air instruments": Mimicry of sound-producing gestures by novices and experts. In: Gibet, S., Courty, N., Kamp, J.-F. (eds.) GW 2005. LNCS (LNAI), vol. 3881, pp. 256–267. Springer, Heidelberg (2006)

15. Godøy, R.I., Haga, E., Jensenius, A.R.: Exploring Music-Related Gestures by Sound-Tracing: A Preliminary Study. In: Ng, K. (ed.) Proceedings of the COST287-ConGAS 2nd International Symposium on Gesture Interfaces for Multimedia Systems, pp. 27–33. Leeds College of Music, Leeds (2006)
16. Godøy, R.I., Jensenius, A.R., Nymoen, K.: Chunking in Music by Coarticulation. Acta Acustica United with Acustica 96(4), 690–700 (2010)
17. Godøy, R.I., Leman, M. (eds.): Musical Gestures: Sound, Movement, and Meaning. Routledge, New York (2010)
18. Haga, E.: Correspondences between Music and Body Movement. PhD Thesis, University of Oslo. Unipub, Oslo (2008)
19. Haken, H., Kelso, J.A.S., Bunz, H.: A theoretical model of phase transitions in human hand movements. Biological Cybernetics 51(5), 347–356 (1985)
20. Hansen, K.F., Bresin, R.: The Skipproof virtual turntable for high-level control of scratching. Computer Music Journal 34(2), 39-50 (2010)
21. Hardcastle, W., Hewlett, N. (eds.): Coarticulation: theory, data and techniques. Cambridge University Press, Cambridge (1999)
22. Haueisen, J., Knösche, T.R.: Involuntary Motor Activity in Pianists Evoked by Music Perception. Journal of Cognitive Neuroscience 13(6), 786–792 (2001)
23. Jensenius, A.R., Wanderley, M.M., Godøy, R.I., Leman, M.: Musical Gestures Concepts and Methods in Research. In: Godøy, R.I., Leman, M. (eds.) Musical Gestures: Sound, Movement, and Meaning, pp. 12–35. Routledge, New York (2010)
24. Kelso, J.A.S., Fuchs, A., Lancaster, R., Holroyd, T., Cheyne, D., Weinberg, H.: Dynamic cortical activity in the human brain reveals motor equivalence. Nature 392(23), 814–818 (1998)
25. Leman, M.: Embodied Music Cognition and Mediation Technology. The MIT Press, Cambridge (2008)
26. Liberman, A.M., Mattingly, I.G.: The Motor Theory of Speech Perception Revised. Cognition 21, 1–36 (1985)
27. Miranda, E.R., Wanderley, M.M.: New Digital Musical Instruments: Control and Interaction Beyond the Keyboard. A-R Editions, Inc, Middleton (2006)
28. Pöppel, E.: A Hierarchical model of time perception. Trends in Cognitive Science 1(2), 56–61 (1997)
29. Porcello, T.: Speaking of sound: language and the professionalization of sound-recording engineers. Social Studies of Science 34(5), 733–758 (2004)
30. Risset, J.-C.: Timbre Analysis by Synthesis: Representations, Imitations and Variants for Musical Composition. In: De Poli, G., Piccialli, A., Roads, C. (eds.) Representations of Musical Signals, pp. 7–43. The MIT Press, Cambridge (1991)
31. Rocchesso, D., Fontana, F. (eds.): The Sounding Object. Edizioni di Mondo Estremo, Firenze (2003)
32. Rodet, X., Lefevre, A.: The Diphone program: New features, new synthesis methods and experience of musical use. In: Proceedings of ICMC 1997, September 25-30, Thessaloniki, Grece (1997)
33. Schaeffer, P.: Traité des objets musicaux. Éditions du Seuil, Paris (1966)
34. Schaeffer, P., Reibel, G., Ferreyra, B.: Solfège de l'objet sonore. INA/GRM, Paris (1988) (first published in 1967)

35. Schleidt, M., Kien, J.: Segmentation in behavior and what it can tell us about brain function. Human Nature 8(1), 77–111 (1997)
36. Schneider, A.: Music and Gestures: A Historical Introduction and Survey of Earlier Research. In: Godøy, R.I., Leman, M. (eds.) Musical Gestures: Sound, Movement, and Meaning, pp. 69–100. Routledge, New York (2010)
37. Verfaille, V., Wanderley, M.M., Depalle, P.: Mapping strategies for gestural and adaptive control of digital audio effects. Journal of New Music Research 35(1), 71–93 (2006)
38. Vines, B., Krumhansl, C., Wanderley, M., Levitin, D.: Cross-modal interactions in the perception of musical performance. Cognition 101, 80–113 (2005)
39. Wanderley, M., Battier, M. (eds.): Trends in Gestural Control of Music. Ircam, Paris (2000)

Chapter 3
Automatic Music Transcription: From Monophonic to Polyphonic

Fabrizio Argenti, Paolo Nesi, and Gianni Pantaleo

Abstract. Music understanding from an audio track and performance is a key problem and a challenge for many applications ranging from: automated music transcoding, music education, interactive performance, etc. The transcoding of polyphonic music is a one of the most complex and still open task to be solved in order to become a common tool for the above mentioned applications. Techniques suitable for monophonic transcoding have shown to be largely unsuitable for polyphonic cases. Recently, a range of polyphonic transcoding algorithms and models have been proposed and compared against worldwide accepted test cases such as those adopted in the MIREX competition. Several different approaches are based on techniques such as: pitch trajectory analysis, harmonic clustering, bispectral analysis, event tracking, nonnegative matrix factorization, hidden Markov model. This chapter analyzes the evolution of music understanding algorithms and models from monophonic to polyphonic, showing and comparing the solutions, while analysing them against commonly accepted assessment methods and formal metrics.

3.1 Introduction

Music Information Retrieval (MIR) is a multidisciplinary research field that has revealed a great increment in academic interest in the last fifteen years, although yet barely comparable to the commercial involvement grown around speech recognition. It must be noticed that music information is much more complex than

Fabrizio Argenti · Paolo Nesi · Gianni Pantaleo
Department of Systems and Informatics
University of Florence
Via S. Marta 3, Florence 50139, Italy
e-mail: nesi@dsi.unifi.it
http://www.disit.dsi.unifi.it

J. Solis and K. Ng (Eds.): Musical Robots and Interactive Multimodal Systems, STAR 74, pp. 27–46.
springerlink.com

speech information, both from a physical (range of frequency analysis) and a semantic (big number, high complexity and many abstraction levels of the possible queries) point of view. Automatic transcription is a specific task within MIR, and it is considered one of the most difficult and challenging problems. It is defined here as the process of both analyzing a musical recorded signal, or a musical performance, and converting it into a symbolic notation (a musical score or sheet) or any equivalent representation concerning note parameters such as pitch, onset time, duration and intensity.

A musical signal is generally understood as composed by a single or a mixture of approximately periodic, locally stationary acoustic waves. According to the Fourier representation, any finite energy signal is represented as the sum of an infinite number of sinusoidal components weighted by appropriate amplitude coefficients. A musical sound is a particular case where, ideally, frequency values of single harmonic components are integer multiples of the first one, called fundamental frequency (defined as *F0*, which is the perceived pitch). Many real instruments, however, produce sounds which do not have exactly harmonically spaced partials. The phenomenon is called *partial inharmonicity*, and it was analytically described by Fletcher and Rossing [18], and brought to the attention of music transcription research community by Klapuri [33].

A major distinctive cue in music transcoding is given by the number of voices a music piece consists of: there can be only one voice playing at each time; these cases are treated as a *monophonic transcription* task. On the contrary, if several voices are played simultaneously, we deal with a *polyphonic transcription* process. The former is currently considered a resolved problem, while the latter is still far from being successfully settled, and additional difficulties arise in the presence of multi-instrumental contexts. In Figure 3.1, some examples of the spectral content of typical audio signals, are shown.

Difficulties arise in polyphonic music transcription since two or more concurrent sounds may contain partials which share the same frequency values. This generates the well known problem of partials overlapping, which is one of the main reasons why simple amplitude spectral analysis is considered inadequate, if not joined to other signal processing techniques or *a priori* knowledge resources.

Retaining the parallel between speech and music, music notation is mainly a set of instructions for a musical performance, rather than a representation of a musical signal; in the same way, written text is to be considered as the equivalent for speech. The main difference is that music information is much more multi-faceted, since it includes many different levels of information (note pitch, harmonic and rhythmic information, indications for expression, dynamics, ...). Besides, modularity is a similar aspect observed also in the human brain [33], [47]. The human auditory system (the inner ear together with the part of the brain appointed to music cognition) results to be the most reliable acoustic analysis tool [33].

Many efforts have been made to realize exhaustive reviews of automatic transcription methods. Remarkable works are the ones by Rabiner [52] for monophonic transcription, and by Bello [2], Klapuri [33], [34], Brossier [6] and Yeh [63] also for polyphonic transcription. However, it is difficult to categorize music transcription methods according to any single taxonomy, since human capability to achieve the comprehension of music transcription is understood as the sum of two different attitudes: bottom-up and top-down processing. This suggests a first boundary of classification, given by the following approaches:

- Bottom-up processing, or the *data-driven model*, starts from low level elements (the raw audio samples) and it uses processing blocks to analyze and cluster these elements in order to gather the required information.
- Top-down processing, or the *prediction-driven model*, starts from information at a higher level (based on external knowledge) and it uses such information to understand and explain elements at lower hierarchy levels (*physical stimuli*).

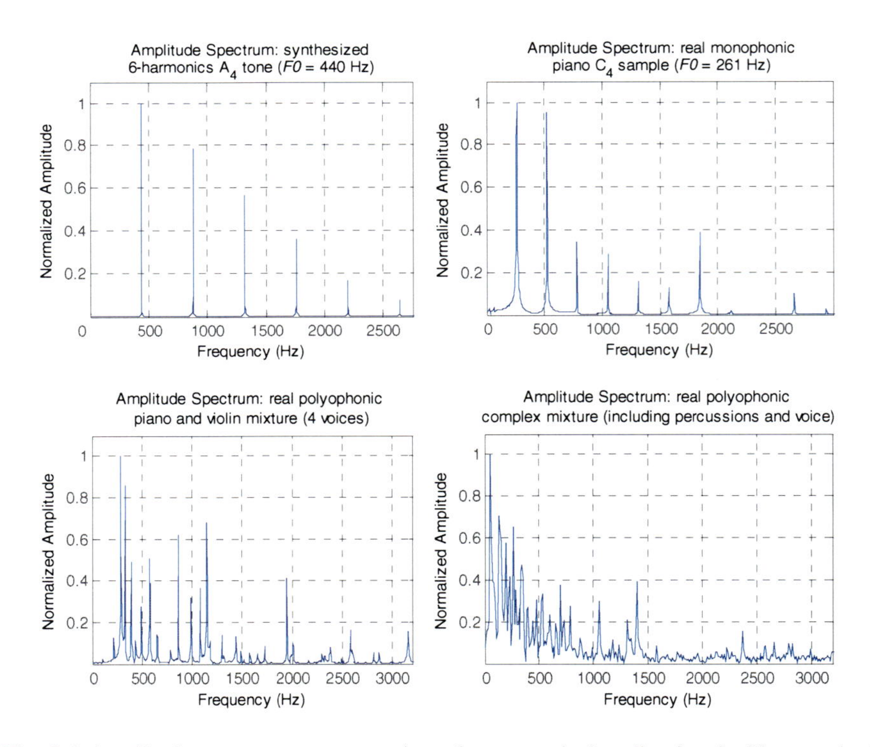

Fig. 3.1 Amplitude spectrum representation of some typical audio signals. Noteworthy is the increasing complexity of the spectral content, as the number of concurrent playing voices increases.

We have considered this, reported by Bello [2], as the most general categorization criterion for the music transcription problem, since these two approaches are non-mutually-exclusive, and ideally contain all the other fields of codification we intend to review in the following.

There are many reviews of automatic music transcription methods in literature, most of them present their own criteria, upon which the different front ends, used to obtain a useful mid-level representation of the audio input signal, are grouped together. One of the most commonly used criterion, adopted by Gerhard [20], Brossier [6] and Yeh [63], is based on a differentiation at signal analysis level:

- Time domain analysis: systems belonging to this category process the audio waveform in order to obtain information about pitches (periodicities of the audio signal) or onset times. In general, this family of methods is suitable for monophonic transcription.
- Frequency domain analysis: methods belonging to this class vary from spectral analysis (FFT, cepstrum, multi-resolution filtering, Wavelet transform and related variants) to auditory models developed first in the 90s within the Computational Auditory Scene Analysis (CASA) framework [58], [16], [43], as well as many spectral matching or spectral features extraction techniques.

Another classification concept is reported by Yeh [63], for whom music transcription methods can be catalogued into two different approaches:

- Iterative estimation: such principle refers to all the methods which iteratively estimate predominant *F0*, and subsequently cancel the residual harmonic pattern of estimated notes from the observed spectrum, processing the residual until a stop criterion is met; usually, a condition related to residual energy is adopted. The block diagram of this architecture is shown in Figure 3.2.

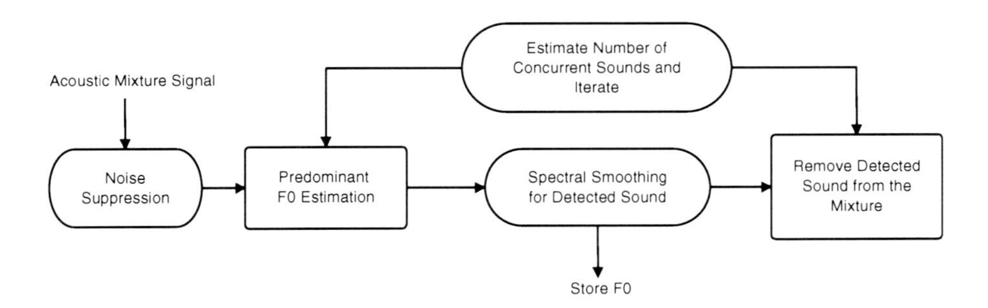

Fig. 3.2 Iterative *F0* estimation and harmonic cancellation architecture, according to the system proposed by Klapuri [32].

- Joint estimation: under this approach we find algorithms that jointly evaluate many hypotheses on *F0* estimation, without involving any cancellation. These solutions include the use of salience functions or other knowledge source, in order to facilitate spectral peak-picking, and other frameworks like Martin's Blackboard architecture [41]. This framework is a problem-solving model, which integrates knowledge from different sources and allows the interaction of different parts of the model. An expert musical knowledge, integrated with signal processing and other physical, engineering or mathematical frameworks, is considered useful to accomplish the task of automatic music transcription. Another sub-group belonging to the Joint Estimation category is the spectral matching by parametric/non-parametric models, like Non-negative Matrix Approaches (NMA) including Non-negative Matrix Factorization (NMF), frequently used in recent literature [62], [11], [61].

Another categorization regards statistical versus non statistical framework. The statistical-inference approach generally aims at jointly performing *F0* estimation and tracking of temporal parameters (onsets and durations) from a time-frequency representation of the input signal. In these models, the quantities to be inferred are considered as a set of hidden variables. The probabilistic model relates these variables to the observation variable sequence (the input signal or a mid-level representation) by using a set of properly defined parameters. Statistical frameworks frequently used for automatic music transcription are Bayesian networks [30], [8] or Hidden Markov Models (HMM) [56], [9].

Finally, another pivotal aspect is the evaluation of the transcription systems proposed so far.

3.1.1 State of the Art

In literature, a large variety of methods for both monophonic and polyphonic music transcription has been realized. Some of these methods were based on time-domain techniques like Zero Crossing Rate [44], or on autocorrelation function (ACF) in the time-domain [53], as well as parallel processing [23] or Linear Predictive Coding (LPC) analysis [38]. First attempts of performing polyphonic music transcription started in the late 1970s, with the pioneering work of Moorer [45] and Piszczalski and Galler [49]. As time went by, the commonly-used frequency representation of audio signals as a front-end for transcription systems has been developed in many different ways, and several techniques have been proposed. Klapuri [32], [35] performed an iterative predominant *F*0 estimation and a subsequent cancelation of each harmonic pattern from the spectrum; Nawab [46] used an iterative pattern matching algorithm upon a constant-Q spectral representation. In the early 1990s, other approaches, based on applied psycho-acoustic models and also known as *Computational Auditory Scene Analysis (CASA)*, from the work by Bregman [5], began to be developed. This framework was focused on the idea of formulating a computational model of the human inner ear system, which is known to work as a frequency-selective bank of passband filters; techniques based on this model, formalized by Slaney and Lyon [58], were proposed by Ellis [16], Meddis and O'Mard [43], Tolonen and Karjalainen [60] and Klapuri [36]. Marolt [39], [40] used the output of adaptive oscillators as a training set for a bank of neural networks to track partials of piano recordings. A systematic and collaborative organization of different approaches to the music transcription problem is the mainstay of the idea expressed in the *Blackboard Architecture* proposed by Martin [41]. More recently, physical [4] and musicological models, like average harmonic structure (AHS) extraction in [13], as well as other *a priori* knowledge [27], and possibly temporal information [3] have been joined to the audio signal analysis in the frequency-domain to improve transcription systems performances. Other frameworks rely on statistical inference, like hidden Markov models [55], [56], [9], Bayesian networks [30], [8] or Bayesian models [22], [14]. Others systems were proposed, aiming at estimating the bass line [57], or the melody and bass lines in musical audio signals [24], [25]. Currently, the approach based on non-negative matrix approximation [54], in different versions like nonnegative matrix factorization of spectral features [59], [62], [11], [61], has received much attention within the music transcription community.

3.2 Methods Overview and Comparison

In this section, a comparative review of some of the most important and cited music transcription systems is proposed. This review is not meant as an exhaustive and omni-comprehensive work, although it covers a large part of the literature, starting from the first pioneering methods, realized at the end of the 70s, until nowadays. In Figure 3.3, a functional block diagram related to the general architecture of an automatic music transcription system, is shown.

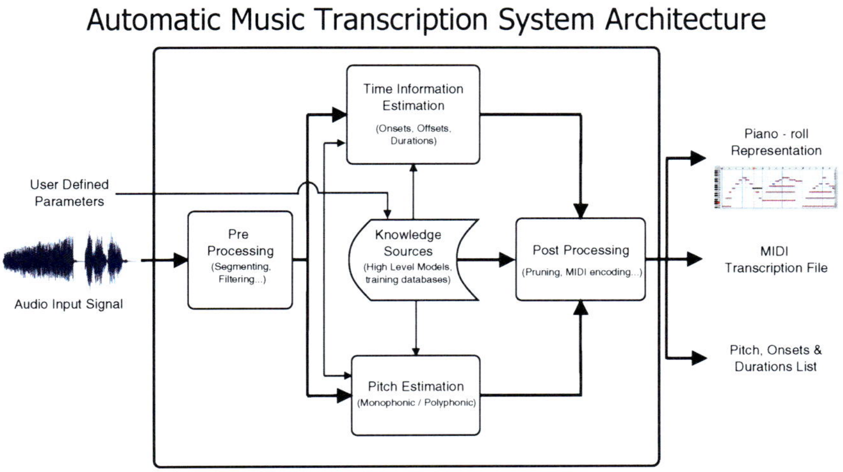

Fig. 3.3 General architecture of an automatic music transcription system.

A **Pre-Processing** module is generally assigned to segment the input signal into frames, and to compute the mid-level representation (spectral analysis, auditory model based representation etc...). The retrieval of pitch information and note temporal parameters is usually performed by dedicated modules, referred to as **Pitch Estimation** and **Time Information Estimation** in Figure 3.3. To achieve better transcription accuracies, additional **Knowledge Sources** (harmonic/instrumental models, training databases etc...) are often implemented in transcription systems, for many different purposes. Finally, a **Post-Processing** module groups all the detected note information and converts it into an appropriate output format (MIDI file, piano-roll or note parameters list).

In the following, a multi-field classification is proposed through the use of a set of parameters which can be helpful to highlight the main characteristics and peculiarities of different algorithms, without forcing a strict categorization, not even focusing on specific parts of the processing framework. The comparison summary is reported in Table 3.2. They are defined as follows:

- **Reference:** this field contains the reference to the authors of each system. In past years, longer-term projects have been undertaken by Stanford university (Centre for Computer Research in Music and Acoustics, CCRMA in the Table 3.2), University of Michigan (U-M), University of Tokyo (UT), National Institute of Advanced Industrial Science and Technology (AIST), Massachusetts Institute of Technology (MIT), Queen Mary University of London (QMUL), University of Cambridge (CAM), Tampere/Helsinki University of Technology (TUT/HUT), and by the Institut de Recherche et Coordination Acoustique/ Musique (IRCAM) of Paris, France. Other names and abbreviations, refer either to the name of the research projects, or to the commercial development of such systems (e.g., KANSEI, SONIC, YIN).
- **Year:** the year of publication of the referenced papers.
- **System Input / Output:** this field contains specifications, if they exist, on the input audio file, and it reports also the output format of the transcription process, whether described in the referenced papers.
- **Pre-Processing and Mid-Level:** a list of the signal processing techniques, used to obtain a useful front end.
- **Real time / Offline:** this field specifies whether the system operates in real time or not.
- **Source Availability:** this specifies if the source code is available, directly or web-linked.
- **Mono / Poly:** this field shows if the system is mainly dedicated to monophonic or polyphonic transcription.
- **Time / Frequency:** indicates if the signal processing techniques used by the algorithm (which are listed in the *Pre-Processing and Mid-Level* categories described above) operates either in the time or in the frequency domain.

- **Pitch Estimation Knowledge:** a brief description about the approaches and the knowledge used to extract pitch information.
- **Rhythm Info Extraction:** in this field the techniques used to retrieve temporal information of estimated F0s (where this task is performed) are summarized. It is divided into two sub-fields: **Onsets** and **Durations**, as they are often estimated with different strategies.
- **Evaluation Material:** this section shortly reports, where described, the type of the dataset used for evaluation and the number of test files / samples. Evaluation results are omitted. MIREX results are reported, for all those algorithms which participated in the past editions. The transcription output (MIDI file or piano-roll usually) is compared with a reference ground truth of the audio source data. For the evaluation of music transcription algorithms, MIREX tasks are defined as follows for Multiple F0 Estimation on a frame by frame basis. Performance measures are defined for this task: *Precision*, which is the portion of correct retrieved pitches for all the pitches retrieved for each frame, *Recall*: it is the ratio of correct pitches to all the ground truth pitches for each frame, *Accuracy*: it is an overall measure of the transcription system performance, and the classical *F-measure* to assess the balance between false positives and false negatives.
- **Additional Notes:** under this entry, any further noteworthy information, which cannot be classified according to the defined categories, is recalled.

When the value of a certain parameter is missing, or information about one of the defined fields is not available in the referenced paper, the abbreviation N.A. is used in Table 3.2. In Table 3.1, other acronyms used in Table 3.2 are defined.

Table 3.1 Definition of acronyms used in Table 3.2.

Acc	Accuracy	IHC	Inner Hair Cell
ACF	Autocorrelation Function	IIR	Infinite Impulse Response filter
AHS	Average Harmonic Structure	MCMC	Markov Chain Monte Carlo
DFT	Discrete Fourier Transform	MF0E	Multiple F0 Estimation MIREX task
F0	Fundamental Frequency	NN	Neural Network
FFT	Fast Fourier Transform	NT	Note Tracking MIREX task
FIR	Finite Impulse Response filter	PCM	Pulse Code Modulation
fs	Sampling Frequency	RWC	Real World Computing database
HMM	Hidden Markov Models	STFT	Short Time Fourier Transform
HTC	Harmonic Temporal Clustering	SVM	Support Vector Machine
HWR	Half Wave Rectification		

Table 3.2 Comparison of Automatic Music Transcription Systems.

Reference (Group)	Year		System Input/Output	Pre-Processing & Mid-Level	Real time	Source	Mono	Time	Pitch Estimation Knowledge	Time Info		Addiotional Notes	Evaluation Material
					Offline	Avail.	Poly	Freq.		Onest	Duration		
(Moorer 1977) (CCRMA)	1975 1977	I	N.A.	Optimum comb filter bank	N.A.	No	Poly	F	Time periodicity research by detecting sinusoidal components	No	No	Max # voices: 2 Limited freq. range F0s ratio can't be an integer number	Synthesized violin, real guitar duets
		O	N.A.	Short - time FFT									
(Piszczalski Galler 1977) (U-M)	1979	I	N.A.	Spectral equalization to enhance partials	Offline	No	Poly	F	Evaluation of harmonic relations among spectral peaks	No	No	Robust with missing F0 and inharmonic partials	Synth. and real signals (carillon bells)
		O	F0s and amplitudes list										
(Friedman 1979)	1979	I	Speech, fs=10 kHz	Band-Pass FIR Lo-Freq emphasis	Suitable for R.T.		Poly	T	Zero-crossing rate on processed waveform	No	No	Pitch estimation for speech	3-second speech sample
		O	24 ms pitch frames										
(Chafe and Jaffe 1986) (CCRMA)	1986	I	Digital recordings	Bounded Q Frequency Transform	N.A.	No	Poly	F	Grouping partials in sinusoidal analysis	Changes in spectral energy over time	No	Knowledge of source acoustic applied	N.A.
		O	Hihg-level MIDI										
(Katayose and Inokuchi 1989) (KANSEI)	1989	I	N.A.	Time - frequency map obtained with interpolation using complex spectra.	N.A.	No	Poly	F	Peaks extraction in the frequency domain and matching procedure	No	No	Heuristic rules implemented to group detected frequency peaks into notes.	Test developed for piano, guitar and shamisen. Results are not reported.
		O	N.A.										
(Slaney and Lyon 1990)	1990	I	N.A.	Correlogram: cochlear model, 2nd order filter bank, HWR	N.A.	Partial-ly	Poly	F	Periodicities research in the correlogram by use of the autocorrelation function	No	No	—	Acoustical Society of America Database Qulitative results
		O	Time-frequency pitch representation										
(Maher 1990)	1990	I	Digital audio signals < 20 s	Short-time FFT (512 and 1024 samples)	Offline	No	Poly	F	McAulay-Quatieri & two-way mismatch	No	No	Limited to duets, non ovrlapping frequency ranges; nearly harmonic sounds	Synthetic and real signals (basson/clarinet, trumpet/tuba) Qualitative results
		O	Chains of peaks for partials tracking	Hi-frequency pre-emphasys					Strategies to resolve colliding sinusoidal partials				

Table 3.2 *(continued)*

Reference (Group)	Year	System Input/Output	Pre-Processing & Mid-Level	Real time Offline	Source Avail.	Mono Poly	Time Freq.	Pitch Estimation Knowledge	Time Info Onsets	Time Info Duration	Addiotional Notes	Evaluation Material
(Kashnio and Tanaka 1992) (UT)	1992	I Monaural signals O Multi-ch. MIDI	A/D conversion and spectrogram representation	N.A.	No	Poly	F	Peaks - peaking in STFT (segregation); statistic rules for partials grouping (integration)	No	No	Timbre models to detect different instruments	Synthesized Vivaldi Concerto (op. 3, no. 6)
(Kashnio and Tanaka 1993) (UT)	1993	I Monaural signals (48 kHz / 16 bits) O Multi-ch. MIDI	Frequency analysis: band-pass 2nd order IIR filters	N.A.	No	Poly	F	Frequency content extraction by pinching planes thresholds and bottom-up clustering	No	No	Automatic timbre modelling based on perceptual rules	Synthesized chords. Good recognition up to 3 voices
(Hawley93) (MIT)	1993	I N.A. O N.A.	Short-time spectral analysis	N.A.	No	Poly	F	Spectral comb filtering for note identification	Hi-freq. content; bilinear time-domain filtering	No	—	Bach piano excerpts Non extensive tests
(Kashino et al. 1995) (UT)	1995	I Monaural signals O Multi-ch. MIDI	STFT	N.A.	No	Poly	F	Frequency content extraction by pinching planes and Bayesian networks	No	No	Many knowledge sources applied (timbre, chord type)	2-3 voices synth. MIDI chords with real samples
(Martin 1996a) (MIT)	1996	I CD quality audio input O Transcription in counterpoint style	STFT Blackboard architecture front-end	Offline	No	Poly	F	Knowledge-based source (KS) applied to sinusoidal track extraction	Peaks picking on squared and low-pass filtered signal energy	No	—	Bach Chorales Many octave errors Good recognition in B2 - A4 interval
(Martin 1996b) (MIT)	1996	I N.A. O MIDI, symbolic score or piano-roll	Log-lag correlogram (Auditory model of pitch perception)	N.A.	No	Poly	F	Periodicities research in the correlogram; Knowledge source applied (Blackboard framework)	Energy maxima signal envelope	No	Advantages of auditory models in detecting octave intervals	Monophonic & polyphonic tests on Bach piano pieces. Qualitative results
(Fernández-Cid and Casajús-Quirós 1998)	1998	I N.A. O MIDI file	Multi scale sinusoidal model; Constant-Q filter bank	Not true real time	No	Poly	F	Prominent harmonic pattern search in synth. spectrum (ampltudes of peaks are set after a *quality-of-fit* measure)	No	No	Source models. Masking effect test Post processing to kill short notes.	Not specified dataset. High error rate for typical musical signals are revealed

Table 3.2 (*continued*)

Reference (Group)	Year	System Input/Output	Pre-Processing & Mid-Level	Real time Offline	Source Avail.	Mono Poly	Time Freq.	Pitch Estimation Knowledge	Time Info Onsets	Time Info Duration	Addiotional Notes	Evaluation Material
(Tolonen and Karjalainen 2000) (HUT)	2000	I N.A. O N.A.	Pre-whitening filter 2 channels - filter bank (Hp and LP, crossover @ 1 kHz)	Real time	No	Poly	ACF Transf. domain	Periodicity estimation by *summary auto-correlation* function on both channels	No	No	F0 estimation examples available on Web	2-4 Clarinet tones mixed to form various chords
(Goto 2000) (Goto 2004) (AIST)	2000 2004	I 16-bit PCM signal fs = 16 kHz O N.A.	STFT obtained with a multi-rate filter bank Band-pass filter; two freq. regions spectrum	Real time	No	Poly	F	Frequency-to-instantaneous freq. mapping Maximize probability function for F0 candidates	No	No	Melody and bass line detection from real-world audio signals	10 excerpts from commercial CD recordings.
(Marolt 2001) (SONIC)	2001	I PCM, fs=44.1 kHz O MIDI file	Auditory model, 200 *gammatone* filters	N.A.	Yes	Poly	F	Network of adaptative oscillators	Multy-layer perceptron	Observation of NN activity	Piano transcription Note range: A1-C8	120 synthesized piano pieces
(Cheveigné and Kawahara 2002) (YIN)	2002	I N.A. O N.A.	Autocorrelation (ACF) Method	Suitable for real time	No	Mono	F	Difference function (similar to autocorrelation) computed via FFT, to find singal periodicities	No	No	Cumulative mean function and parabolic interpolation to reduce sub-harmonic errors	Speech databases Informal evaluation on music
(Raphael 2002)	2002	I N.A. O MIDI file	Fourier analysis of input audio signal	N.A.	No	Poly	F	Generative HMM (Baum-Welch training)	"Burstiness" measures for attack, steady & silence		Method for piano music transcription	Mozart's piano sonata 18, K570
(Godsill and Davy 2003) (CAM)	2003	I PCM, fs=22.05 kHz O Frame by frame F0 list	Sinusoidal model differentiated for mono/poly	N.A.	No	Mono	T	Bayesian Models; MCMC harmonic inference	Frame by frame F0 tracking		Audio examples on the Web	Solo flute extract (Debussy's *Syrinx*)
(Klapuri 2003) (TUT)	2003	I 16-bit PCM signal fs = 44.1 kHz O Frame by frame F0	DFT on Hamming windowed signal frames Signal/Noise model	N.A.	Algorithm only	Poly	F	Analysis of harmonic relationships between partials on 18 overlapping bands	No	No	Iterative estimation and harmonic pattern cancellation	Mixed samples: McGill, Iowa and IRCAM database
(Bruno and Nesi 2005)	2005	I PCM, 44.1 kHz (mono or multi-ch.) O List of note features	Patterson-Meddis auditory model	N.A.	No	Both	F	Neural Network tracking of pitches detected by the onset detection algorithm	Peak-Picking on signal envelope	Offset detected by Neural Networks	Different instrument models are used. Training mode available	Piano, guitar and violin samples Bach chorales

Table 3.2 *(continued)*

Reference (Group)	Year	System Input/Output	Pre-Processing & Mid-Level	Real tim Offline	Source Availabl	Mon Poly	Tim Freq	Pitch Estimation Knowledge	Time Info Onsets	Time Info Durations	Addiotional Notes	Evaluation Material
(Ryynänen Klapuri 2005) (TUT)	2005	**I** PCM stereo fs = 44.1 kHz **O** MIDI file	70 Channels band pass filter; HWR and STFT for all bands	N.A.	No	Poly	F	Comb filters bank estimates periodicity in frequency domain	Positive changes in F0s energy	Notes tracking by HMM	MIREX 2007-08: Acc.≈61% (MF0E) F-measure≈34% (NT)	Excerpts from RWC Database
(Bello et al. 2006) (QMUL)	2006	**I** PCM files fs = 22.05 kHz **O** N.A.	STFT & smoothing; Signal modeled as the sum of an internal database piano waveforms	N.A.	No	Poly	F	F0 hypotheses for relevant amplitude partials Heuristic rules for partials grouping	Temporal parameters estimated integratin frame estimations over time		Transcription of recorded piano music. Hybrid method combining time-freq. info	Disklavier played piano MIDI files. Error rate increases with number of voices
(Cemgil and Kappen 2006)	2006	**I** N.A. **O** Piano-roll of note parameters	Sinusoidal model (state space form) to obtain a piano-roll like representation	Real time	No	Poly	T	Bayesian networks, switching Kalman filters, generative model to estimate note parameters.	Onsets and offsets detected by transitions (*mute-sound* states) of the generators		The approach used allows to remove the frame by frame assumption for audio analysis	Own recordings of 2-3 voices chords. Qualitative results, many offset errors
(Poliner and Ellis 2007)	2007	**I** 8 kHz audio (sampled from MIDI) **O** N.A.	STFT	N.A.	No	Poly	F	87 One-versus-all SVM classifier for piano notes trained with *Sequential Minimal Optimization*	Note tracking by two state (on/off) HMM	No	Method for piano music transcription; evaluation tests are available	Synthesized, recorded and Disklavier played MIDI files
(Kameoka et al. 2007) (UT)	2007	**I** PCM files fs = 44.1 kHz **O** F0s, onsest & offsets list	Multi-resolution power spectrum obtained via Gabor wavelet transform	Offline	No	Poly	F	Harmonic temporal clustering (HTC) model for source separation	Joint estimation by using HTC model		Evaluated in MIREX 2007, 2008 and 2009: Acc. ≈ 49% (MF0E) F-measure ≈ 32% (NT)	Excerpts from RWC-Classical and RWC-Jazz databases
(Klapuri 2008) (TUT)	2008	**I** N.A. **O** N.A.	Auditory (*gammatone*) filter bank (2 types of 2nd order IIR resonators); IHC model, HWR	Real time	No	Poly	F	Periodicities search in the *Summary spectrum*. Detect F0 as peaks of a salience function	Energy peak detection on signal envelope	No	Iterative estimation and harmonic pattern cancellation from the summary spectrum	Mixed samples from McGill, Iowa and IRCAM database

Table 3.2 (*continued*)

Reference (Group)	Year	System Input/Output	Pre-Processing & Mid-Level	Real time Offline	Source Avail.	Mono Poly	Time Freq.	Pitch Estimation Knowledge	Time Info Onsets	Time Info Duration	Addiotional Notes	Evaluation Material
(Duan et al. 2008)	2008	**I** PCM audio signals **O** Piano-roll (F0 tracking)	STFT	N.A.	No	Poly	F	Maximum Likelihood F0 estim. Average Harmonic Structure (AHS) extraction for source separation	No	No	The system performs bad F0 recognition for inharmonic sounds	Synth., real instruments, singing voice Non standard metrics used for evaluation
(Pertusa and Iñesta 2008)	2008	**I** PCM mono signals fs = 44.1 kHz **O** Sequence of MIDI notes	STFT Gaussian smoothing of spectral patterns	N.A.	No	Poly	F	Candidate F0s with best salience function, calculated from partial amplitude and spectral smoothness	No	No	Evaluated in MIREX 2007 and 2008; Acc.≈62% (08 MF0E) F-measure≈25% (08 NT)	4000 chords; random mixtures of various samples (1 to 4 voices polyphony)
(Vincent et al. 2008)	2008	**I** N.A. **O** N.A.	ERB - scale time / frequency representation (similar to STFT)	Offline	No	Poly	F	NMF with harmonic/inharmonic constraints on the basis spectra with fixed/adaptative tuning	Thresholding amplitude sequence of detected pitches	Same as for onsets	Evaluated in MIREX 2007 - 2008 Acc.≈54% (08 MF0E) F-measure≈20% (08 NT)	43 Disklavier 30 seconds excerpts
(Chang et al. 2008) (IRCAM)	2008	**I** N.A. **O** N.A.	Spectral analysis based on sinusoidal and noise model	N.A.	No	Poly	F	Spectral matching, spectral smoothing and synchronous amplitude evolution of single sources	F0 tracking using a high-order HMM model with two states: attack and sustain		Evaluated in MIREX 2007 - 2008- 2009; Acc.≈69% (09 MF0E) F-measure≈36% (08 NT)	Samples from McGill, Iowa and IRCAM database
(Duan et al. 2009)	2009	**I** PCM audio signals fs = 44.1 kHz **O** N.A.	Spectral analysis Spectrum divided into *peaks* and non-peaks regions	N.A.	No	Poly	F	Maximum Likelihood parameter estimation in the freq. domain, using also neighborframes estimates	Build pitch trajectories by constraint clustering problem with 2 classes: *must-link* /*cannot-link*		Evaluated in MIREX 2009 Acc.≈57% (MF0E) F-measure≈22% (NT)	10 real music performances (4-parts Bach's chorales)
(Argenti et al. 2009)	2009	**I** PCM mono/stereo 16 bits, fs=44.1 kHz **O** MIDI file; pitches, onsets & offsets list	Joint constant-Q and Bispectral (higher order spectra) analysis	Offline	No	Poly	F	Iterative 2D harmonic pattern matching (in the bispectrum domain) and pattern cancellation	Peaks-picking in spectral Kullback-Leibler divergence	Tracking of note events on STFT	MIREX 2009, 1st ranked in piano NT task Acc.≈48% (MF0E) F-measure≈23% (NT)	Excerpts from RWC Classical datbase;

3.3 Review of Some Music Transcription Systems

Goto (2000 and 2004): It was one of the first who proposed a transcription system (*PreFEst*, from "Predominant F0 Estimation") for real-world audio signals [24], [25], characterized by complex polyphony, presence of drum or percussion, and singing voice also. To achieve such a goal, the music scene description and the signal analysis are carried out at a more specific level, focusing on the transcription of the melody and the bass line in musical fragments. The front end extracts instantaneous frequency components by using a STFT multi-rate filter bank, thus limiting the frequency regions of the spectrum with two band-pass filters. A probability density function is then assigned to each filtered frequency component; this function is a weighted combination of different harmonic-structure tone models. An Expectation-Maximization (EM) algorithm then estimates the model parameters. The frequency value that maximizes the probability function is detected as a predominant F0. Finally, a multi-agent architecture is used to sequentially track F0 peak trajectories, and to select the most stable ones; this operation is carried out by a salience detection and a dynamic thresholding procedures.

Ryynänen and Klapuri (2005): This system [56] uses a probabilistic framework, a hidden Markov Model (HMM), to track note events. The multiple F0 estimator front end is based on auditory model: a 70-channel bandpass filter bank splits the audio input into sub-band signals which are later compressed, half-wave rectified and low-pass filtered with a frequency response close to $1/f$. Short time Fourier Transform is then performed across the channels, and the obtained magnitude spectra are summed together into a summary spectrum. Predominant F0 estimation, and cancelation from the spectrum of the harmonic set of detected F0 is performed iteratively. Onset detection is also performed by observing positive energy variation in the amplitude of detected F0 values. The output of F0 estimator is further processed by a set of three probabilistic models: a HMM note event model tracks the likelihood for each single detected note; a silence model detects temporal intervals where no notes are played; finally, a musicological model controls the transitions between note event and silence models.

Bruno and Nesi (2005): The proposed system [7] processes the input audio signal through a Patterson-Meddis auditory model. A partial tracking module extracts the harmonic content, which is analyzed to estimate active pitches. Onset detection is performed by using a peak-picking algorithm on the signal envelope. Pitch tracking is carried on, for each note by a bank of neural networks. This network can be trained by a set of parameters describing several instrument models (concerning partial amplitude weights, frequency range etc.).

Vincent, Bertin and Badeau (2008): They have proposed a system based on Non-negative Matrix Factorization (NMF) [61]. By using this technique, the observed signal spectrogram (Y) is decomposed into a weighted sum of basis spectra (contained in H) scaled by a matrix of weighting coefficients (W): $Y = WH$.

Since the elements of Y are non-negative by nature, the NMF method approximates it as a product of two non-negative matrixes, W and H.

The system uses a family of constrained NMF models, where each basis spectrum is a sum of narrow-band spectrum containing partials at harmonic or inharmonic frequencies. This ensures that the estimated basis spectra are pitched at known fundamental frequencies; such a condition is not always guaranteed if standard NMF models are applied without any of these constraints. The input signal is first pre-processed to obtain a representation similar to the Short-time Fourier Transform, by performing an ERB-scale representation. Then, the parameters of the models are adapted by minimizing the residual loudness after applying the NMF model. Pitches, onsets and offsets of detected notes are transcribed by simply thresholding the amplitude sequence. The system has been evaluated in the MIREX 2007 framework: the two submitted versions reached average accuracies of 46.6% and 54.3% in the task 1 (multi-F0 estimation over 10 ms frames) and an average F-measure of 45.3% and 52.7% in the task 2 (note tracking).

Chang, Su, Yeh, Roebel and Rodet (2008): In method [9], instantaneous spectra are obtained by FFT analysis. A noise level estimation algorithm is applied to enhance the peaks generated by sinusoidal components (produced by an unknown number of audio sources) with respect to noise peaks. Subsequently, a matching between a set of hypothetical sources and the observed spectral peaks is made, by using a score function based on the following three assumptions: *spectral match with low inharmonicity*, *spectral smoothness* and *synchronous amplitude evolution*. These features are based on physical characteristics generally showed by the partials generated by a single source. Musical notes tracking is carried out by applying a high order hidden Markov model (HMM) having two states: *attack* and *sustain*. This is a probabilistic framework aimed at describing notes evolution as a sequence of states evolving on a frame by frame basis. The goal is to estimate optimal note paths and the length of each note trajectory. Finally, the source streams are obtained by pruning the candidate trajectories, in order to maximize the likelihood of the observed polyphony. The system has been evaluated within the MIREX 2007 framework, and improved versions were submitted to MIREX 2008 and MIREX 2009 contests. Best multiple F0 estimation accuracy of 69% has been achieved in 2009 running (1st ranked in task 1): this is currently the highest accuracy reached in all the MIREX editions for the first task. Best performance in the note tracking task was reached in 2008 edition, with an F-measure of 35.5% (1st ranked).

Argenti, Nesi and Pantaleo (2009): This transcription method [1] has been proposed by the authors of the present chapter. It has an original front-end: a constant-Q bispectral analysis is actually applied to the input signal. The bispectrum belongs to the class of higher-order spectra (HOS), or polyspectra. They are defined as the Fourier Transform of corresponding order cumulants, which are strictly related to statistical moments. The bispectrum, in particular, is also known as the third-order spectrum: it is a bivariate frequency function, $B(f_1, f_2)$, capable of detecting nonlinear activities like phase or frequency coupling, for example amongst the partials of a sound, or a mixture of sounds. Pitch estimation is performed by harmonic pattern matching procedure in the bispectrum domain. In the spectrum domain, a monophonic musical signal is described as a comb-pattern of amplitude peaks, located at integer multiple values of the fundamental frequency. In the bispectrum domain, a monophonic sound composed of T partials generates a 2D pattern characterized by peaks positions. Two sounds presenting some colliding partials generate spectral overlapping patterns; this is a well known problematic situation that leads to detection errors in a pattern matching/correlation based method; besides, in an iterative pitch estimation and subtraction algorithm. The geometry of bispectral 2D pattern is useful in preserving information about overlapping partials. This is demonstrated by evaluation results, made on excerpts from the RWC database: a comparison between a spectral based and a bispectral based transcription system (both performing an iterative F0 estimation and harmonic pattern extraction procedure) shows that the latter outperforms the former, with average F-measures of 72.1% and 57.8%, respectively. Onset detection are estimated using the Kullback-Leibler divergence, thus highlighting energy variations which are expected to be found at onset times. Note durations are estimated by thresholding the spectrogram envelope. The system has been evaluated in the MIREX 2009 framework: it has reached a 48.8% frame by frame F0 estimation accuracy (task 1); it has been 3rd ranked in the mixed set note tracking (task 2a, with an F-measure of 22.7%), and 1st ranked in the piano-only tracking note task (task 2b).

3.4 Discussion and Conclusions

From this review work some general aspects, concerning automatic music transcription systems can be gathered. Automatic transcription of polyphonic music is one of the most challenging task in the MIR research field; in fact, this is to be considered as a conjunction of several tasks, which can be accomplished jointly or by using dedicated procedures. From this point of view, a modular architecture seems to be the most robust approach for a problem solution. Such a construct perfectly matches with Martin's idea of blackboard architecture [41].

While the human perceptual approach to music has been successfully studied and implemented through the *Computational Auditory Scene Analysis* (*CASA*), it is more difficult to code knowledge at higher levels of into a computational framework, since it must be consistent with experience, and it often needs training to avoid misleading or ambiguous decisions. Such knowledge is commonly represented by all those models which aim at reproducing human capabilities in features extraction and grouping (e.g., harmony related models, musical key finding etc…). The experience of a well-trained musician can be understood as a greatly flexible and deep network of state-machine like hints, as well as complex matching procedures.

Review of music transcription systems in literature suggests that time-frequency representation (usually performed through short-time Fourier transform) of the signal is the most used front end, upon which pitch estimation and onset/offset detection strategies can be applied. Multi resolution spectrogram representation (obtained by using constant-Q or wavelet transform) seems to be, in our opinion, the most suitable, since it fits properly the exponential spacing of note frequencies, and it also reduces computational load to achieve the desired time/frequency resolution. Auditory model based front ends have been largely studied and applied in the 90s; however, the interest toward this approach has decreased. Time domain techniques are becoming more and more infrequent, since they have provided poor performances in polyphonic contexts.

With regards to pitch estimation strategies, the largely adopted class of spectral content peak-picking based algorithms has revealed to be not sufficient to achieve satisfactory transcription accuracies. Actually, amplitude thresholding in the spectrum domain, as well as simple harmonic pattern matching, leads to frequent false positive detection, if no other knowledge is applied. A large variety of models has been proposed to spectral analysis, and it is not easy to find out which is the best approach. The most used techniques in recent literature are: Nonnegative Matrix Factorization [59], [62], [61], Hidden Markov Models [55], [56], [9], Bayesian models [30], [21], [22], [14], generative harmonic models [8], and the use of jointed frequency and time information.

Onset detection is often devolved upon detecting rapid spectral energy over time. Techniques such as the phase-vocoder based functions, applied to audio spectrogram, seem to be more robust with respect to peak-picking algorithms performed upon the signal envelope. Offset detection is still considered as of less perceptual importance. Statistical frameworks offer an interesting perspective in solving discontinuities in joint time-pitch information, typically yielded by lower processing levels techniques. On the contrary, other devices that usually reach a deep level of specialization, like neural networks, are more suitable for particular areas or subsets of automatic transcription; actually this kind of tool is often trained at recognizing specific notes or at inferring particular instrumental models [39].

References

[1] Argenti, F., Nesi, P., Pantaleo, G.: Automatic Transcription of Polyphonic Music Based on Constant-Q Bispectral Analysis for MIREX 2009. In: Proc. of 10th ISMIR Conference (2009)

[2] Bello, J.P.: Towards the Automated Analysis of Simple Polyphonic Music: A Knowledge-based Approach. PhD Thesis (2003)

[3] Bello, J.P., Daudet, L., Sandler, M.B.: Automatic piano transcription using frequency and time-domain information. IEEE Transactions on Audio, Speech, and Language Processing 14(6), 2242–2251 (2006)

[4] Ortiz-Berenguer, L.I., Casajús-Quirós, F.J., Torres-Guijarro, S.: Multiple piano note identification using a spectral matching method with derived patterns. Journal of Audio Engineering Society 53(1/2), 32–43 (2005)

[5] Bregman, A.: Auditory Scene Analysis.The MIT Press, Cambridge (1990)

[6] Brossier, P.M.: Automatic Annotation of Musical Audio for Interactive Applications. PhD Thesis, Centre for Digital Music Queen Mary, University of London (2006)

[7] Bruno, I., Nesi, P.: Automatic Music Transcription Supporting Different Instruments. Journal of New Music Research 34(2), 139–149 (2005)

[8] Cemgil, A.T., Kappen, H.J., Barber, D.: A Generative Model for Music Transcription. IEEE Transaction on Audio, Speech and Language Processing 14(2), 679–694 (2006)
[9] Chang, W.C., Su, A.W.Y., Yeh, C., Roebel, A., Rodet, X.: Multiple F0 Tracking Based on a High Order HMM Model. In: Proc. of the 11th Int. Conference on Digital Audio Effects, DAFx 2008 (2008)
[10] Chafe, C., Jaffe, D.: Source separation and note identification in polyphonic music. In: Proc. of IEEE Int. Conf. on Acoustics, Speech and Signal Processing (ICASSP 1986), vol. 11, pp. 1289–1292 (1986)
[11] Cont, A., Shlomo, D., Wessel, D.: Realtime multiple-pitch and multiple-instrument for music signals using sparse non-negative constraints. In: Proc. of 10th Int. Conference of Digital Audio Effects, DAFx 2007 (2007)
[12] De Cheveigné, A., Kawahara, H.: YIN, a fundamental frequency estimator for speech and music. The Journal of the Acoustical Society of America 111(4), 1917–1930 (2002)
[13] Duan, Z., Zhang, Y., Zhang, C., Shi, Z.: Unsupervised single-channel music source separation by average harmonic structure modeling. IEEE Transactions on Audio, Speech and Language Processing 16(4), 766–778 (2008)
[14] Dubois, C., Davy, M.: Joint detection and tracking of time-varying harmonic components: a general bayesian framework. IEEE Transactions on Audio, Speech and Language Processing 15(4), 1283–1295 (2007)
[15] Duan, Z., Han, J., Pardo, B.: Harmonically Informed Multi-pitch Tracking. In: Proc. of 10th International Society for Music Information Retrieval Conference, ISMIR 2009 (2009)
[16] Ellis, D.P.W.: Prediction-driven Computational Auditory Scene Analysis. PhD Thesis, Massachusetts Institute of Technology (1996)
[17] Fernández-Cid, P., Casajús-Quirós, F.J.: Multi-pitch estimation for Polyphonic Musical Signals. In: Proc. Of IEEE Int. Conf. on Acoustics Speech and Signal Processing (ICASSP 1998), vol. 6, pp. 3565–3568 (1998)
[18] Fletcher, N.F., Rossing, T.D.: The physics of musical instruments, 2nd edn. Springer, New York (1998)
[19] Friedman, D.H.: Multichannel Zero-Crossing-Interval Pitch Estimation. In: Proc. of IEEE Int. Conf. on Acoustics, Speech and Signal Processing (ICASSP 1979), vol. 4, pp. 764–767 (1979)
[20] Gerhard, D.: Pitch Extraction and Fundamental Frequency: History and Current Techniques. Tech. Report TR-CS 2003-06, Dep. of Computer Science, University or Regina, Canada (2003)
[21] Godsill, S.J., Davy, M.: Bayesian Harmonic Models for Musical Signal Analysis. Bayesian Statistics 7, 105–124 (2003)
[22] Godsill, S.J., Davy, M., Idier, J.: Bayesian analysis of polyphonic western tonal music. Journal of the Acoustical Society of America 119(4), 2498–2517 (2006)
[23] Gold, B., Rabiner, L.R.: Parallel Processing Techniques for Estimating Pitch Periods of Speech in the Time Domain. Journal of Acoustic Society of America 46(2), 442–448 (1969)
[24] Goto, M.: A robust predominant-f0 estimation method for real-time detection of melody and bass lines in cd recordings. In: Proc. of IEEE Int. Conf. on Acoustics, Speech and Signal Processing (ICASSP 2000), Istanbul, Turkey, vol. 2, pp. 757–760 (2000)
[25] Goto, M.: A real-time music-scene-description system: Predominant-F0 estimation for detecting melody and bass lines in real-world audio signals. Speech Communication - ISCA Journal 43(4), 311–329 (2004)

[26] Hawley, M.: Structure out of sound. Ph.D. thesis, MIT Media Laboratory, Cambridge, Massachusetts (1993)
[27] Kameoka, H., Nishimoto, T., Sagayama, S.: A multipitch analyzer based on harmonic temporal structured clustering. IEEE Transactions on Audio, Speech, and Language Processing 15(3), 982–994 (2007)
[28] Kashino, K., Tanaka, H.: A Sound Source Separation System Using Spectral Features Integrated by Dempster's Law of Combination. Annual Report of the Engineering Research Institute, vol. 52. University of Tokyo (1992)
[29] Kashino, K., Tanaka, H.: A Sound Source Separation System with the Ability of Automatic Tone Modeling. In: Proc. of International Computer Music Conference (ICMC 1993), pp. 248–255 (1993)
[30] Kashino, K., Nakadai, K., Kinoshita, T., Tanaka, H.: Application of Bayesian Probability Network to Music Scene Analysis. In: Computational Auditory Scene Analysis Workshop (IJCAI 1995), pp. 32–40 (1995)
[31] Katayose, H., Inokuchi, S.: The KANSEI Music System. Computer Music Journal 13(4), 72–77 (1989)
[32] Klapuri, A.P.: Multiple fundamental frequency estimation based on harmonicity and spectral smoothness. IEEE Transactions on Speech and Audio Processing 11(6), 804–816 (2003)
[33] Klapuri, A.P.: Signal Processing Methods for the Automatic Transcription of Music. PhD thesis, Tampere University of Technology (2004)
[34] Klapuri, A.P.: Automatic Music Transcription as We Know it Today. Journal of New Music Research 2004 33(3), 269–282 (2004)
[35] Klapuri, A.P.: A perceptually motivated multiple-f0 estimation method. In: Proc. of IEEE Workshop on Applications of Signal Processing to Audio and Acoustics, pp. 291–294 (2005)
[36] Klapuri, A.P.: Multipitch analysis of polyphonic music and speech signals using an auditory model. IEEE Transactions on Audio, Speech, and Language Processing 16(2), 255–266 (2008)
[37] Maher, R.C.: Evaluation for a Method for Separating Digitized Duet Signals. Journal of Acoustic Engineering Society 38(12), 956–979 (1990)
[38] Markel, J.D.: The SIFT Algorithm for Fundamental Frequency Estimation. IEEE Transactions on Audio and Electroacoustics 16, 367–377 (1972)
[39] Marolt, M.: SONIC: Transcription of polyphonic piano music with neural networks. In: Workshop on Current Research Directions in Computer Music, Barcelona, Spain, pp. 217–224 (2001)
[40] Marolt, M.: Networks of adaptive oscillators for partial tracking and transcription of music recordings. Journal of New Music Research 33(1), 49–59 (2004)
[41] Martin, K.D.: A blackboard system for automatic transcription of simple polyphonic music. Perceptual Computing Technical Report 385, MIT Media Lab (1996)
[42] Martin, K.D.: Automatic Transcription of Simple Polyphonic Music: Robust Front End Processing. Technical Report #399, MIT Media Lab, Perceptual Computing Section, The MIT Press (1996)
[43] Meddis, R., O'Mard, L.: A Unitary Model of Pitch Perception. The Journal of the Acoustical Society of America 102(3), 1811–1820 (1997)
[44] Miller, N.J.: Pitch detection by data reduction. IEEE Transaction on Audio, Speech and Language Processing 23(1), 72–79 (1975)
[45] Moorer, J.: On the transcription of musical sound by computer. Computer Music Journal 1(4), 32–38 (1977)

[46] Nawab, S.H., Ayyash, S.A., Wotiz, R.: Identification of musical chords using constant-q spectra. In: IEEE Proc. on Acoustic, Speech and Signal Processing (ICASSP 2001), vol. 5, pp. 3373–3376 (2001)
[47] Peretz, I., Coltheart, M.: Modularity of music processing. Nature Neuroscience 6(7), 688–691 (2003)
[48] Pertusa, A., Iñesta, J.M.: Multiple Fundamental Frequency estimation using Gaussian smoothness. In: Proc. of the IEEE Int. Conf. on Acoustics, Speech, and Signal Processing, ICASSP 2008, Las Vegas, USA, pp. 105–108 (2008)
[49] Piszczalski, M., Galler, B.: Automatic music transcription. Computer Music Journal 1(4), 24–31 (1977)
[50] Piszczalski, M., Galler, B.: Automatic music transcription. Computer Music Journal 66(3), 710–720 (1979)
[51] Poliner, G.E., Ellis, D.P.W.: A Discriminative Model for Polyphonic Piano Transcription. IEEE Transaction on Audio, Speech and Language Processing 14(4), 1247–1256 (2007)
[52] Rabiner, L.R.: On the Use of Autocorrelation Analysis for Pitch Detection. IEEE Transaction on Acoustics, Speech and Signal Processing 25(1), 24–33 (1977)
[53] Rabiner, L.R.: A Comparative Performance Study of Several Pitch Detection Algorithms. IEEE Transaction on Acoustics, Speech and Signal Processing 24(5), 399–418 (1977)
[54] Raczynksi, S., Ono, N., Sagayama, S.: Multipitch analysis with harmonic nonnegative matrix approximation. In: Proc. of the 8th International Conference on Music Information Retrieval (ISMIR 2007), pp. 381–386 (2007)
[55] Raphael, C.: Automatic transcription of piano music. In: Proc. on 3rd Int. Conf. on Music Information Retrieval, pp. 15–19 (2002)
[56] Ryynänen, M.P., Klapuri, A.P.: Polyphonic Music Transcription Using Note Event Modeling. In: Proc. of 2005 IEEE Workshop on Applications of Signal Processing to Audio and Acoustics, New Paltz, NY, October 16-19, pp. 319–322 (2005)
[57] Ryynänen, M.P., Klapuri, A.P.: Automatic bass line transcription from streaming polyphonic audio. In: Proc. of IEEE International Conference on Acoustics, Speech and Signal Processing (ICASSP 2007), vol. 4, pp. 1437–1440 (2007)
[58] Slaney, M., Lyon, R.F.: A Perceptual Pitch Detector. In: Proc. of IEEE Int. Conf. on Acoustics Speech and Signal Processing (ICASSP 1990), vol. 1, pp. 357–360 (1990)
[59] Smaragdis, P., Brown, J.C.: Non-negative matrix factorization for polyphonic music transcription. In: IEEE Workshop on Applications of Signal Processing to Audio and Acoustics, New Paltz (NY), pp. 177–180 (2003)
[60] Tolonen, T., Karjalainen, M.: A computationally efficient multipitch analysis model. IEEE Transactions on Speech and Audio Processing 8(6), 708–716 (2000)
[61] Vincent, E., Bertin, N., Badeau, R.: Harmonic and Inharmonic Nonnegative Matrix Factorization for Polyphonic Pitch Transcription. In: Proc. of IEEE Int. Conf. on Acoustics, Speech and Signal Processing (ICASSP 2008), pp. 109–112 (2008)
[62] Virtanen, T.: Monaural sound source separation by nonnegative matrix factorization with temporal continuity and sparseness criteria. In: Proc. of IEEE Int. Conf. on Computational Intelligence for Measurement Systems and Applications, vol. 15(3), pp. 1066–1074 (2007)
[63] Yeh, C.: Multiple Fundamental Frequency Estimation of Polyphonic Recordings. PhD Thesis, Université Paris VI (2008)

Chapter 4
Multimodal Analysis of Expressive Gesture in Music Performance

Antonio Camurri and Gualtiero Volpe

Abstract. This chapter focuses on systems and interfaces for multimodal analysis of expressive gesture as a key element of music performance. Research on expressive gesture became particularly relevant in recent years. Psychological studies have been a fundamental source for automatic analysis of expressive gesture since their contribution in identifying the most significant features to be analysed. A further relevant source has been research in the humanistic tradition, in particular choreography. As a major example, in his Theory of Effort, choreographer Rudolf Laban describes the most significant qualities of movement. Starting from these sources, several models, systems, and techniques for analysis of expressive gesture were developed. This chapter presents an overview of methods for the analysis, modelling, and understanding of expressive gesture in musical performance. It introduces techniques resulted from the research developed over the years by the authors: from early experiments of human-robot interaction in the context of music performance up to recent set-ups of innovative interfaces and systems for active experience of sound and music content. The chapter ends with an overview of possible future research challenges.

4.1 Introduction

This chapter presents methods and techniques for multimodal analysis of expressive gesture in music performance. The chapter adopts an historical perspective, taking as reference and as a source for examples the research developed over years

Antonio Camurri
Casa Paganini – InfoMus, DIST, University of Genova,
Viale Causa 13, 16145 Genova, Italy
e-mail: antonio.camurri@unige.it

Gualtiero Volpe
Casa Paganini – InfoMus, DIST, University of Genova,
Viale Causa 13, 16145 Genova, Italy
e-mail: gualtiero.volpe@unige.it

J. Solis and K. Ng (Eds.): Musical Robots and Interactive Multimodal Systems, STAR 74, pp. 47–66.
springerlink.com

at the formerly InfoMus Lab in Genova, Italy, nowadays the Casa Paganini – InfoMus Research Centre. Starting from some pioneering studies and interfaces for expressivity in musical robots, it reviews models, techniques and interfaces developed along the years in several research projects, and it ends with an overview of topics and challenges for future research.

Research on expressive gesture became particularly relevant in recent years (e.g. see the post-proceedings of Gesture Workshops 2003 [6], 2005 [22], 2007 [39], and 2009 [26]). Several definitions of gesture exist in the literature. The most common use of the term is with respect to natural gesture, which is defined as a support to verbal communication. McNeill [34], in his well-known taxonomy, divides the natural gestures generated during a discourse into four different categories: iconic, metaphoric, deictic, and beats. In a wider perspective Kurtenbach and Hulteen [28] define gesture as "a movement of the body that contains information". A survey and a discussion of existing definition of gesture can be found in [3]. In artistic contexts and in particular in the field of music and of performing arts, gesture is often not intended to denote things or to support speech as in the traditional framework of natural gesture, but the information it contains and conveys is related to the affective/emotional domain. Starting from this observation Camurri and colleagues [10, 12] considered gesture to be "expressive" depending on the kind of information it conveys: expressive gesture carries what Cowie et al. [18] call "implicit messages", and what Hashimoto [23] calls KANSEI. That is, expressive gesture is responsible for the communication of information called expressive content. Expressive content is different and in most cases independent from, even if often superimposed to, possible denotative meaning. Expressive content concerns aspects related to feelings, moods, affect, and intensity of emotional experience. In this framework, expressive gesture is understood in a broad perspective, which is only partially related to explicit body movement. Expressive gesture is considered as the result of a juxtaposition of several dance, music, and visual gestures, but it is not just the sum of them, since it also includes the artistic point of view of the artist who created it, and it is perceived as a whole multimodal stimulus by human spectators. In the particular case of music performance, expressive gestures include explicit non-functional gestures of performers, but also embodied experience of expressive sound and music content.

Psychology has been a fundamental source for automatic analysis of expressive gesture since it identified the most significant features for analysis [2, 19, 49]. A further relevant source has been research in the humanistic tradition, in particular choreography. As a major example, choreographer Rudolf Laban described in his Theory of Effort the most significant qualities of movement [29, 30].

4.2 Pioneering Studies

Pioneering studies on analysis of expressive gesture in complex music and dance performance scenarios date back to the Eighties. Taking as example the research carried out at InfoMus Lab (which was born in 1984), a significant pioneering work consisted of the interactive systems developed for the music theatre production *Outis* by the famous composer Luciano Berio (Teatro alla Scala, Milan, 1996):

starting from the research on interactive systems originated at InfoMus Lab in late eighties [5, 7], and from a fruitful collaboration with Nicola Bernardini and Alvise Vidolin, a real-time system was developed for on-stage gesture detection and the control of live electronics. Simplifying for sake of brevity, the request by the composer was to obtain a perfect synchronisation between the orchestra, the movement of mimes on stage, and a few sections of live electronics, including the real-time activation and processing of synthetic percussion sounds based on the movements of the mimes. The solution was to give the orchestra conductor the responsibility to conduct also the gestures of the mimes on stage. Such gestures were detected and associated to specific activations and processing events on the live electronics. Specific sensor systems were also developed (e.g. floor sensitive areas on the stage) and custom electronics to connect such sensor signals to computers placed at a long distance (e.g. electronics to transmit MIDI signals on a few hundred meters distance cables). These electronics were also provided to Nicola Bernardini and Alvise Vidolin at Salzburg Festival in 1999 to set-up a remote live electronics for a music theatre opera by the composer Adriano Guarnieri.

Another important exploitation of the research work at InfoMus Lab and, at the same time, a relevant source of inspiration for subsequent research (indeed, it originated the first consolidated public version of the software EyesWeb, about one year later) was the participation to the music theatre production *Cronaca del Luogo* again by Luciano Berio (Salszburg Festival, 1999). In this framework, InfoMus Lab developed the interaction mechanisms for one of the major actors of the piece, Nino, played by David Moss. Nino is a character with a double personality: one intelligent, crazy, and aggressive, and the other a sort of repository of knowledge, wisdom, and calmness. These two personalities are modelled by two different live-electronics contexts on his own voice, controlled by his movement. David Moss, according to his movement from one part to the other of the stage, had the possibility to change smoothly in real-time his voice from one character to the other. This was obtained with small sensors on his costume (FSRs), connected to a Wireless-Sensor-to-Midi system developed by the Lab, exploiting wireless microphone channels (modulating and demodulating MIDI signals). Further, a camera placed over the stage detected his overall movement, to recognise how he occupied the stage. This was used to change the degrees of intervention of the live electronics between the two contexts corresponding to the two characters. Therefore, the actor had the possibility to manage a sort of morphing between the different characters, creating an interesting dialogue between the two Nino personalities (see Figure 4.1). This was obtained with the first version of EyesWeb, which was refined and extended for this occasion. This experience was also a precious occasion to gather novel requirements for the subsequent version of the system.

In this artistic project too, custom hardware systems were developed in order to achieve the needed robustness and effectiveness. A big problem was the tracking from a video-camera of the movement of the main character Nino on a large stage (about 30 m by 15 m, the Festspielhouse, Salzburg Festival), with huge lighting and scenery changes, and with a high number of other actors on the stage (see Figure 4.1a). InfoMus Lab developed, thanks to the collaboration with Matteo Ricchetti, an original method to solve the problem: the helmet he was wearing (a part of his on stage costume) was endowed with a cluster of infrared LEDs flashing in synch and at half the rate of the infrared video-camera used for the motion tracking. Therefore, an algorithm was implemented in EyesWeb to detect a small blob which is present in one frame and not in the next one, and again present in the next one, and so on. This was an original and very effective technique to obtain the position of Nino on stage in the case of such a noisy environment.

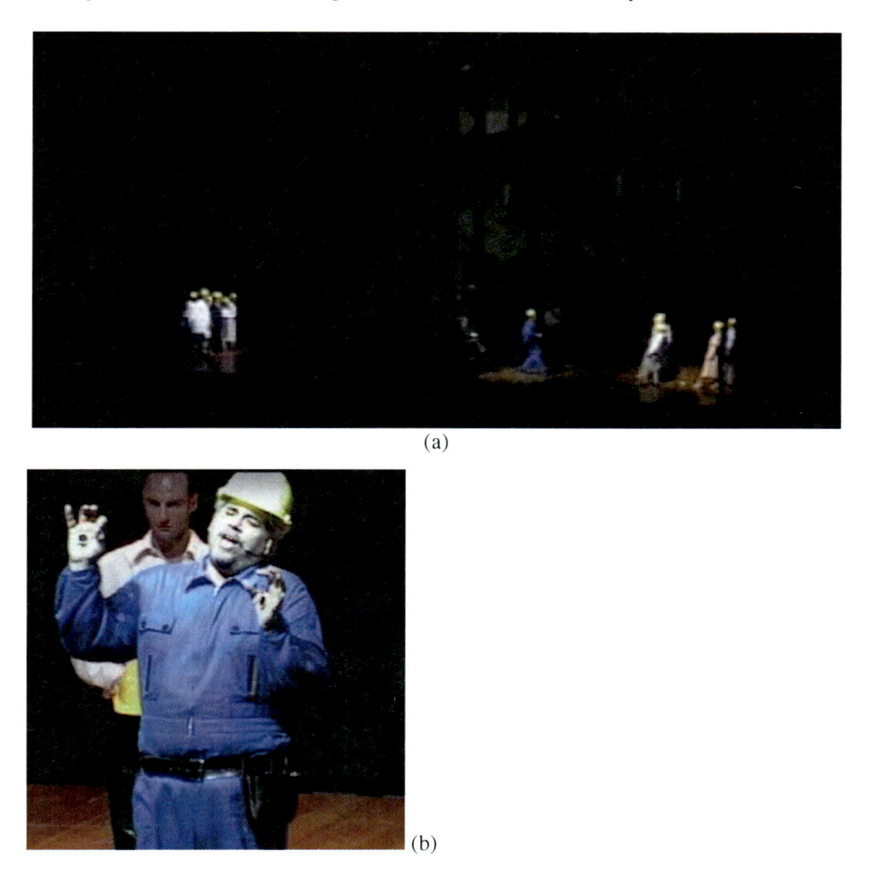

Fig. 4.1 David Moss during the rehearsals of *Cronaca del Luogo* (composer L. Berio, Salzburg Festival, 1999). (a) Moving from one side to the other of the stage he changes his personality. (b) The sensors on his hands allow him to control in real-time the processing of his voice by the movements of his hands.

These are a few examples, among the many that InfoMus Lab experienced and still is experiencing at present days, of how the participation to artistic projects can stimulate novel developments in scientific and technological research.

In the late Nineties, expressive gesture was also studied in the framework of human-robot interaction [8]. In an experimental set-up, a "theatrical machine" was developed for the performance of the music piece *Spiral*, by K. Stockhausen, for one singer or player (in this performance, one trombone) endowed with a short wave radio. The radio, audio amplifier, and loudspeakers were installed on board of a robot navigating on the stage, thus creating effects of physical spatialisation of sound due to the movement of the robot during the performance (trombone: Michele Lo Muto, live electronics: Giovanni Cospito, Civica Scuola di Musica, Sezione Musica Contemporanea, Milano, June 1996). The movements of the robot depended on sound parameters and on the gesture of the trombone performer: for example, a high "energy" content in the trombonist's gesture and a high sound spectral energy were stimuli for the robot to move away from the performer. Smooth and calm phrasing and movements were stimuli attracting the robot near and around the performer. Further, the robot sound and music outputs were part of the interaction process, i.e. the expressive gesture nuances and the sound produced by the performer influenced the robot, and vice-versa.

For *L'Ala dei Sensi* (see Figure 4.2), a multimedia performance about human perception (director: Ezio Cuoghi, choreographer and dancer: Virgilio Sieni, Ferrara, Italy, November 1998), two episodes were developed concerning interactive dance/music performance and making use of a small mobile robotic platform (a Pioneer 2 from Stanford Research Institute). The robot was equipped with sensors, an on-board video projector, and a video-camera. Sensors allowed the robot to avoid collisions with the scenery and the dancers. In the main episode, initially the on-board video projector and the video-camera were directly controlled in real-time by the director (off-stage). He also used the images coming from the robot (the robot's point of view) to mix them in real-time on a large screen. The director controlled the movements of the robot too. That is, the robot was a sort of passive companion of the dancer. At a certain point, the dancer plugged off the electric power cable of the robot. This was a specially important gesture: the robot came to life and a deeper dialogue with the dancer started. The dancer was equipped with two sensors on the palms of his hands. By acting on the first one, he was allowed to influence the robot towards one of two different styles of movement: a kind of "ordered" movement (aiming at a direct, constant speed movement) and a "disordered" type of movement. Through the second sensor the movement could be stopped and restarted. The dancer was also observed by a video-camera and his expressive gestures were a further stimulus for the robot, which was able to react by changing (morphing) its style of moving.

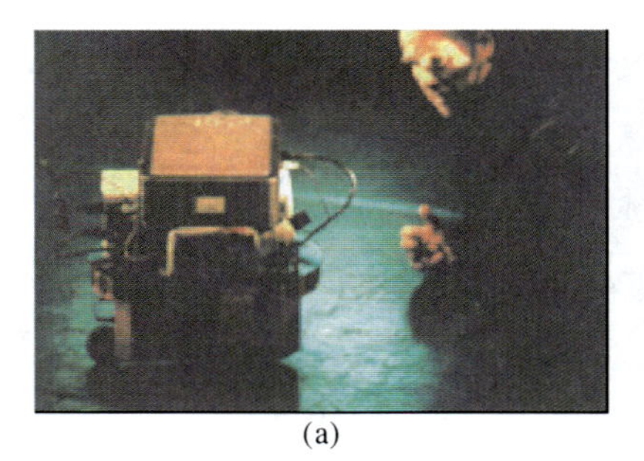
(a)

(b)

Fig. 4.2 Collaborative human-robot interaction in the multimedia performance *L'Ala dei Sensi* (Ferrara, Italy, November 1998): (a) robot/dancer interaction, (b) example of visual feedback.

4.3 Multimodal Expressive Gesture Analysis toward Consolidation

After these pioneering studies, in the framework of the European Project MEGA (Multisensory Expressive Gesture Applications, FP5, November 2000 – October 2003) multimodal analysis of expressive gesture consolidated toward maturity. The concept of expressive gesture was better defined and a conceptual model for multimodal expressive gesture processing was developed [10, 12]. In the model, analysis of expressive gesture is understood as a process involving several layers of abstraction, and a multi-layered architecture is envisaged in which analysis is carried out by progressively extracting higher-level information from lower-level signals. A platform, EyesWeb (www.eyesweb.org), was developed as well as applications in several domains, including music.

4.3.1 A Conceptual Model

The multilayered conceptual model defined in MEGA consists of four layers (see Figure 4.3) and is considered under a multimodal perspective, i.e. it aims at integrating analysis of audio, video, and sensor signals. Integration is tighter as far as analysis moves from lower to upper layers.

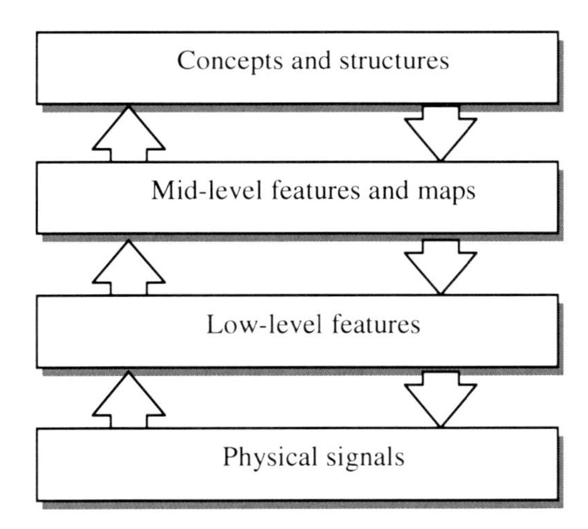

Fig. 4.3 The conceptual framework for expressive gesture processing designed in the European Project MEGA (2000 – 2003).

Layer 1 (*Physical Signals*) receives as input information captured by the sensors of a computer system. Physical signals may have different formats strongly dependent on the kind of sensors that are used. For example, they may consist of sampled signals from tactile, infrared sensors, signals from haptic devices, video frames, sampled audio signals, MIDI messages. In this context the word "sensors" is often related to both the physical sensors employed and to the algorithm used to extract a given set of low-level data. It is therefore possible to speak of "virtual sensors" or "emulated sensors". For example, in the case of analysis of movement through video-cameras, a CCD camera can be an example of physical sensor, while the optical flow, the motion templates, or the positions of certain points in the frame sequence are examples of data extracted from "virtual sensors" implemented by the cited algorithms. Layer 1 applies pre-processing, filtering, signal conditioning, and audio and video analysis techniques to the incoming rough data to obtain cleaner data and further signals derived from the rough input. For example, in the case of video analysis of human movement, usually the following steps are needed: a background subtraction step for identifying the moving subjects and separating them from the background and a motion tracking step to follow the moving subjects across frames. Two types of output are generated: pre-processed images (e.g. the silhouettes of the moving subjects) and geometric data, such as trajectories of body parts (e.g. centre of gravity, limbs), the bounding rectangle (i.e. the smallest rectangle including the silhouette of a moving subject), the contour, and the convex hull (i.e. the smallest polygon including the silhouette).

Layer 2 (*Low-level features*) gets as input the pre-processed signals coming from Layer 1 and applies algorithms to extract a collection of low-level features. The employed techniques range from computer vision algorithms, to signal processing, to statistical techniques. The extracted low-level descriptors are features that psychologists, musicologists, researchers on music perception, researchers on human movement, and artists deemed important for conveying expressive content. In the case of analysis of expressive gesture in human movement, two significant examples [11] are (i) the amount of movement detected by the video-camera (Motion Index, formerly Quantity of Motion), computed as the normalised area (i.e. number of pixels) of a Silhouette Motion Image (SMI), i.e. an image carrying information about the variations of the silhouette shape and position in the last few frames, and (ii) the amount of contraction/expansion (Contraction Index), computed as the ratio of the area of the bounding rectangle and the area of the silhouette. Important features are those related to the Effort dimensions described in Rudolf Laban's Theory of Effort [29, 30], e.g. directness (computed as the ratio of the length of the straight line joining the first and last point of the trajectory followed by a moving subject and the length of the trajectory itself), impulsivity, and fluidity (both computed from the shape of the velocity profiles). In the case of music, low-level features are related to tempo, loudness, pitch, articulation, spectral shape, periodicity, dynamics, roughness, tonal tension, and so on: a similar conceptual framework and a taxonomy of audio features can be found in [31, 32]. Notice that analogies can be found among features in movement and in music, e.g. amount of motion – loudness, contraction/expansion – melodic contour or spectral width, bounded, hesitant movement – roughness.

Layer 3 (*Mid-level features and maps*) has two main tasks: segmenting expressive gestures and representing them in a suitable way. Such a representation would be the same (or at least similar) for gestures in different channels, e.g. for expressive gestures in music and dance. Data from several different physical and virtual sensors are therefore likely to be integrated in order to perform such a step. Each gesture is characterised by the measures of the different features extracted in the previous step (e.g. speed, impulsiveness, directness, etc. for movement, loudness, roughness, tempo, etc. for music). Segmentation is a relevant problem at this level: the definition of expressive gesture does not help in finding precise boundaries. For example, a motion stroke (i.e. the time window from the start of a movement to its end) can be considered as an expressive gesture (and segmentation can be performed on the basis of the detected amount of motion). In fact, this is quite an arbitrary hypothesis: sub-phases of a motion phase (e.g. the phase of motion preparation) could also be considered as expressive gestures as well as sequences of motion and pause phases. Several possibilities are open for the common representation Layer 3 generates as its output. For example, an expressive gesture can be represented as a point or a trajectory in a feature space. Clustering algorithms may then be applied in order to group similar gestures and to distinguish different ones. Another possible output is a symbolic description of the observed gestures along with measurements of several features describing them. For example, in the *Expressive HiFi* application [12] a bi-dimensional feature space was adopted having the Motion Index and Fluidity as axes.

Layer 4 (*Concepts and structures*) extracts high-level expressive content from expressive gestures. It can be organised as a conceptual network mapping the extracted features and gestures into (verbal) conceptual structures. For example, the focus can be on emotional labels (e.g. the basic emotions anger, fear, grief, and joy) or on dimensional approaches, such the well-known valence-arousal space, or other dimensional models [38, 43], or those especially developed for analysis and synthesis of expressive music performance, e.g. [16, 24, 48]. Other outputs include, for example, the Laban's types of Effort such as "pushing", "gliding", etc [29, 30]. Machine learning techniques can be used at Layer 4, including statistical techniques like multiple regression and generalized linear techniques, fuzzy logics or probabilistic reasoning systems such as Bayesian networks, various kinds of neural networks (e.g. classical back-propagation networks, Kohonen networks), support vector machines, decision trees. For example, in [10] decision trees were used to classify expressive dance gestures according to emotional labels.

The conceptual architecture sketched above is conceived for analysis (upward arrows). A similar structure can be employed also for synthesis (downward arrows). Consider for example Layer 4: it may consist of a network in which expressive content is classified in terms of the four basic emotions anger, fear, grief, and joy, depending on current measures of low and mid-level features. If, instead of considering the framework from a bottom-up perspective, a top-down approach is taken, an emotion a virtual performer wants to convey can be translated by a similar network structure in values of low and mid-level features to be applied to generated audio and/or visual content.

4.3.2 Implementation: The EyesWeb Platform

The EyesWeb open platform ([9, 14], www.eyesweb.org) was designed at InfoMus Lab as a development and prototyping environment for both research and applications on multimodal analysis and processing of non-verbal expressive gesture, including music performance. EyesWeb consists of a number of integrated hardware and software modules that can be easily interconnected and extended in a visual environment. It includes a development environment and a set of libraries of reusable software components ("blocks") that can be assembled by the user in a visual language to build applications ("patches") as in common computer music languages (e.g. Max/MSP, pd).

EyesWeb implements the above-sketched conceptual model and supports the user in experimenting computational models of non-verbal expressive communication and in mapping, at different levels, gestures from different modalities (e.g. human full-body movement, music) onto real-time generation of multimedia output (e.g. sound, music, visual media, mobile scenery). It allows fast development and experiment cycles of interactive performance setups. EyesWeb supports the integrated processing of different streams of (expressive) data, such as music audio, video, and, in general, gestural information. EyesWeb is an open platform, since users can extend it by building modules and use them in patches.

EyesWeb has been recently enhanced and re-engineered. A first collection of new features was introduced in version 4. These include the distinction between the kernel engine and the patch editor (i.e. the kernel does not rely on the user interface for its proper working, thus allowing different interfaces to be provided to edit as well as to run patches), a new concept of clock for enhanced performance and synchronisation of multimodal data streams, a new layer of abstraction for devices that adds the possibility to map the available physical hardware resources to virtual devices used by blocks, the availability of kernel objects for optimising the most crucial operations, the availability of a sub-patching mechanism. The current version (XMI, for eXtended Multimodal Interaction) enhances the support to processing of synchronised streams at different sampling rates (e.g. audio, video, biometric data). It has been ported to Linux and has been extended to work as server in complex distributed applications where several clients and servers collaborate in distributed analysis and processing of expressive content.

A collection of libraries is integrated with or connected to EyesWeb. They include, for example, the EyesWeb Expressive Gesture Processing Library for extraction of expressive cues and analysis of expressive content in human movement [11], modules for extraction of audio cues (including a real-time auditory model), modules for processing of time-series (e.g. both sensor signals or higher level expressive features). The EyesWeb XMI libraries have been recently extended with software modules for real-time analysis of affective social interaction, included in the novel EyesWeb Social Signal Processing Library [46]. The library operates on multidimensional time-series consisting of the temporal sequence of expressive feature vectors, and computes features such as a phase synchronisation index and a leadership index (see also Section 2.4).

4.3.3 Applications and Validation of Research on Interactive Systems

The conceptual model and the techniques described above have been exploited in several applications to music and dance performance, as well as in applications for active experience of pre-recorded sound and music content. Different interfaces (e.g. full-body or tangible interfaces) and different techniques (e.g. either for full-body gesture analysis or focusing on single body parts) have been used.

For example, in the music piece *Allegoria dell'opinione verbale* (composer Roberto Doati, actress Francesca Faiella, live electronics Alvise Vidolin, writings Gianni Revello; see also [12]) performed for the first time in September 2001 at the opening concert of the season of Teatro La Fenice, Venice, the focus was on the analysis of the lips of the actress. The actress is on stage, seated on a stool, turned towards the left backstage (the audience sees her profile). A big screen projects her face in frontal view. A video-camera is placed hidden in the left part of the backstage, and is used both to project her face on the big screen and to capture her lips and face movements. The audience can therefore observe the movements of the actress' face while listening to the piece and thus perceiving the overlapping and interaction of her movement with sound changes from the loudspeakers. The actress plays a text in front of the camera while the movements of her lips and face are processed (see Figure 4.4), in order to obtain expressive features used to record and process in real-time her voice and diffuse spatialised electroacoustic music on four loudspeakers placed at the four corners of the auditorium in a standard electroacoustic music setup. The signals reproduced by the loudspeakers are only derived by the actress' voice: former recordings of her voice, real-time recordings, and post-processing in real-time.

In the music theatre production *Un avatar del diavolo* (composer Roberto Doati), performed at La Biennale di Venezia in September 2005 [13], a wooden chair (see Figure 4.5) was sensorised to transform it into a sensitive object. The interface is the whole chair. Sensorisation was based on Tangible Acoustic Interfaces (TAI) technology. A TAI exploits the propagation of sound in a material to get information about where (i.e. location) and how (i.e. which expressive intention, e.g. softly, aggressively) an object is touched. A specific vocabulary of touching gestures by the actor on stage was defined with the composer and the director of the music theatre opera. In particular gestures were localised and analysed in real-time: the position where the chair is touched, the way it is touched (e.g. with hard and quick tapping-like movements or with light and smooth caress-like movements) were detected. The chair could be touched through sitting on it (e.g. caressing the sides and legs), or while standing or moving around it and touching, tapping, caressing it in all its parts (see Figure 4.5a). The recognised positions, qualities, and types of tactile interaction were used to control in real-time sound generation and processing, as well as video-projections. The voice of an actress is the main sound produced and processed in real-time, depending on the high-level gestural information obtained from the actor. Figure 4.5b shows a snapshot of the live performance.

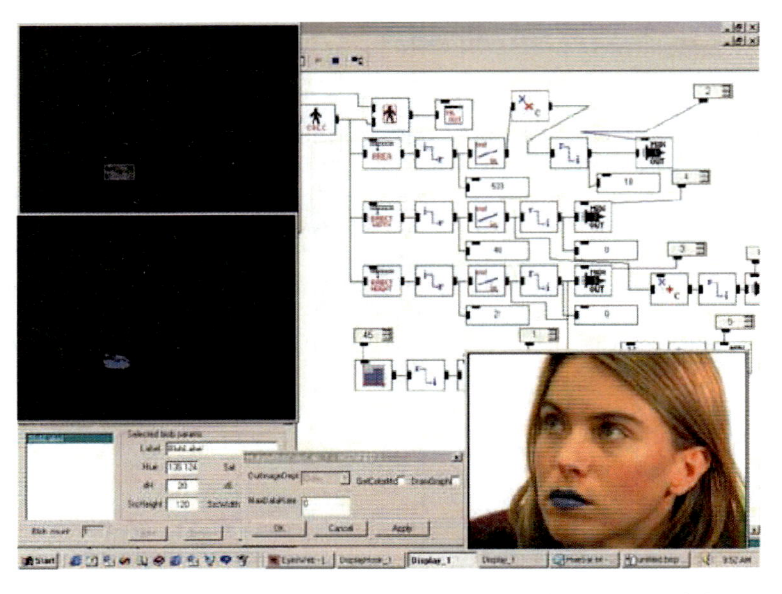

Fig. 4.4 The EyesWeb patch for the concert *Allegoria dell'opinione verbale* (composer Roberto Doati, actress Francesca Faiella). On the left, the lips of the actress as extracted by EyesWeb (bounding box in the top window). The movement of the lips of the actress is analysed in real-time and mapped onto the real-time processing of her voice.

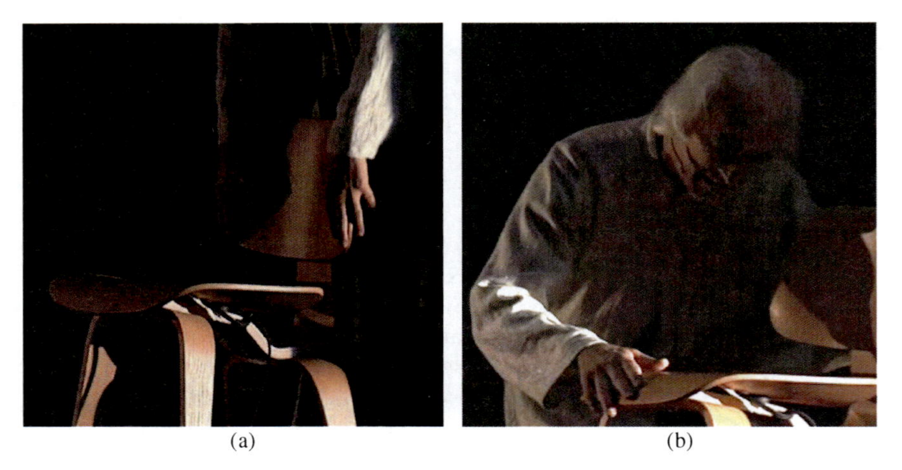

Fig. 4.5 The sensorised chair for the music theatre production *Un avatar del diavolo*: location and quality of the touch gestures are analysed and used for controlling sound processing in real-time.

In a more recent application – the interactive collaborative installation *Mappe per Affetti Erranti*, presented at the Festival della Scienza of Genova [15] – the expressive gestures performed by the users were analysed in order to control in real-time the reproduction of four expressive performances of the same music piece (a choral by J.S. Bach). The four expressive performances corresponded to the following expressive intentions: Happy/Joyful, Solemn, Intimate/Shy, and Angry/Aggressive. These were associated to the same four expressive intentions classified from the users' expressive gestures. Analysis of expressive gesture was performed by means of twelve expressive features: Motion Index (i.e. motor activity), computed on the overall body movement and on translational movement only; Impulsiveness; vertical and horizontal components of velocity of peripheral upper parts of the body; speed of the barycentre; variation of the Contraction Index; Space Occupation Area; Directness Index; Space Allure (inspired by the Pierre Schaeffer's Morphology); Amount of Periodic Movement; and Symmetry Index. Such features were computed in real-time for each single user (a maximum of four users could experience the installation at the same time). Further features were also computed on the whole group of users, such as, for example, the contraction/expansion of the group. This perspective corresponds to Rudolf Laban's General Space [30]. Classification was performed following a fuzzy-logic like approach. Such an approach had the advantage that it did not need a training set of recorded movement and it was also flexible enough to be applied to the movement of different kinds of users (e.g. adults, children, and elder people). Figure 4.6 shows *Mappe per Affetti Erranti* experienced by a group of users and a couple of dancers during a dance performance.

(a) (b)

Fig. 4.6 *Mappe per Affetti Erranti* experienced by a group of users (a) and by two dancers (b).

4.3.4 Scientific Experiments

Besides the developments and applications involving the EyesWeb platform, the consolidation of the approaches to analysis of expressive gesture in music performance also passed through the foundational research carried out with a collection of experiments aiming at assessing and validating models and techniques on the one hand, and at investigating the mechanisms of expressive content communication in music performance on the other hand.

For example, Castellano and colleagues [4] proposed an approach for affect recognition based on the dynamics of movement expressivity and validated it in the framework of music performance. The approach was inspired by theories from psychology [40] claiming that emotional expression is reflected to a greater extent in the timing of the expression than in absolute or average measures. By focusing on the dynamics of expressive motion cues, this approach addressed how motion qualities vary at different temporal levels. A mathematical model that allows for the extraction of information about the dynamics of movement expressivity was developed. The idea behind this model is to use information about temporal series of expressive movement features in a format that is suitable for feature vector-based classifiers. The model provides features conveying information about the dynamics of movement expressivity, i.e. information about fluctuations and changes in the temporal profiles of features. Based on this model, the upper-body and head gestures in musicians expressing emotions were analysed. Results showed that features related to the timing of expressive motion features such as the Motion Index and velocity were more effective for the discrimination of affective states than traditional statistical features such as the mean or the maximum. For example, in an experiment a pianist was asked to play the same excerpt with different emotional expressive intentions: personal, sad, allegro, serene, and over-expressive. The main aim was to verify whether these expressions could be distinguished based solely on the motor behaviour and which motion features are more emotion-sensitive. Analyses were done through an automatic system capable of detecting the temporal profile of two motion features: Motion Index of the upper body and velocity of the head. Results showed that these motion features were sensitive to emotional expression, especially the velocity of the head. Further, some features conveying information about their dynamics over time varied among expressive conditions allowing an emotional discrimination.

Another experiment, carried out earlier in the MEGA Project, investigated the mechanisms responsible for the audience's engagement in a musical performance, by analysing the expressive gesture of the music performer(s) [44]. The aim of this experiment was twofold: (i) individuating which auditory and visual features are mostly involved in conveying the performer's expressive intentions, and (ii) assessing the conceptual model and the techniques by comparing their performance to spectators' ratings of the same musical performances. The research hypotheses combined hypotheses from Laban's Theory of Effort with hypotheses stemming from performance research [17, 36] and from research on the intensity of emotion and tension in music and dance [27, 41, 42].

A professional concert pianist (Massimiliano Damerini) performed Etude Op.8 No.11 by Alexandr Scriabin on a Yamaha Disklavier at a concert that was organised for the experiment's purpose. He performed the piece first without audience in a normal manner and in an exaggerated manner, and then with the audience in a normal concert manner. Exaggerated means in this case with an increased emphasis in expressivity consistent with the style of performance of early 20th Century pianist style. The Scriabin's Etude is a slow and lyrical piece in a late Romantic style that has a considerable number of modulations. The pianist performed on a grand coda piano (Yamaha Disklavier), which made it possible to register MIDI information of the performance. In addition, video was recorded from four sides (see Figure 4.7) and audio with microphones both near the instrument and in the environment. The video recordings from the left were presented to the participants of the experiment. The participants saw the performances on a computer screen and heard them over high-quality loudspeakers twice. At the first hearing, they indicated the phrase boundaries in the music by pressing the button of the joystick. At the second hearing, they indicated to what extent they were emotionally engaged with the music by moving a MIDI-slider up and down.

Fig. 4.7 The frontal, left, right, and top view of a pianist playing a Scriabin's Etude. Analysis of expressive gesture from such videos was correlated with audience engagement measures.

The analyses of the performance data suggested an opposite relation between emotional intensity and the performer's posture. The pianist leaned forward for softer passages and backward for intensive passages. In addition it suggested a differentiation in expressive means with tempo on one side and key-velocity and movement velocity on the other. When relating the performers data to the listeners' data, this differentiation in expressive means was confirmed. Tempo communicates phrase boundaries, while dynamics are highly predictive for the intensity of felt emotion. Hardly any evidence was found for movement features to influence listeners' ratings. The sound seemed the primary focus of the participants and vision only subsidiary. The local phrase-boundaries indicated by tempo did not lead to release of emotional intensity. The modulation of dynamics over a larger time-span communicates the overall form of the piece and, at that level, intensity did increase and decrease within phrases.

4.4 A Shift of Paradigm and Novel Research Challenges

Besides the work presented here, the importance of analysis of expressive gesture for the scientific community has grown in recent years and several systems were developed able to classify expressive qualities of gesture (see for example [1, 25]). Most of such systems, however, classify gestures according to basic emotion categories (e.g. the basic emotions) or simple dimensional approaches (e.g. the valence-arousal space), whereas the subtlest and more significant emotional expressions, such as empathy and emotional engagement, are still neglected. The last experiment discussed above goes in this direction: towards the analysis of the emotional engagement of an audience.

Moreover, almost all of the existing systems are intended for a single user, whereas social interaction is neglected. Nevertheless, social interaction still is one of the most important factors in music performance (e.g. interaction between performers, between performers and conductor, between performers and audience). Research on social interaction is thus a very promising direction. Current research, however, does not focus on the high-level emotional aspects, but rather on group cohesion and decision-making. In this framework, pioneering studies by Pentland [37] developed techniques to measure social signals in scenarios like salary negotiation and friendship. Particular attention was also directed to the recognition of functional roles (e.g. the most dominant people) played during small-group meetings [21]. These works are often based on laboratory experiments and do not address the subtlest aspects such as empathy. Empathy, in fact, has been studied mainly in the framework of synthesis of (verbal) dialogues by virtual characters and embodied conversational agents (e.g. [20, 35]). Recently, the EU-ICT project SAME (www.sameproject.eu) developed techniques for social active listening to music by mobile devices, i.e. for allowing a group of users to mould collaboratively a pre-recorded music piece they are listening to.

Music performance, indeed, is an ideal test-bed for the development of models and techniques for measuring creative social interaction in an ecologically valid framework. Music is widely regarded as the medium of emotional expression par excellence. Moreover, ensemble performance is one of the most closely synchronised activities human beings engage in: it is believed that this ability from individuals and groups to entrain to music is unique only to humans and that, unlike speech, musical performance is one of the few expressive activities allowing simultaneous participation.

According to such novel research challenges, during the last three years research at Casa Paganini - InfoMus focused on the analysis of famous string quartets and on duos of violin players. The ensembles Cuarteto Casal, Quartetto di Cremona, and Quartetto Prometeo have been involved initially in feasibility studies (e.g. to study and understand which multimodal features can explain their expressive social behaviour) and in experiments in occasion of their concerts at the Opera House of Genova. In addition, in collaboration with Ben Knapp

(SARC, Queen's University, Belfast) and Carol Krumhansl measurements were carried out on duos of violinists participating in the International Violin Competition Premio Paganini in 2006, in the framework of the EU Summer School of the HUMAINE Network of Excellence. More recently, again in collaboration with the SARC colleagues, multimodal synchronised recordings of the Quartetto di Cremona were performed. Figure 4.8 shows snapshots from the experiments.

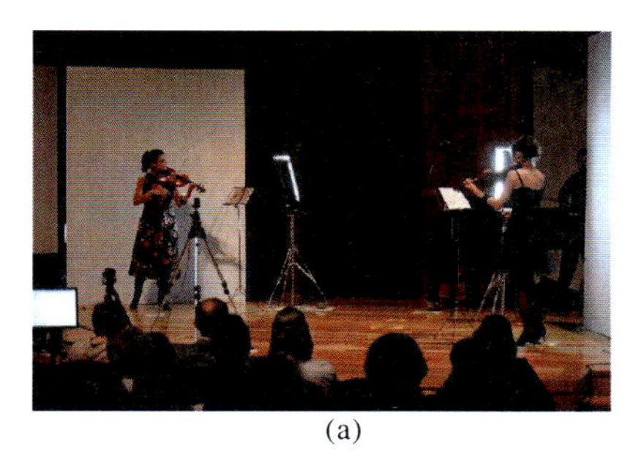

(a)

(b)

Fig. 4.8 Experiments on joint music performance: (a) music duo performance in which the two musicians can communicate, also exchanging visual information, (b) the Quartetto di Cremona during the experiment. Each musician wears a white hat including a green passive marker and a 3-axis accelerometer, plus a 3-axis accelerometer on the back, and physiological sensors.

Using approaches based on Recurrence Quantification Analysis (RQA) and analysis of Phase Synchronisation (PS) [33, 50], several results emerged: for example, in the case of a music duo performance, it was possible to evaluate how the visual and acoustic channels affect the exchange of expressive information during the performance and how positive emotion can affect the emergence of synchronisation [45]. Moreover, foundations for a criterion to distinguish between parallel and reactive empathic outcomes were defined. Measures of the direction of PS confirmed the hypothesis on egalitarian distribution of dominance in a duo performance.

Preliminary results from the analysis of string quartets highlighted how the induction of a positive emotion in one of the musicians of the group resulted in an increased synchronisation among musicians (in terms of heads movement), with respect to the no emotion induction condition. In the same experiment, the SARC colleagues found high physiological synchronisation with the structural changes in the music. Moreover, measures relating to performer mistakes, and the perceived difficulty of the music were found, which also strongly affect both intra- and interpersonal synchronisation.

A real-time implementation of these techniques resulted in the EyesWeb XMI Social Signal Processing Library [46], which was employed to develop applications for social active music listening experiences. For example, the *Sync'n'Move* application prototype, based on EyesWeb XMI and its extensions to Nokia S60 mobile phones, enables users to experience novel forms of social interaction based on music and gesture [47]. Users move rhythmically (e.g. dance) while wearing their mobiles. Their PS is extracted from their gesture (e.g. using the accelerometer data from the mobiles) and used to modify in real-time the performance of a pre-recorded music. More specifically, every time users are successful in synchronising among themselves, music orchestration and rendering is enhanced; while in cases of low synchronisation, i.e. poor collaborative interaction, the music gradually corrupts, looses sections and rendering features, until it becomes a very poor audio signal.

The research challenges related to the analysis of subtle expressive qualities in a social dimension, such as, for example, empathy, are the focus of a project: the EU-ICT-FET Project SIEMPRE (Social Interaction and Entrainment using Music PeRformance Experimentation), started on May 2010. We believe that research on such topics will bring on the one hand new significant insights on the emotional mechanisms underlying expressive music performance and, on the other hand, will enable a novel generation of applications, exploiting the social impact of music performance in a broad range of application scenarios: entertainment, edutainment, museum and cultural applications, therapy and rehabilitation, learning, social inclusion (young people, elderly).

Acknowledgments. We thank our colleagues at Casa Paganini – InfoMus for their precious contributions to research. Recent work has been carried out with the partial support of the EU-ICT Project SAME (www.sameproject.org).

References

1. Bernhardt, D., Robinson, P.: Detecting affect from non-stylised body motions. In: Paiva, A.C.R., Prada, R., Picard, R.W. (eds.) ACII 2007. LNCS, vol. 4738, pp. 59–70. Springer, Heidelberg (2007)
2. Boone, R.T., Cunningham, J.G.: Children's decoding of emotion in expressive body movement: The development of cue attunement. Developmental Psychology 34, 1007–1016 (1998)
3. Cadoz, C., Wanderley, M.: Gesture – Music. In: Wanderley, M., Battier, M. (eds.) Trends in Gestural Control of Music, Ircam, Paris (2000)
4. Castellano, G., Mortillaro, M., Camurri, A., Volpe, G., Scherer, K.: Automated Analysis of Body Movement in Emotionally Expressive Piano Performances. Music Perception 26(2), 103–120 (2008)
5. Camurri, A.: Interactive Dance/Music Systems. In: Proc. Intl. Computer Music Conference ICMC 1995, The Banff Centre for the Arts, pp. 45–252. ICMA-Intl. Comp. Mus. Association, Canada (1995)
6. Camurri, A., Volpe, G. (eds.): Gesture-based Communication in Human-Computer Interaction. LNCS(LNAI), vol. 2915. Springer, Heidelberg (2004)
7. Camurri, A., Catorcini, A., Innocenti, C., Massari, A.: Music and Multimedia Knowledge Representation and Reasoning: the HARP System. Computer Music Journal 18(2), 34–58 (1995)
8. Camurri, A., Coletta, P., Ricchetti, M., Volpe, G.: Expressiveness and Physicality in Interaction. Journal of New Music Research 29(3), 187–198 (2000)

9. Camurri, A., Hashimoto, S., Ricchetti, M., Trocca, R., Suzuki, K., Volpe, G.: EyesWeb – Toward Gesture and Affect Recognition in Interactive Dance and Music Systems. Computer Music Journal 24(1), 57–69 (2000)
10. Camurri, A., Mazzarino, B., Ricchetti, M., Timmers, R., Volpe, G.: Multimodal analysis of expressive gesture in music and dance performances. In: Camurri, A., Volpe, G. (eds.) GW 2003. LNCS (LNAI), vol. 2915, pp. 20–39. Springer, Heidelberg (2004)
11. Camurri, A., Mazzarino, B., Volpe, G.: Analysis of Expressive Gesture: The EyesWeb Expressive Gesture Processing Library. In: Camurri, A., Volpe, G. (eds.) GW 2003. LNCS (LNAI), vol. 2915, pp. 460–467. Springer, Heidelberg (2004)
12. Camurri, A., De Poli, G., Leman, M., Volpe, G.: Toward Communicating Expressiveness and Affect in Multimodal Interactive Systems for Performing Art and Cultural Applications. IEEE Multimedia Magazine 12(1), 43–53 (2005)
13. Camurri, A., Canepa, C., Drioli, C., Massari, A., Mazzarino, B., Volpe, G.: Multimodal and cross-modal processing in interactive systems based on Tangible Acoustic Interfaces. In: Proc. International Conference Sound and Music Computing 2005, Salerno, Italy (2005)
14. Camurri, A., Coletta, P., Demurtas, M., Peri, M., Ricci, A., Sagoleo, R., Simonetti, M., Varni, G., Volpe, G.: A Platform for Real-Time Multi-modal Processing. In: Proc. International Conference Sound and Music Computing, Lefkada, Greece, pp. 354–358 (2007)
15. Camurri, A., Canepa, C., Coletta, P., Ferrari, N., Mazzarino, B., Volpe, G.: The Interactive Piece The Bow is bent and drawn. In: Proc. 3rd ACM International Conference on Digital Interactive Media in Entertainment and Arts (DIMEA 2008), Athens, Greece, pp. 376–383 (2008)
16. Canazza, S., De Poli, G., Drioli, C., Rodà, A., Vidolin, A.: Audio Morphing Different Expressive Intentions for Multimedia Systems. IEEE Multimedia Magazine 7(3), 79–83 (2000)
17. Clarke, E.F., Davidson, J.W.: The body in music as mediator between knowledge and action. In: Thomas, W. (ed.) Composition, Performance, Reception: Studies in the Creative Process in Music, pp. 74–92. Oxford University Press, Oxford (1998)
18. Cowie, R., Douglas-Cowie, E., Tsapatsoulis, N., Votsis, G., Kollias, S., Fellenz, W., Taylor, J.: Emotion Recognition in Human-Computer Interaction. IEEE Signal Processing Magazine 1 (2001)
19. De Meijer, M.: The contribution of general features of body movement to the attribution of emotions. Journal of Nonverbal Behavior 13, 247–268 (1989)
20. de Rosis, F., Cavalluzzi, A., Mazzotta, I., Novielli, N.: Can embodied conversational agents induce empathy in users? In: Proc. of AISB 2005 Virtual Social Character Symposium (2005)
21. Dong, W., Lepri, B., Cappelletti, A., Pentland, A., Pianesi, F., Zancanaro, M.: Using the influence model to recognize functional roles in meetings. In: Proc. 9th Intl. ACM Conf. on Multimodal Interfaces, pp. 271–278 (2007)
22. Gibet, S., Courty, N., Kamp, J.F.: Gesture in Human-Computer Interaction and Simulation. In: Gibet, S., Courty, N., Kamp, J.-F. (eds.) GW 2005. LNCS (LNAI), vol. 3881, Springer, Heidelberg (2006)
23. Hashimoto, S.: KANSEI as the Third Target of Information Processing and Related Topics in Japan. In: Camurri, A. (ed.) Proc. of the International Workshop on KANSEI: The technology of emotion, pp. 101–104. AIMI (Italian Computer Music Association) and DIST-University of Genova (1997)

24. Juslin, P.N.: Cue utilization in communication of emotion in music performance: relating performance to perception. Journal of Experimental Psychology: Human Perception and Performance 26(6), 1797–1813 (2000)
25. Kapur, A., Kapur, A., Virji-Babul, N., Tzanetakis, G., Driessen, P.F.: Gesture-based affective computing on motion capture data. In: Tao, J., Tan, T., Picard, R.W. (eds.) ACII 2005. LNCS, vol. 3784, pp. 1–7. Springer, Heidelberg (2005)
26. Kopp, S., Wachsmuth, I.: Gesture in Embodied Communication and Human-Computer Interaction. In: Kopp, S., Wachsmuth, I. (eds.) GW 2009. LNCS, vol. 5934. Springer, Heidelberg (2010)
27. Krumhansl, C.L.: Can dance reflect the structural and expressive qualities of music? A perceptual experiment on Balanchine's choreography of Mozart's Divertimento No. 15. Musicae Scientiae 1, 63–85 (1997)
28. Kurtenbach, G., Hulteen, E.: Gestures in Human Computer Communication. In: Laurel, B. (ed.) The Art and Science of Interface Design, pp. 309–317. Addison-Wesley, Reading (1990)
29. Laban, R.: Modern Educational Dance. Macdonald & Evans Ltd., London (1963)
30. Laban, R., Lawrence, F.C.: Effort. Macdonald & Evans Ltd., London (1947)
31. Leman, M., et al.: Correlation of gestural musical audio cues and perceived expressive qualities. In: Camurri, A., Volpe, G. (eds.) GW 2003. LNCS (LNAI), vol. 2915, pp. 40–54. Springer, Heidelberg (2004)
32. Lesaffre, M., Leman, M., Tanghe, K., De Baets, B., De Meyer, H., Martens, J.P.: User-Dependent Taxonomy of Musical Features as a Conceptual Framework for Musical Audio-Mining Technology. In: Proc. Stockholm Music Acoustics Conference SMAC 2003, KTH, Stockholm, Sweden (2003)
33. Marwan, N., Romano, M.C., Thiel, M., Kurths, J.: Recurrence plots for the analysis of complex systems. Physics Reports 438, 237–329 (2007)
34. McNeill, D.: Hand and Mind: What Gestures Reveal About Thought. University Of Chicago Press, Chicago (1992)
35. McQuiggan, S.W., Lester, J.C.: Modeling and evaluating empathy in embodied companion agents. Intl. J. Human-Computer Studies 65(4), 348–360 (2007)
36. Palmer, C.: Music Performance. Annual Review of Psychology 48, 115–138 (1997)
37. Pentland, A.: Social signal processing. IEEE Signal Processing Magazine 24(4), 108–111 (2007)
38. Russell, J.A.: A circumplex model of affect. Journal of Personality and Social Psychology 39, 1161–1178 (1980)
39. Sales Dias, M., et al.: Using hand gesture and speech in a multimodal augmented reality environment. In: Sales Dias, M., et al. (eds.) GW 2007. LNCS (LNAI), vol. 5085, pp. 175–180. Springer, Heidelberg (2009)
40. Scherer, K.R.: Appraisal considered as a process of multi-level sequential checking. In: Scherer, K.R., Schorr, A., Johnstone, T. (eds.) Appraisal processes in emotion: Theory, Methods, Research, pp. 92–120. Oxford University Press, New Yark (2001)
41. Scherer, K.R.: Why music does not produce basic emotions: pleading for a new approach to measuring the emotional effects of music. In: Proc. Stockholm Music Acoustics Conference SMAC 2003, KTH, Stockholm, Sweden, pp. 25–28 (2003)
42. Sloboda, J.A., Lehmann, A.C.: Tracking performance correlates of changes in perceived intensity of emotion during different interpretations of a Chopin piano prelude. Music Perception 19(1), 87–120 (2001)

43. Tellegen, A., Watson, D., Clark, L.A.: On the dimensional and hierarchical structure of affect. Psychological Science 10(4), 297–303 (1999)
44. Timmers, R., Marolt, M., Camurri, A., Volpe, G.: Listeners' emotional engagement with performances of a Skriabin etude: An explorative case study. Psychology of Music 34(4), 481–510 (2006)
45. Varni, G., Camurri, A., Coletta, P., Volpe, G.: Emotional Entrainment in Music Performance. In: Proc. 8th IEEE International Conference on Automatic Face and Gesture Recognition (FG 2008), Amsterdam, The Netherlands (2008)
46. Varni, G., Camurri, A., Coletta, P., Volpe, G.: Toward Real-time Automated Measure of Empathy and Dominance. In: Proc. 2009 IEEE International Conference on Social Computing (SocialCom 2009), Vancouver, Canada (2009)
47. Varni, G., Mancini, M., Volpe, G., Camurri, A.: Sync'n'Move: social interaction based on music and gesture. In: Proc: 1st Intl. ICST Conference on User Centric Media, Venice, Italy (2009)
48. Vines, B.W., Krumhansl, C.L., Wanderley, M.M., Ioana, M.D., Levitin, D.J.: Dimensions of Emotion in Expressive Musical Performance. Ann. N.Y. Acad. Sci. 1060, 462–466 (2005)
49. Wallbott, H.G.: Bodily expression of emotion. Eu. J. of Social Psychology 28, 879–896 (1998)
50. Zbilut, J., Webber Jr., C.L.: Embeddings and delays as derived from quantification of recurrence plots. Phys. Lett. A 5, 199–203 (1992)

Chapter 5
Input Devices and Music Interaction

Joseph Malloch, Stephen Sinclair, Avrum Hollinger, and Marcelo M. Wanderley

Abstract. This chapter discusses some principles of digital musical instrument design in the context of different goals and constraints. It shows, through several examples, that a variety of conditions can motivate design choices for sensor interface and mapping, such as robustness and reliability, environmental constraints on sensor technology, or the desire for haptic feedback. Details of specific hardware and software choices for some DMI designs are discussed in this context.

5.1 What Is a DMI?

Simply stated, a digital musical instrument (DMI) is tool or system for making music in which sound is synthesized digitally using a computer and the human interface is formed using some type of sensor technology. Whether the computer is physically embedded in the interface or a stand-alone general-purpose PC, a defining characteristic of DMIs is that the user interface and the sound production mechanism are conceptually (and often physically) separate; control and sound synthesis parameters must be artificially associated by *mapping* [25]. The synthesized sound is available to be experienced by an audience, but it also forms a channel of *feedback* for the performer/interactor. Visual displays (not discussed in this chapter) may also be used to provide additional feedback channels for the performer/interactor. Vibration or force actuators can be used to provide *haptic* feedback similar to the intrinsic vibrations produced by acoustic instruments, or expanded feedback which an acoustic system could not produce. The combination of audio and haptic feedback can even allow natural, nuanced interaction with virtual models of acoustic systems.

There are a great many motivations for creating DMIs, but here we will focus on three main areas: DMIs for musical performance, DMIs for studying music cognition in special environments, and DMIs comprising haptic interfaces and

Joseph Malloch · Stephen Sinclair · Avrum Hollinger · Marcelo M. Wanderley
Input Devices and Music Interaction Laboratory (IDMIL), Centre for Interdisciplinary Research in Media Music and Technology (CIRMMT), McGill University,
555, Sherbrooke Street West, H3A 1E3 Montreal, Qc, Canada

J. Solis and K. Ng (Eds.): Musical Robots and Interactive Multimodal Systems, STAR 74, pp. 67–83.
springerlink.com

virtual models of acoustic systems. Some case studies from the Input Device and Music Interaction Laboratory (IDMIL) at McGill University are provided as examples.

5.2 DMIs for Musical Performance

Digital Musical Instruments allow us to perform music using sounds and types of musical control impossible with acoustic instruments, as well as enabling individuals who may not be able to use traditional instruments due to physical or mental handicaps to engage in live music-making. As can be seen from the interfaces presented at new music conferences every year, many different ideas and perspectives can form the inspiration for a new DMI. For example:

- A specific posture, gesture, or series of gestures.
- An activity, such as dance, or playing chess.
- A type of sensing.
- An object or collection of objects.
- A particular musical composition.
- An algorithm.
- A question or problem.

Any of these inspirations could result in an interesting and successful musical interface, but there are a number of issues which should be considered when conceiving and developing a new interface for making music. We will discuss a few of them below.

5.2.1 Immersion vs. Physical Embodiment

Fitzmaurice, Ishii and Buxton define a design space for "graspable user interfaces" [11] and articulate advantages of tangible interfaces over traditional WIMP interfaces[1]: graspable interfaces encourage bimanual interaction, and typically allow parallel input by the user; they leverage well-developed prehensile manipulation skills, and allow a variety of ways to perform the same task; and finally they supply "tactile confirmation" and "interaction residue."

Although these points are aimed at the general human-computer interaction (HCI) tangible user interface (TUI) community, they are completely relevant for the design of digital musical instruments. It is interesting that the NIME (New Interfaces for Musical Expression) community would think many of the points completely self-evident, since this community has followed a decidedly different path to the conceptualization and development of manipulable interfaces. There have been several classic (i.e. "objects on a table") TUI projects in music (Musical Trinkets [30], the Reactable [17]) but for the most part the musical interface design problem has been approached from the perspective of augmenting or

[1] Windows, Icons, Mouse, Pointer—referring to the standard graphical operating system desktop on personal computers.

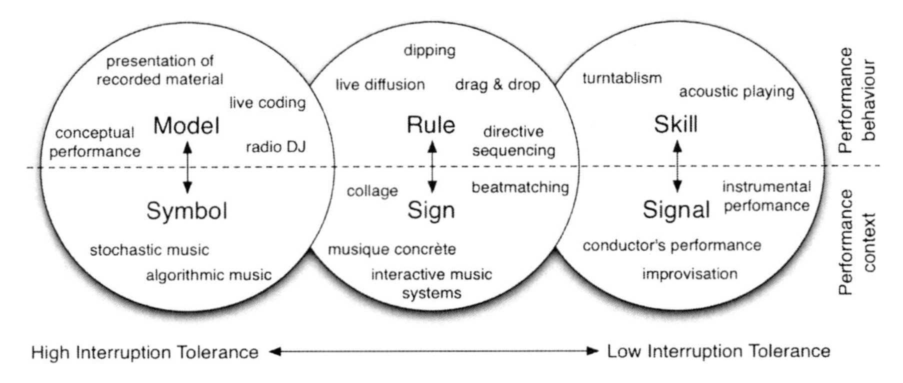

Fig. 5.1 Model visualization based on Rasmussen's typology of human information processing [21]. From left to right, the systems represented are less and less tolerant of interruption of the channels of control.

emulating existing acoustic instruments, which obviously are already tangible, physically-embodied, single-purpose artifacts. In fact, much of the seminal work in DMI conceptualization and design has focused on *reducing* the traditional identification of interface as "object," with Theremins, gloves, body sensors, etc., believing that one can "relieve the performer of the tedium of manipulation" [34].

Rovan and Hayward, among others, advocated using haptic feedback to bring back the tactile experience of playing an acoustic, vibrating musical instrument [34] while performing open-air gestures. This helps, but the user is still deprived of all the other TUI gains mentioned above, such as the leveraging of prehensile skill, embodiment of I/O, and "visual interaction residue." The use of haptic interfaces for tangible interaction with virtual musical instruments is discussed in section 5.3.3.

5.2.2 Mapping and Metaphor

One issue faced by both designer and performer of gestural controllers is that connections between gesture and sound must be *designed*, since in the DMI idiom these relationships do not emerge from inherent properties of the building materials and playing technique [35]. While this allows unprecedented flexibility in terms of musical instrument design, the task of designing an instrument that will please performers and audiences can be somewhat daunting. Wessel and Wright proposed using explicitly chosen metaphors as a method for organizing and guiding interaction with computer-based instruments [39]. Metaphors discussed include different spatial pitch-arrangements, "timbre space," "drag and drop," "scrubbing," "dipping," and "catch and throw." The authors make the important point that metaphors like these determine how we perceive and interact with the world around us and naturally should be considered carefully when designing new systems for control.

This approach can be seen as parallel to the embedding of physical behaviour into audio and visual systems in virtual reality or games. There are not actually physical laws at work inside the simulation, but modeling behaviour familiar to the user from their experience in the "real world" is logical and provides a basic level of environmental knowledge that the user can leverage to explore the system.

5.2.3 Musical Control

There is a huge variety of situations and interactions which might be termed "Musical Performance," ranging from classical "virtuosic" performance on acoustic musical instruments to turntablism, mixing or live diffusion, to live coding of musical processes and sound synthesis. It is obvious that these different musical interactions present very different needs in terms of interface design, so we have found it essential to differentiate between the intentions of the creator, performer, and audience in order to establish contexts for discussing, designing, or evaluating DMIs.

In particular we use a paradigm of interaction and musical context based on Rasmussen's model of human information processing [33], previously used to aid DMI design in [8]. In Rasmussen's model, interaction behaviours are described as being *skill-*, *rule-*, or *model*-based. Skill-based behaviour is defined as a real-time, continuous response to a continuous signal, whereas rule-based behaviour consists of the selection and execution of stored procedures in response to cues extracted from the system. Model-based behaviour refers to a level yet more abstract, in which performance is directed towards a conceptual goal, and active reasoning must be used before an appropriate action (rule or skill-based) is taken. Each of these modes is linked to a category of human information processing, distinguished by their human interpretation; that is to say, during various modes of behaviour, environmental conditions are perceived as playing distinct roles, which can be categorized as signals, signs, and symbols.

Figure 5.1 shows a visualization we have developed for comparing and discussing musical devices based on Rasmussen's framework [21]. Performance behaviours are represented on the top row, and performance contexts on the bottom row. Since these contexts and behaviours may be blurred or mixed, we have also included "interruption tolerance" as a horizontal axis, meaning the tolerance of the system to interruption of the channels of control between user and machine. For example, if the performer stops "playing" and leaves to get coffee, will the system be affected immediately, after some length of time, or not at all?[2]

Skill-based behaviour is identified by [7] as the mode most descriptive of musical interaction, in that it is typified by rapid, coordinated movements in response to continuous signals. Rasmussen's own definition and usage is somewhat broader, noting that in many situations a person depends on the experience of previous attempts rather than real-time signal input, and that human behaviour is very seldom restricted to the skill-based category. Usually an activity mixes rule and skill-based behaviour, and performance thus becomes a sequence of automated (skill-based) sensorimotor patterns. Instruments that belong to this

[2] This idea has also been represented as "granularity of control" and later as "balance of power in performance" [29]; we feel that "interruption tolerance" is less subject to value-judgements and conflicting interpretations.

mode of interaction have been compared more closely in several ways. The "entry-fee" of the device [39], allowance of continuous excitation of sound after an onset [20], and the number of musical parameters available for expressive nuance [9] may all be considered.

During rule-based performance the musician's attention is focused on controlling a process rather than a signal, responding to extracted cues and internal or external instructions. Behaviours that are considered to be rule-based are typified by the control of higher-level processes and by situations in which the performer acts by selecting and ordering previously determined procedures, such as live sequencing, or using "dipping" or "drag and drop" metaphors [39]. Rasmussen describes rule-based behaviour as goal-oriented, but observes that the performer may not be explicitly aware of the goal. Similar to the skill-based domain, interactions and interfaces in the rule-based area can be further distinguished by the rate at which a performer can effect change and by the number of task parameters available as control variables.

The model domain occupies the left side of the visualization, where the amount of control available to the performer (and its rate) is determined to be low. It differs from the rule-based domain in its reliance on an internal representation of the task, thus making it not only goal-oriented but goal-controlled. Rather than performing with selections among previously stored routines, a musician exhibiting model-based behaviour possesses only goals and a conceptual model of how to proceed. She must rationally formulate a useful plan to reach that goal, using active problem-solving to determine an effectual course of action. This approach is thus often used in unfamiliar situations, when a repertoire of rule-based responses does not already exist.

By considering their relationship with the types of information described by Rasmussen, performance context can also be distributed among the interaction domains. The signal domain relates to most traditional instrumental performance, whether improvised or pre-composed, since its output is used at the signal-level for performance feedback. The sign domain relates to sequenced music, in which pre-recorded or pre-determined sections are selected and ordered. Lastly, the symbol domain relates to conceptual music, which is not characterized by its literal presentation but rather the musical context in which it is experienced. In this case, problem solving and planning are required; for example, conceptual scores may lack specific "micro-level" musical instructions but instead consist of a series of broader directives or concepts to be actively interpreted by the performer [6].

5.2.4 From the Lab to the Stage: Context, Repertoire, and Pedagogy

A major issue in the use of new instruments is the lack of cultural context surrounding the instrument, since there is no pre-existing repertoire or performance practice. Distribution of audio and video recordings can help expose potential audiences to the new interface, but it traditionally takes decades or even centuries for a musical instrument to accumulate cultural baggage.

If the DMI is to be used in the context of performer-centric/weighted music (rather than interface-centric music such as glitch), the resilience of the interface under the demands of countless hours of practice and rehearsal must be considered. Interaction metaphors cultivated by the instrument-designer—or emergent metaphors based on the behaviour of the working system—may be detrimentally affected by breakage, or repair downtime, especially if the continuity of the metaphor depends on active feedback. For the performer to achieve maximal rapport with the DMI it needs to be very robust, since we want the performer to concentrate on gesture and sound rather than sensors and computers; "musical performance" rather than "laboratory experiment" [23]. The history of the development of the Continuum Fingerboard provides an interesting example of this search for robustness [13, 25].

A perennial temptation when writing pre-composed music for new DMIs is to notate the gestures required rather than the musical result desired. This seems to stem from two sources: the fact that new users/performers of the instrument will likely not know how to produce a notated sound or musical process, and the distraction of using new technology. We advocate using traditional notation whenever possible, however, as it allows vital room for performer interpretation, as well as avoiding overloading the performer with instructions. Adequate training and practice time will allow the performer to interpret the composer's intentions correctly, aided by instructional material (documentation, video, audio) included with the score.

5.3 Case Studies

Following the above general remarks, the rest of this chapter will be devoted to providing some examples of digital musical instrument research and development carried out by the authors.

5.3.1 Case Study: The T-Stick

The T-Sticks are a family of digital musical instruments being developed in the IDMIL, in collaboration with performers and composers as part of the McGill Digital Orchestra Project [31]. Nearly twenty T-Stick prototypes have been completed, in tenor, alto, and soprano versions, allowing collaboration with multiple performers and composers, and use of the instrument in ensemble performances. The T-Stick project has several main motivations:

- The T-Sticks are intended to form a family analogous to the orchestral string instruments, in which the basic construction, user interface, and interaction design are the same, but each subclass of T-Stick differs from its siblings in size, weight, and register. Seeing and hearing multiple T-Sticks in ensemble works will help "parallelize" the task of providing context to the audience.

Fig. 5.2 Percussionist Fernando Rocha investigating the first prototype of the T-Stick, before the protective outer covering has been applied.

- The physical interface should be robust enough that it could be practiced and performed on for hours every day without breaking or crashing. It is vital that the performers feel that they can work intensively with the interface without fear of breaking it.
- The DMI should be comparable in simplicity to an electric guitar in terms of set-up time and electronic knowledge required. By doing this, the performer will hopefully think of their DMI as they would a traditional musical instrument, rather than as a "lab experiment."
- Excitation of sound should require physical energy expenditure by the performer.
- Sensing and mapping of the performers gestures should be done in an integral, interrelated way such that any manipulation of the physical interface will affect the sound in an intuitive way (as excitation, modification, or selection [2]). Novice interaction should be guided by these relationships, as well as the appearance, feel, and weight of the interface, so that the performer can quickly construct a mental model of the control system.
- The DMI should be designed for the use of expert users, rather than novices. To this end, more emphasis should be put on extending any "ceiling on virtuosity" than on lowering the "entry-fee" [39]. New users

should be able to produce sound from the DMI, but not necessarily what they would judge to be musical sound. In the context of the framework mentioned above, the DMI should fit into the skill/signal category, exhibit low interruption tolerance, low latency, and maximize the transfer of information (including sensor information) as signals rather than signs, events, or triggers.

- The design of the system should aim to keep the performer's focus on the sound and its relation to the entire physical object, rather than individual sensors. Technological concerns should be subsumed under performance or musical concerns (or appear to be such to the performer). As discussed by Rasmussen, the ability to focus on the task rather than the interface may improve with practice.

5.3.1.1 Hardware

The T-Stick is constructed using cylindrical plastic pipe as a structural base, with all sensors and electronics hidden inside. Its appearance to the performer is a simple cylinder, 60 to 120 cm long depending on the family member, and 5 cm in diameter. Some models require a USB cable for power and data communication to be plugged in to one end of the cylinder; others are wireless.

An effort was made to sense all of the affordances of the interface. The entire length of the tube is equipped with capacitive multitouch sensing, allowing the interface to sense where and how much of its surface is being touched over time. A long force-sensing resistor (FSR) or custom paper force sensor [18] is positioned along one side of the tube and covered with thin compressible foam to suggest a squeezing affordance and provide proprioceptive feedback for an essentially isometric force sensor. Three-axis accelerometers are used to sense tilting, rolling, swinging and shaking; a piezoelectric contact microphone is bonded to the inside of the tube to sense deformation due to tapping, hitting, or twisting. A third revision of the hardware produced the "SpaT-Stick," which includes a 3-axis magnetometer for direction sensing and was used to control sound spatialization in live performance.

5.3.1.2 Mapping and Synthesis

Mapping for the T-Stick was developed using the Digital Orchestra Tools (DOT), a communication protocol and suite of software tools for plug-and-play gesture signal processing and connection negotiation [22]. The parameter-space exposed for mapping is structured according to the *gesture description interchange format* (GDIF) proposal [16]; raw and preprocessed versions of the sensor data are included alongside extracted higher-level parameters corresponding to postures and gestures.

Several software tools have been used to synthesize the voice of the T-Stick over the years, including granular and physical modeling techniques implemented in Cycling 74's Max/MSP and the Sculpture synthesizer from Apple's Logic Pro.

5.3.1.3 Performance Technique

Based on the typology of musical gesture proposed in Cadoz and Wanderley [5], the various performance techniques used for the T-Stick can be broken down as follows:

- *excitation gestures*: tapping, brushing, jabbing, swinging, and twisting
- *damping gestures*: touching and holding, sometimes in a specific area (depending on mapping version)
- *modification gestures*: tilting, rolling, touch location, surface coverage
- *selection gestures*: choosing which pressure pad to activate
- *biasing gestures*: specific grips, for example cradling the instrument with one arm.

5.3.1.4 Repertoire and Performances

Through its use in interdisciplinary university seminars and the three-year-long McGill Digital Orchestra Project [31], the T-Stick has already achieved rare progress in the transformation from technology demonstration to "real" instrument, with its own associated pedagogy and composed repertoire. To date, the T-Stick has been performed in concerts and demonstrations in Canada, Brazil, Portugal, Italy, and the USA. In total, six performers are preparing to perform or have already performed using the T-Stick, and three composers have now used the DMI in six written works, including a trio for two T-Sticks and another DMI and a piece for two T-Sticks with chamber ensemble. As a testament to the success of the design goals (specifically robustness and technical simplicity), several partner institutions are also using the T-Stick for artistic and research projects, including Casa da Música in Portugal, Universidade Federal de Minas Gerais in Brazil, and Aalborg University in Denmark.

5.3.2 Case Study: An MRI-Compatible Musical Instrument

Music performance involves coordinating motor activity while integrating auditory, visual, haptic, and proprioceptive feedback; the underlying neuropsychological mechanisms for this complex task are still poorly understood. Studies that investigate the neural activity of musicians and non-musicians as they perform musical tasks, such as audition and pitch discrimination or performance and motor learning, help neuroscientists better understand the function of specific brain areas and their connectivity. A prominent technique available to neuroscientists is functional magnetic resonance imaging (fMRI), which provides 3D images of the brain, enabling the correlation of action and thought with localized neural activity. Brain scans need to be correlated with measured behavioural changes if one is to link, for instance, motor learning and activation of specific neural centres.

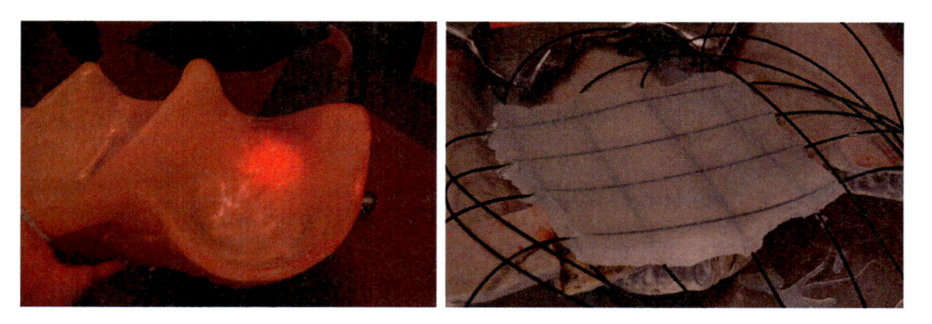

Fig. 5.3 Initial interface prototype for the Ballagumi (left), showing optical sensor grid (right).

The behavioural data or performance metrics of a musical task need to be electronically synchronized with the image acquisition for off-line analysis and real-time feedback, be it auditory or visual—thus the need for a DMI. The intense magnetic and radio-frequency (RF) fields generated by the scanner can adversely affect electronic devices, and conversely, electronic devices can interfere with image acquisition. In order to overcome these issues posed by ferromagnetic materials and current-carrying conductors, care must be taken to ensure MRI compatibility and special sensing methods are employed.

As part of an ongoing collaborative research project at the IDMIL, CIRMMT, BRAMS, and the MNI[3], several electronic MRI-compatible musical controllers were commissioned. These consisted of a piano keyboard [14], a cello-like interface (currently under development), and a novel flexible interface called the *Ballagumi*.

5.3.2.1 The Ballagumi: A Flexible Interface

The Ballagumi [15] consists of a flexible physical interface cast in silicone with embedded fibre optic flexion and pressure sensors and a self-contained hardware platform for signal acquisition, mapping, and synthesis. Its design was approached with a firm set of constraints and goals:

- The instrument should be composed of a physical interface generating analogue signals with the purpose of controlling a physical modeling synthesizer.
- The interface should be sensed in a holistic fashion encouraging a complex mapping with nuanced control of acoustic properties.
- The instrument should be self-contained and not require an external computer for mapping and audio synthesis.
- The physical interface should be compatible with magnetic resonance imaging (MRI) scanners in order for it to be used in neuroimaging studies.

[3] IDMIL: Input Devices and Music Interaction Laboratory, CIRMMT: Centre for Interdisciplinary Research in Music Media and Technology, BRAMS: International Laboratory for Brain, Music and Sound Research, MNI: Montreal Neurological Institute.

The novelty of this instrument for MRI studies is quite exciting as the subject has no preconceived notions of how to interact with the Ballagumi nor the auditory feedback presented, thus providing a totally unique opportunity to study motor learning of musical tasks.

The Ballagumi was to have nuanced control of timbre, pitch, and amplitude, and to that end many sensors are used to collect information about the deformation of the instrument as a whole. Optical fibres are used to sense flexion and pressure and are distributed throughout the Ballagumi. While redundant sensing leads to coupled sensor responses, the coupling of acoustic properties of the sound is inherently coupled within the dynamical system, making the mapping section an equally important part of the equation. In order to draw users to the instrument, an organic form and a viscerally exciting material was necessary. While the winged bat or ray-like form is an idiosyncratic design, it provides the performer with affordances for striking, pushing, pulling, plucking, twisting, and stretching gestures. It is meant to sit on the lap to enable bi-manual control and allows the legs to provide support and leverage for manual manipulation. The silicone rubber material, as noted by Morris et al. [26], provides an engaging and continuous sensation of haptic feedback owing to its viscoelastic properties.

5.3.2.2 Mapping and Synthesis

Although a few different synthesis algorithms have been implemented, mapping strategies are still under development. A mapping system was designed to enable rapid prototyping of different mappings. The mapping structure takes sensor data from an on-board PSoC and ARM's on-chip ADC, performs signal conditioning (ambient light compensation and signal difference operations), cubic scaling (for linearization or otherwise) for each signal independently, and from there synthesis parameters are computed as a linear combination of cooked data.

For development purposes, the ARM can communicate with a PC over USB to query the namespace of the mapping sources (interface control signals) and destinations (synthesis parameters) using a command-line interface. Other commands allow the user to set scaling and mapping values as well as check sensor input values, assert calibration mode and save a mapping to memory.

Three physical modeling synthesizers were implemented: a single neuron FitzHugh-Nagumo (FHN) model, a two-dimensional grid of FHN neurons connected with a Laplacian diffusion model, and the Ishizaka-Flanagan (IF) vocal fold model [15]. All of these models were first tested in Matlab, then Max/MSP externals were created. Finally, the C-code was implemented on the ARM microcontroller. While static parameters have been tested, and have yielded interesting results—especially in the case of the IF model—dynamically-controlled mappings have yet to be explored fully. As a many-to-many mapping is likely required, some implicit mapping using computational neural networks are in progress, as is a PC-side communication layer for using the device with the IDMIL's graphical mapping tools.

5.3.3 Case Study: Haptic Interaction with Virtual Models

5.3.3.1 Virtual Musical Instruments

A mapping concept that is sometimes used for digital musical instruments is the *virtual* musical instrument. Mulder [28] proposed that there may be advantages to using a representation of an object in virtual space for interaction with an input device, whose dynamic properties can in turn be used for control of sound.

This concept brings to mind classical ideas of immersive virtual reality, in which the input device is not used to directly modulate input signals to a synthesis system, but is rather used for interaction with an environment. Sound synthesis is then generated based on this environment. In short, the design of virtual instruments can be seen as the design of digital musical instruments constructed in a virtual space and subject to the possibilities afforded by computer modeling. It is even possible, for example, to swap input devices entirely, keeping the virtual instrument constant.

Since the virtual instrument may have a number of degrees of freedom that are more or less than the input device used to interact with it, the VMI can be seen as an M-to-N mapping layer. If the VMI implements physical dynamics, this mapping layer may even be time-variant, non-linear, and potentially quite complex. The ability for a musician to deal with a virtual instrument will be dependent on the input device used to interact with it and the demands the particular choice of VMI behaviour impose. On the other hand, the idea of representing mapping as an "object" can provide a player with a palpable metaphor which may help to understand and internalize otherwise abstract relationships (Fig. 5.4) [28].

Fig. 5.4 Abstract functional gesture-sound mapping (left) vs. mapping based on a virtual instrument layer (right).

The concept of *object* in virtual instruments naturally brings forward an idea of tangibility. One wants to reach out and touch these virtual objects which are only manipulable through the restricted channels of the given input device. One can conceptually "push" on an object using a force sensor, for example, but not feel its inertia. A way of realising a more intimate and immediate connection with virtual objects is to add *haptic feedback*.

Haptic technology, meaning the sense of touch, provides a means to synthesize vibrations and real forces by using actuators, such as DC motors, in response to human input. Using a high speed digital feedback loop that implements a simulation, it is possible to create the impression of walls, textures, and other effects. See for example [19] for a summary of such techniques.

5.3.3.2 Physical Models

The virtual musical instrument as described by Mulder was a geometric shape parameterized by signals generated at an input device [27]. However, just as Cook pointed out that any physical object can be used as the basis for musical instrument design [10], so too can any virtual object modeled on the computer. A more or less obvious candidate is the physical model.

In a physical model, a numerical simulation of a physical system is integrated through time. Physical models are exploited readily in computer graphics, for example, to create convincing visual effects, but it is also possible to use them for controlling or generating sound and haptic feedback. In fact, Cadoz proposed that modeling physical principles is the only way to maintain a coherent energetic relationship during simulated instrumental interaction [3].

Acoustic models can be simulated using techniques such as mass-spring interaction [4], modal synthesis [1, 12] or the digital waveguide [37], allowing for realistic sound and haptic feedback. However, for control over *arbitrary* synthesis parameters—that is, to use physics as a general-purpose mapping layer—we have found it interesting to make use of rigid body simulation.

5.3.3.3 Rigid Body Virtual Instruments

Rigid bodies can be a useful tool for modeling collections of solid objects. They are often used in video games, for example, to create stacks of boxes or brick walls which can be destroyed. A typical rigid body simulator allows a variety of geometric shapes or possibly arbitrary shapes based on triangle meshes, and can model hard constraints between them. Hard constraints allow stable stacking behaviour, modeling of various joint types, and contact handling that avoids inter-penetration [40].

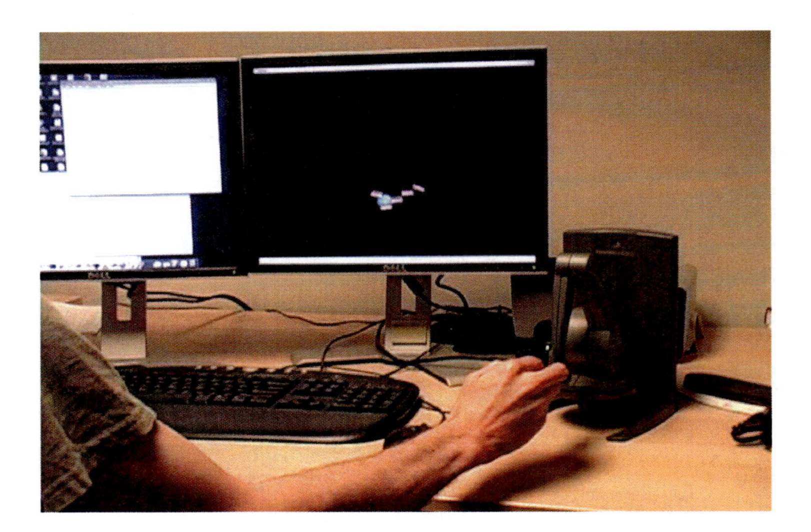

Fig. 5.5 A user interacting through a haptic device with a set of mutually constrained shapes in DIMPLE. Each object controls a frequency modulation synthesizer. The user is manipulating a Phantom Desktop haptic device from SensAble Technologies.

Using such a simulator, one can construct mechanical systems composed of rigid bodies that can be manipulated by an input device or a haptic interface. One can picture this process as designing a machine out of solid parts based on hinges, axles, and levers. Movement of the bodies and their intercollisions can then be used as continuous or event-based control signals for sound synthesis.

We have designed a software system, called DIMPLE[4] based on this idea which can be used to construct real-time virtual mechanisms to be probed with a haptic device, and communicates with several popular audio synthesis software packages [36].

Since haptics has different timing requirements than audio, it is typical in software systems to have a concurrent operating system thread or process called the *servo loop* which runs with high priority and whose sole responsibility is to calculate feedback forces and handle communication with the haptic device. The idea behind DIMPLE was to provide this loop as a real-time programmable rigid body engine. Objects and their properties can be created and accessed entirely using the Open Sound Control communication protocol [41], which is accessible to many popular audio programming environments.

To date DIMPLE has been used with Pure Data [32], Max/MSP (Cycling '74), SuperCollider [24], and ChucK [38]. Initially, it presents a window containing an empty background and a small 3D sphere representing the position of the haptic device handle. This object is called the *cursor*, or sometimes the *proxy* object. The user program sends messages to the DIMPLE process instructing it to instantiate shapes and constraints.

These shape objects are immediately "touchable" by the haptic device, and respond to forces. This then provides some tools necessary for building mechanical contraptions that have a desired interactive behaviour. The audio environment can then send requests to DIMPLE for a given object property, such as position, velocity, applied forces and torques, or collisions, and use this data as control for synthesis parameters.

5.4 Conclusions

We have presented some motivations and design principles of digital musical instruments, and provided examples under various configurations of design constraints. Firstly, to explore the possibility of using sensor technology to fully capture the affordances of an object; secondly, to deal with the difficulties of strict design constraints in uncooperative environments; thirdly, to use actuated interfaces to manifest physical interaction with virtual objects.

A topic that has not been fully covered here is the complexity of choices in mapping. As mentioned, for DMIs it is necessary to explicitly provide the connection to sound, and there are a myriad of ways to accomplish this goal. There may be no "best" answer to this problem for any particular situation, but ongoing research attempts to solidify some principles for mapping based on what

---

[4] The Dynamically Interactive Musically PhysicaL Environment, available at `http://idmil.org/software/dimple`

we know about human interaction, design, ergonomics, and, of course, the needs of musical composition and aesthetics. We encourage the interested reader to follow up the references presented in this chapter to get a more complete overview of these topics.

Thanks to the Natural Sciences and Engineering Research Council of Canada and the Canadian Foundation for Innovation for funding of this work.

References

[1] Adrien, J.-M.: Physical model synthesis: The missing link. In: Poli, G.D., Piccialli, A., Roads, C. (eds.) Representations of Musical Signals, pp. 269–297. MIT Press, Cambridge (1991)

[2] Cadoz, C.: Instrumental Gesture and Musical Composition. In: Proc. of the 1988 International Computer Music Conference, pp. 1–12. International Computer Music Association, San Francisco, Calif (1988)

[3] Cadoz, C.: Retour au réel: le sens du feedback. In: Rencontres Musicales Pluridisciplinaires: Le Feedback dans la Création Musicale (2006)

[4] Cadoz, C., Luciani, A., Florens, J.-L.: CORDIS-ANIMA: A Modeling and Simulation System for Sound and Image Synthesis - The General Formalism. Computer Music J. 17(1), 19–29 (1993)

[5] Cadoz, C., Wanderley, M.M.: Gesture-Music. In: Trends in Gestural Control of Music, Ircam - Centre Pompidou (2000)

[6] Cage, J.: Silence: Lectures and Writings. Wesleyan University Press, Middletown (1961)

[7] Cariou, B.: Design of an Alternate Controller from an Industrial Design Perspective. In: Proc. of the 1992 International Computer Music Conference, pp. 366–367. International Computer Music Association, San Francisco (1992)

[8] Cariou, B.: The aXiO MIDI Controller. In: Proc. of the 1994 International Computer Music Conference, pp. 163–166. International Computer Music Association, San Francisco (1994)

[9] Clarke, E.F.: Generative processes in music: the psychology of performance, improvisation, and composition, ch. 1, pp. 1–26. Clarendon Press, Oxford (1988)

[10] Cook, P.: Principles for designing computer music controllers. In: Proceedings of the 2001 Conference on New Interfaces for Musical Expression, pp. 1–4. National University of Singapore, Singapore (2001)

[11] Fitzmaurice, G.W., Ishii, H., Buxton, W.: Bricks: Laying the foundations for graspable user interfaces. In: Proceedings of ACM CHI 1995, Denver, Colorado, pp. 442–449 (May 1995)

[12] Florens, J.-L.: Expressive bowing on a virtual string instrument. In: Camurri, A., Volpe, G. (eds.) GW 2003. LNCS (LNAI), vol. 2915, pp. 487–496. Springer, Heidelberg (2004)

[13] Haken, L., Tellman, E., Wolfe, P.: An Indiscrete Music Keyboard. Computer Music J. 22(1), 31–48 (1998)

[14] Hollinger, A., Steele, C., Penhune, V., Zatorre, R., Wanderley, M. M.: fMRI-compatible electronic controllers. In: Proceedings of the International Conference on New Interfaces for Musical Expression, pp. 246–249. ACM Press, New York (2007)

[15] Hollinger, A., Thibodeau, J., Wanderley, M.M.: An embedded hardware platform for fungible interfaces. In: Proceedings of the International Computer Music Conference, ICMA, pp. 26–29 (2010)
[16] Jensenius, A.R., Kvifte, T., Godøy, R.I.: Towards a gesture description interchange format. In: Proceedings of the Conference on New Interfaces for Musical Expression, Paris, pp. 176–179. IRCAM – Centre Pompidou (2006)
[17] Jordà, S.: Sonigraphical instruments: from FMOL to the reacTable. In: Proceedings of the Conference on New Interfaces for Musical Expression, Montreal, Canada, pp. 70–76 (2003)
[18] Koehly, R., Curtil, D., Wanderley, M.M.: Paper FSRs and latex/fabric traction sensors: Methods for the development of home-made touch sensors. In: NIME 2006: Proceedings of the 2006 conference on New interfaces for musical expression, Paris, pp. 230–233. IRCAM—Centre Pompidou (2006)
[19] Laycock, S., Day, A.: A survey of haptic rendering techniques. Computer Graphics Forum 26(1), 50–65 (2007)
[20] Levitin, D., McAdams, S., Adams, R.L.: Control parameters for musical instruments: a foundation for new mappings of gesture to sound. Organised Sound 7(2), 171–189 (2002)
[21] Malloch, J., Birnbaum, D., Sinyor, E., Wanderley, M.M.: Towards a new conceptual framework for digital musical instruments. In: Proc. of the Int. Conf. on Digital Audio Effects (DAFx 2006), Montreal, Quebec, Canada, Sept. 18–20 2006, pp. 49–52 (2006)
[22] Malloch, J., Sinclair, S., Wanderley, M.M.: A Network-Based Framework for Collaborative Development and Performance of Digital Musical Instruments. In: Kronland-Martinet, R., Ystad, S., Jensen, K. (eds.) CMMR 2007. LNCS, vol. 4969, pp. 401–425. Springer, Heidelberg (2008)
[23] Malloch, J., Wanderley, M.M.: The T-Stick: From musical interface to musical instrument. Proceedings of the 2007 International Conference on New Interfaces for Musical Expression (NIME 2007), New York City, USA (2007)
[24] McCartney, J.: Rethinking the computer music language: SuperCollider. Computer Music Journal 26, 61–68 (2002)
[25] Miranda, E.R., Wanderley, M.M.: New Digital Instruments: Control and Interaction Beyond the Keyboard. A-R Publications, Middleton (2006)
[26] Morris, G.C., Leitman, S., Kassianidou, M.: SillyTone squish factory. In: Proceedings of the 2004 Conference on New Interfaces for Musical Expression, Hamamatsu, Shizuoka, pp. 201–202 (Japan 2004)
[27] Mulder, A.: Virtual musical instruments: Accessing the sound synthesis universe as a performer. In: Proceedings of the First Brazilian Symposium on Computer Music, pp. 243–250 (1994)
[28] Mulder, A.G.E., Fels, S.S., Mase, K.: Mapping virtual object manipulation to sound variation. IPSJ SIG Notes 122, 63–68 (1997)
[29] Overholt, D.: The musical interface technology design space. Organised Sound 14(2), 217–226 (2009)
[30] Paradiso, J.A., Hsiao, K.Y., Benbasat, A.: Tangible music interfaces using passive magnetic tags. In: Workshop on New Interfaces for Musical Expression - ACM CH 2001, Seattle, USA, pp. 1–4 (April 2001)

[31] Pestova, X., Donald, E., Hindman, H., Malloch, J., Marshall, M.T., Rocha, F., Sinclair, S., Stewart, D.A., Wanderley, M.M., Ferguson, S.: The CIRMMT/McGill digital orchestra project. In: Proc. of the 2009 International Computer Music Conference. International Computer Music Association, San Francisco (2009)
[32] Puckette, M.: Pure Data: another integrated computer music environment. In: Proceedings, Second Intercollege Computer Music Concerts, Tachikawa, Japan, pp. 37–41 (1996)
[33] Rasmussen, J.: Information Processing and Human-Machine Interaction: an Approach to Cognitive Engineering. Elsevier Science Inc., New York (1986)
[34] Rovan, J., Hayward, V.: Typology of tactile sounds and their synthesis in gesture-driven computer music performance. In: Wanderley, M. M., Battier, M. (eds.) Trends in Gestural Control of Music, Paris, pp. 297–320. IRCAM (2000)
[35] Ryan, J.: Some Remarks on Musical Instrument Design at STEIM. Contemporary Music Review 6(1), 3–17 (1991)
[36] Sinclair, S., Wanderley, M.M.: A run-time programmable simulator to enable multi-modal interaction with rigid-body systems. Interact. Comput. 21(1-2), 54–63 (2009)
[37] Smith, J.O.: Physical modeling using digital waveguides. Comp. Mus. J. 16(4), 74–91 (1992)
[38] Wang, G., Cook, P.: ChucK: a programming language for on-the-fly, real-time audio synthesis and multimedia. In: MULTIMEDIA 2004: Proceedings of the 12th Annual ACM International Conference on Multimedia, New York, NY, pp. 812–815 (2004)
[39] Wessel, D., Wright, M.: Problems and prospects for intimate control of computers. Computer Music Journal 26(3), 11–22 (2002)
[40] Witkin, A., Gleicher, M., Welch, W.: Interactive dynamics. SIGGRAPH Comput. Graph. 24(2), 11–21 (1990)
[41] Wright, M., Freed, A.: OpenSoundControl: A new protocol for communicating with sound synthesizers. In: Proceedings of the International Computer Music Conference, ICMA (1997)

Chapter 6
Capturing Bowing Gesture: Interpreting Individual Technique

Diana S. Young

Abstract. Virtuosic bowed string performance in many ways exemplifies the incredible potential of human physical performance and expression. Today, a great deal is known about the physics of the violin family and those factors responsible for its sound capabilities. However, there remains much to be discovered about the intricacies of how players control these instruments in order to achieve their characteristic range and nuance of sound. Today, technology offers the ability to study this player control under realistic, unimpeded playing conditions to lead to greater understanding of these performance skills. Presented here is a new methodology for investigation of bowed string performance that uses a playable hardware measurement system to capture the gestures of right hand violin bowing technique. This measurement system (which uses inertial, force, and electric field position sensors) was optimized to be small, lightweight, and portable and was installed on a carbon fiber violin bow and an electric violin to enable study of realistic, unencumbered violin performances. The application of this measurement system to the study of standard bowing techniques, including *détaché*, *martelé*, and *spiccato*, and to the study of individual players themselves, is discussed.

6.1 Violin Performance

One of the primary reasons that the violin is such a difficult instrument to master is that the playing techniques required of the left hand and the right hand are very different from each other (unlike other instruments, such as the piano, that are played with similar left and right hand techniques). The left hand of a violinist not only controls pitch (by stopping the string against the surface of the violin fingerboard, effectively shortening the string length), but also can control changes in

Diana S. Young
Wyss Institute for Biologically Inspired Engineering
Harvard University
Boston, MA
USA
e-mail: diana.young@wyss.harvard.edu

J. Solis and K. Ng (Eds.): Musical Robots and Interactive Multimodal Systems, STAR 74, pp. 85–103.
springerlink.com

timbre by means of vibrato (small alterations in pitch). Good right hand technique requires an accomplished violinist to be proficient in a vast number of bowing methods (such as *détaché*, *martelé*, and *spiccato*) in order to obtain the range of expression demanded in standard repertoire [3, 12, 13].

Despite the formidable obstacles imposed by the difficult playing interface of the violin, many musicians still devote themselves to its pursuit, committing years to obtain decent sound and decades to achieve mastery. As reward for this dedication, the violin offers tremendous range of tone color and expressive powers that are unsurpassed. But violin playing technique is not just difficult to learn, it is also quite difficult to articulate for many despite important pedagogical work in documenting playing exercises, as well as postures, and physical instructions designed to improve many different techniques (e.g., [2, 10]). Of course, the difficulties of bowing technique find some explanation in the physics that govern the violin bow and strings (discussed in detail in [6]). This bow-string interaction is discussed below.

6.2 The Bowed String

The characteristic sound of a bowed string instrument is due to the phenomenon of Helmholtz motion, depicted in Figure 6.1. When Helmholtz motion is achieved by a player, the string forms a corner that travels in a parabolic path back and forth between the bridge and nut of the violin. When this corner is between the bridge and the bow, the string "slips" and moves in the direction opposite to that of the bow motion. When the corner is between the bridge and the nut, the string "sticks" to the bow hair and therefore moves with the same velocity as the bow.

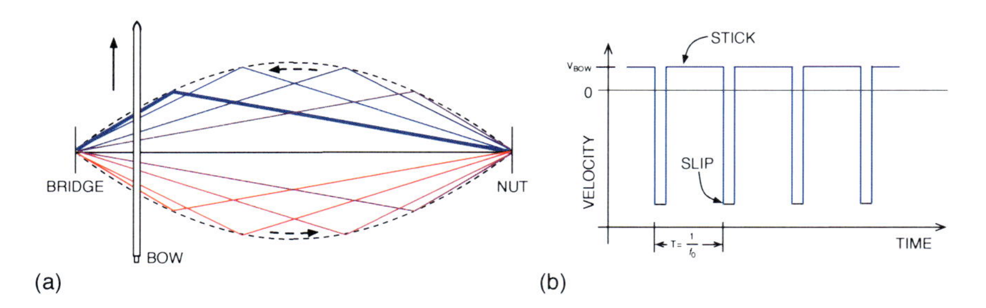

Fig. 6.1 (a) illustrates the string in Helmholtz motion; (b) describes the string velocity during the sticking and slipping intervals, from [9].

This characteristic "slip-stick" behavior, in which the string slips just once in its period, occurs because of the friction component inherent in bow-string interaction. For a given bow-bridge distance, the achievement of Helmholtz motion (the goal of all bowed-string players) depends on the careful negotiation between bow speed and force. As described in [36], for steady bowing, a player must control the bow speed v_{bow}, position β (bow-bridge distance, normalized to the length between the bridge and nut of the violin, assuming an open string), and the normal force between the bow and the string F. If F is too low, the bow will not stick to the string and will produce what is known as "surface sound". If the force is too high, the string does not release when the Helmholtz attack occurs, and the motion becomes raucous. From these two extremes, the range of force for normal play has been analyzed as follows:

$$F_{\text{max}} = \frac{2Z_0 v_{\text{bow}}}{\beta(\mu_s - \mu_d)} \tag{6.1}$$

$$F_{\text{min}} = \frac{Z_0^2 v_{\text{bow}}}{2\beta^2 R(\mu_s - \mu_d)}, \tag{6.2}$$

where μ_s and μ_d are the static and dynamic friction coefficients, respectively. Z_0 is the characteristic impedance of the string and R indicates the equivalent of the rate of energy loss into the violin body.

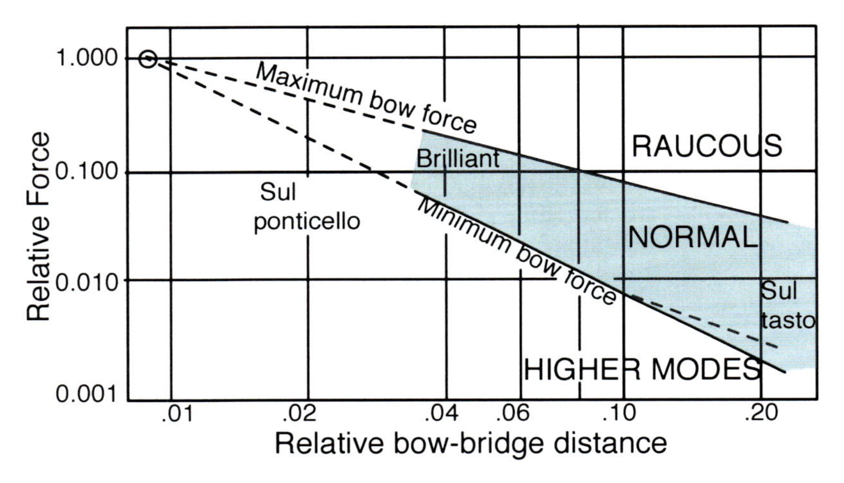

Fig. 6.2 Schelleng diagram showing the region of playability between the minimum and maximum force required for normal violin playing, with respect to bow-bridge distance (bow speed is constant).

For constant bow speed, the dependency between the remaining two parameters, force and position (bow-bridge distance), in creating good tone may be illustrated in a two-dimensional plot. This visualization, first developed by Schelleng [26] and shown in Figure 6.2, outlines a region of "playability" wherein good tone is achieved. Note, this represents a strong idealization, as in any typical musical performance, the bow speed is continually varying.

The key point of this discussion is that the relationship between bowing parameters and the sound they produce is quite complex due to the nonlinear friction mechanism described [11, 14, 37-38]. Unlike the case of the piano playing, in which key velocity is the primary physical input parameter, there is no invertible mapping between inputs and output. In fact, the situation is one of a "many-to-one" mapping. This point concerning the mapping between the input bowing parameters and output sound can be further understood by considering a simple model of a violin (or viola, cello, or double bass), which has a transverse bridge force given by [5]:

$$\hat{F}_b(t) = Z_b \Delta v(t) \tag{6.3}$$

where Z_b is the bridge impedance, and Δv is the difference in velocity of the incoming and outgoing traveling waves in the string at the bridge. In the regime of Helmholtz motion, Δv is (to first approximation) a sawtooth function with period $T = 2l/c$, l is the length of the string including finger input and c is the speed of the wave. The maximum of $\hat{F}_b(t)$ has been determined to be

$$\hat{F}_{b,\max} = \frac{Z_b}{\beta} v_{\text{bow}} \tag{6.4}$$

The transverse bridge force is what drives the body of the instrument to produce the sound (a linear transformation given the frequency response of the body cavity). Equation 4 shows that there are multiple values of bow velocity and bow-bridge distance that can achieve the same transverse bridge force, and hence the Helmholtz motion. Therefore, though the sound may be predicted when the bowing parameters are known, it is not possible to determine the bowing parameters used from inspection of the audio result. (In consideration of this point, audio recordings of bowed string playing can be seen as rather incomplete representations, as they do not contain all of the relevant information to reconstruct these performances.)

6.3 Measurement of Bowing Parameters

Because violin bowing parameters cannot be determined from audio alone, they are the topic of keen research interest [1, 7-8, 27-29, 33, 39]. The primary motivation of the study described herein is to investigate the potential of physical measurement of violin bowing dynamics to inform understanding of bowed string performance, to answer such questions as: *What do musicians do in order to achieve the sonic results they desire? How do they differ in their approaches to the same performance tasks? What are the actual limits of their physical capabilities?*

Accurate and reliable measurement of violin bowing gesture for use in the study of realistic performance presents many challenges. In designing the measurement system used in the bowing technique study presented here, the highest priority was the goal to maintain playability of the measurement system, so that the traditional bowing technique that is of interest in this work remains unimpaired to facilitate related research [43-45]. Therefore, great effort was spent to ensure that the electronics created are as small and light as possible, free from unnecessary wires that would constrain player movement, and that the bow itself remains comfortable to use.

6.3.1 Sensing System

The hardware sensing system shown in Figure 6.3 consists of four types of sensors: force sensors (composed of foil strain gauges), accelerometers, gyroscopes, and an array of four electric field position sensors.

6.3.1.1 Force Sensing

In order to sense the downward bow force (as in [41-42]), foil strain gauges from Vishay® Micro-Measurements [16] were used. These devices were chosen due to their large bandwidth (small capacitance and inductance) and negligible hysteresis. These features were both highly desired, as violin bowing exhibits rapid changes in force that must be accurately recorded.

Two force sensors, each composed of four foil strain gauges, in full Wheatstone Bridge configuration, were installed on the carbon fiber bow stick and calibrated [43] in order to provide "downward" (normal to the string in standard playing position) and "lateral" (orthogonal to the string in standard playing position) force measurements.

6.3.1.2 Acceleration and Angular Velocity Sensing

To enable measurement of bow velocity and bow tilt with respect to the violin (which is often related to the area of bow hair in contact with the string) a combination of 3D acceleration sensing (as in [41-42], similar to that used in [24-25]) and 3D angular velocity sensing, comprising a six degrees of freedom (6DOF) inertial measurement unit (IMU), is implemented within a bow sensing subsystem. In addition to this 6DOF IMU on the bow, an identical one is included in the violin sensing subsystem, as seen in Figure 6.3. Both of these 6DOF IMUs measure tilt with respect to gravity, and so pitch and roll angles between bow and violin can be accurately estimated (see below).

Although the measurement system included two 6DOF IMUs that can be used to estimate the bow-bridge distance and bow velocity measurements, the errors on these two estimates increase quadratically with respect to time and linearly with respect to time, respectively, due to accumulating integration errors. Therefore, an additional position sensor was required to provide both accurate and precise estimates of bow-bridge distance and bow velocity.

6.3.1.3 Position and Velocity Sensing

In order to improve the bow-bridge distance and bow velocity estimates, the original electric field bow position sensor, first designed for use in the Hypercello project and detailed in [22], and adopted in the earlier Hyperbow systems [41-42] was retained. The basic design of this sensor includes a resistive strip (composed of carbon-impregnated plastic from UPM [34]) extending from the frog of the bow to the tip. From either end of this resistive strip, square wave signals are transmitted. These signals are received by an antenna mounted behind the bridge of the violin, and their corresponding magnitudes are measured to estimate bow-bridge distance and tip-frog bow position (x-axis and y-axis, respectively).

6.3.2 Parameter Estimation

Due to the growing errors in time from the drift in the accelerometer and gyroscope data, the linear acceleration and angular velocity estimates are refined using a Kalman filter (via an Extended Kalman Filter, as the update dynamics are nonlinear) [4]. There are two main steps in any Kalman filter algorithm: the prediction step and the update step. In the prediction step, the state of the system and its error covariance matrix are predicted using an *a priori* model. In the update step, the state and the error covariance are corrected using actual measurements and a Kalman gain matrix that minimizes the error covariance matrix.

By applying Kalman Filtering, as well as proper sensor calibration techniques, the primary bowing parameters of bow force, bow velocity, and bow-bridge distance may be estimated, as seen in Figure 6.4. Here these three physical bowing parameters, as well as an approximation of bow tilt, are shown for a performance of *martelé* bowing technique.

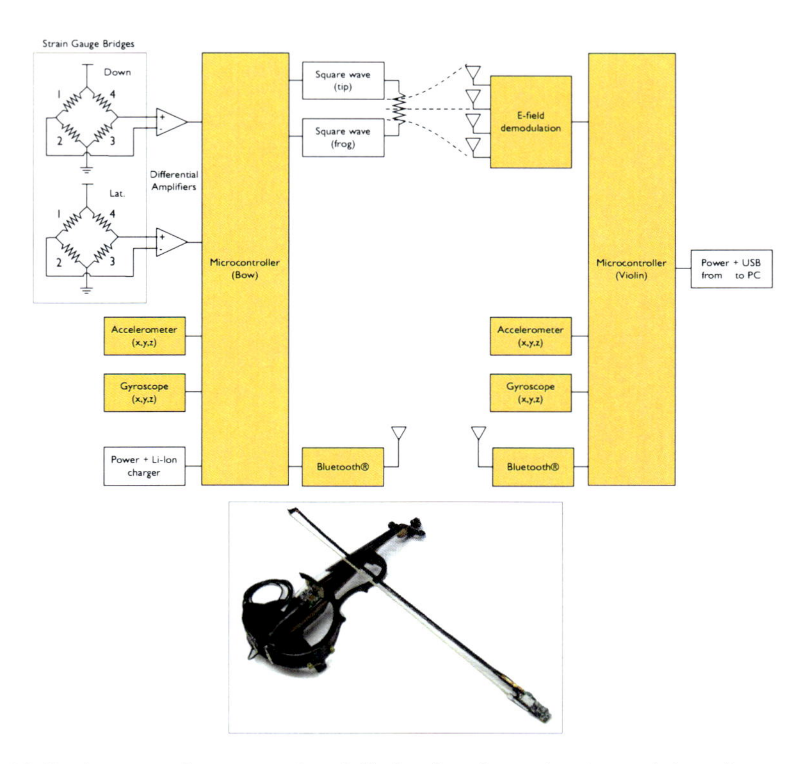

Fig. 6.3 Hardware sensing system for violin bowing (lower inset), consisting of: two strain gauge force sensors to capture both downward and lateral bending in the bow stick; 3D acceleration and 3D angular velocity sensing on both the violin and the bow; an electric field sensor for each of the four violin strings. In this implementation [43], the bow sub-system remains wireless, as a Bluetooth® module is used for data communication to the violin and its power is supplied by a Lithium-ion battery. The violin sub-system is powered via the USB port, which also was used for data communication (of both the bow and violin data) to the computer. The gesture data was synchronized with the audio by recording the gesture data on an independent audio track with the recorded violin audio in the software application Pure Data (Pd).

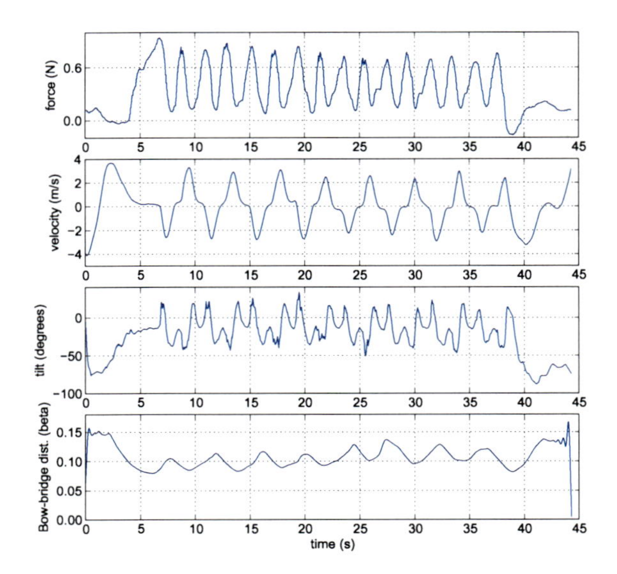

Fig. 6.4 Estimates of bow force, bow velocity, bow tilt angle, and bow-bridge distance parameters produced during a performance of *martelé* bowing technique.

6.4 Studying Bowing Technique

The study of right hand bowing technique is often pursued by using measurement that incorporates an established global coordinate system, such as optical motion capture [7-8, 18-20, 27-28, 30-31, 33, 39]. Because the custom measurement system described above relies on local sensing only, it must be validated. In service of this, a study was conducted to determine if the system is capable of capturing the distinctions between common bowing techniques. In this study, gesture and audio data generated by eight violinists performing six different bowing techniques on each of the four violin strings were recorded for later analysis. The details of the study protocol, experimental setup, and participants are discussed below.

6.4.1 Study Protocol

In this study each of the eight conservatory student participants was asked to perform repetitions of a specific bowing technique originating from the Western classical music tradition. To help communicate the kind of bowstroke desired, a musical excerpt (from a work of the standard violin repertoire) featuring each bowing technique was provided from [3]. In addition, an audio example of the bowing technique for each of the four requested pitches was provided to the player. The bowing technique was notated clearly on a score, specifying the pitch and string, tempo, as well as any relevant articulation markings, for each set of the recordings.

Two different tempos were taken for each of the bowing techniques (on each pitch). First, trials were conducted using a characteristic tempo for each individual bowing technique. Immediately following these, trials were conducted using one common tempo. Though the target trials were actually those that were conducted with the same tempo across all of the bowing techniques, it was found in early pilot testing that requesting performances using the characteristic tempo first enabled the players to perform at the common tempo with greater ease. The tempos required for each bowing technique were provided by a metronome. In some cases, a dynamics marking was written in the musical example, but the participants were instructed to perform all of the bowstrokes at a dynamic level of *mezzo forte*. Participants were instructed to take as much time as they required to either play through the musical example and/or practice the technique before the start of the recordings to ensure that the performances would be as consistent as possible.

Three performances of each bowing technique, comprising one trial, were requested on each of the four pitches (one on each string). During the first preliminary set of recording sessions that were conducted in order to refine the experimental procedure, participants were asked to perform these bowing techniques on the "open" strings. (The rationale for this instruction was that the current measurement system does not capture any information concerning the left hand gestures.) However, it was observed that players do not play as comfortably and naturally on "open" strings as when they incorporate left hand fingering. Therefore, in the subsequent recording sessions that comprise the actual technique study, the participants were asked to perform the bowing techniques on the fourth interval above the open string pitch, with no vibrato.

The bowing techniques included in this study are *accented détaché, détaché lancé*, *louré*, *martelé*, *staccato*, and *spiccato*. The study instructions for the *accented détaché* technique are shown in Figure 6.5. For this technique, the characteristic tempo and the common tempo were the same (this common tempo was later used in for the subsequent bowing techniques, after trials using the characteristic tempo for each technique were recorded).

6.4.2 Experimental Setup

In each trial of the bowing technique study, the physical gesture data were recorded simultaneously with the audio data produced in the performances of each technique. These gesture data were converted into an audio channel and combined into a multi-channel audio track with the recorded violin audio. By ensuring the audio and gesture capture delay was sufficiently small, all data could be maintained on an identical time base with no synchronization events required. The experimental setup was simple: custom violin bowing measurement system installed on a CodaBow® Conservatory™ violin bow [5] and the Yamaha SV-200 Silent Violin [40]; headphones (through which the participants heard all pre-recorded test stimuli and real-time sound of the test violin); M-Audio Fast Track USB audio interface [15]; Apple Macbook with a 2 GHz Intel Core Duo processor (OS X) running PureData (Pd) [23].

Detaché

Franck, *Sonata in A Major* for Violin and Piano (last movement)

Example of *accented détaché*

Fig. 6.5 This figure shows the study instructions for the *accented détaché* bowing technique, including a musical example from [3]. Participants were asked to perform each of the four lines three times (constituting one trial).

6.5 Technique Study Results and Discussion

The primary goal of the technique study was to determine whether the gesture data provided by the measurement system would be sufficient to recognize the different six bowing techniques (*accented détaché*, *détaché lancé*, *louré*, *martelé*, *staccato*, and *spiccato*) played by the eight violinist participants.

To begin this classification exploration, a subset of the gesture data provided by the measurement system was considered for the evaluations. Included in the analyses were data from the eight bow gesture sensors only: the downward and lateral forces; *x*, *y*, *z* acceleration; and angular velocity about the *x*, *y*, and *z* axes.

This data subset was chosen to explore the progress possible with the augmented bow alone (without the additional violin sensing subsystem), as this subsystem may easily be adapted to other instruments if shown to be sufficient.

The k-Nearest-Neighbor (k-NN) algorithm was chosen for data classification, as it is one of the simplest machine learning algorithms and robust for well-conditioned data. Compared to other supervised classification algorithms, such as Neural Networks (NN), Support Vector Machines (SVM) and decision trees, k-NN has the advantages of a simple, clear, heuristic-free geometric interpretation which can be compute efficient for low-dimensional data sets, but the disadvantage of not producing results with as high performance of some of the afore mentioned algorithms (especially SVM). In this study, the dimensionality of the gesture data vector used, 9152 (1144 samples in each time series x 8 gesture channels), was far too high. Therefore, the dimensionality of the gesture data set was first reduced before being input to the classifier.

6.5.1 Data Preparation

Before beginning any data analysis, the bow gesture data was visually inspected to search for obvious outliers. The data from each of the eight bow sensors (lateral force and downward force; x, y, z acceleration; and angular velocity about the x, y, and z axes), as well as the audio waveform produced, were plotted for each of the recorded files that comprise the bowing technique study. In all, there were 576 recorded files (8 players x 6 techniques x 4 strings x 3 performances of each) from the bowing technique study. Each recording included 16 instances of each bow-stroke. For this exercise, the entire time-series length, 1144 samples, of each gesture recording was used. The gesture data for each of the recordings were aligned with the start of the first attack in the audio. (This computation across the full data set was done in Matlab.) With the help of the audio example of each bowing technique and the metronome, the players in this study were able to produce highly repeatable bowing data.

This precision is evidenced in Figure 6.6, which shows gesture data from three performances of one technique on the first individual string by the same player. This visual inspection was conducted for the player's performances of the same technique on each of the three remaining strings. Then, an overlay plot containing the data from all twelve examples (4 strings x 3 performances) of that player's technique was produced. This process was repeated once for each of the five other bowing techniques performed by that player. Then, the same was done for the data from each of the remaining seven players.

The eight channels of bow gesture data as well as the audio waveform of each trial recording were inspected for gross inconsistencies, such as timing errors made by the violinists. Throughout this visual evaluation process, 22 of the original 576 files (approximately 4%) were omitted and replaced with copies of files in the same class (technique, player, string). After performing this inspection process, the raw gesture data (now aligned with the beginning of the first audio attack of each recording and of the same length of 1144 samples), it was now possible to proceed to the dimensionality reduction phase of the analysis.

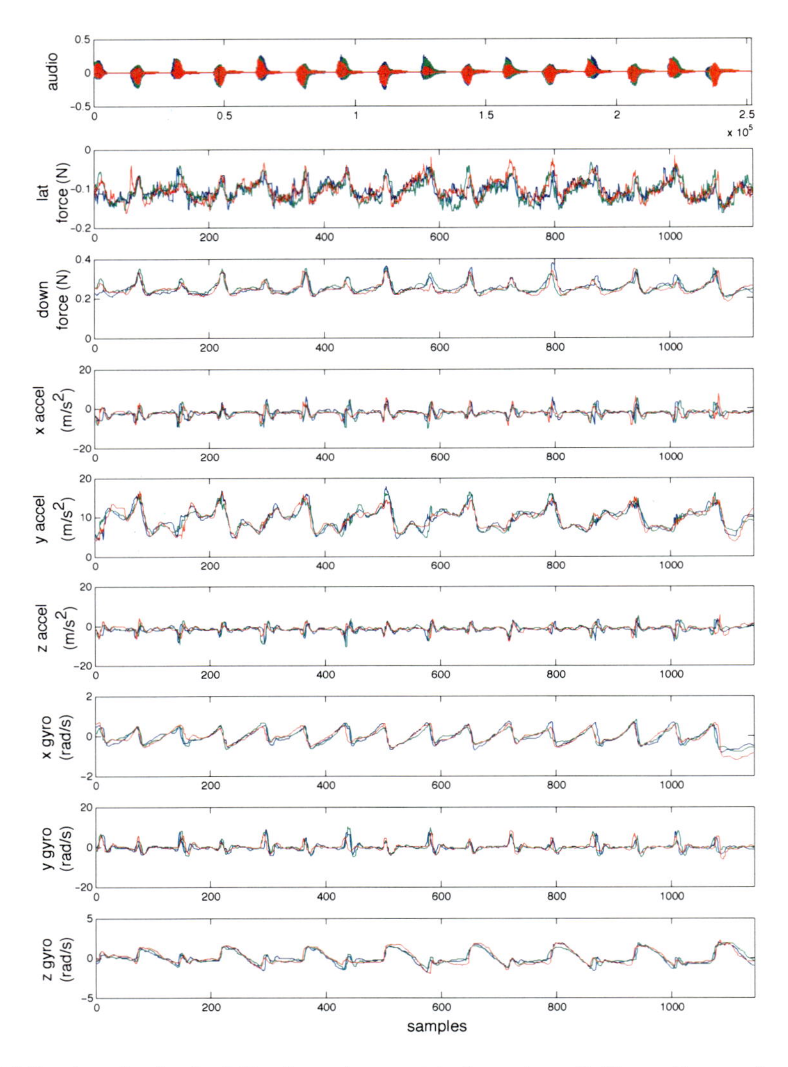

Fig. 6.6 Overlay plot for the 3 E string spiccato trials from player 3. The audio waveform, as well as the data from each of the 8 bow sensors is shown (length of each trial is 5.72 seconds).

6.5.2 *Dimensionality Reduction*

In order to prepare the data for classification, the dimensionality of the raw data was first reduced. Principal Component Analysis (PCA), also known as the Karhunen-Loève transform, is one common technique used for this purpose [32]. PCA is a linear transform that transforms the data set into a new coordinate system such that the variance of the data vectors is maximized along the first coordinate dimension (known as the first principal component). That is, most of the variance is represented, or "explained", by this dimension. Similarly, the second greatest variance is along the second coordinate dimension (the second principal component), the third greatest variance is along the third coordinate dimension (the third principal component), et cetera. Because the variance of the data decreases with increasing coordinate dimension, higher components may be disregarded for similar data vectors, thus resulting in decreased dimensionality of the data set. In this case, because the number of data samples used is smaller than the dimension of the data, the Singular Value Decomposition (SVD) offers an efficient means of calculating the principal components. PCA was chosen because it is the optimal linear transform for minimizing the least squares of each principle component relative to the original data. That is, PCA will determine the best linear features that describe the data.

Figure 6.7 shows one analysis made possible by the above dimensionality reduction algorithm. Here, a scatter plot represents the first 3 principal components corresponding to each of the 6 bowing techniques produced by one player participant. The clear visual separability between bowing techniques demonstrated by this exercise is taken as a strong indicator that technique classification may be successful. One important point is that while PCA will find the principal components of the data to minimize the least square error, it rarely gives insight into what these components represent. As a result, while an effective technique, this prevented this analysis exercise from extracting any conceptually relevant high-level features.

6.5.3 *Classification*

After computing the principal components produced by the SVD method above, the challenge of classifying the data was undertaken using the k-nearest-neighbor classifier. Specifically, Nabney's matlab implementation [17] was employed.

6.5.3.1 Technique Classification

The first classification exercise applied to the technique study bow data was an inspection of the performance of each individual player.

After reducing the dimensionality of two-thirds of the data from a *single player* by computing the principal components using SVD, the k-NN algorithm was trained on the same two-thirds of the data in order to classify the remaining one-third of the player's data. The number of nearest neighbors was maintained at one, while the number of principal components was increased from one to ten (the results given by using small numbers of components are only included to illustrate the trends in the results). Three-fold cross validation was performed by rotating the test/predict data three times. For each number of principal components, the mean and standard deviation of the cross-validation trials were computed to determine the overall success of the technique classification for this intra-player case.

This whole procedure was repeated for each of the other eight players. Good prediction rates for each player's gesture data were achieved for even low numbers of principal components. The results of this analysis for each individual player are shown in Figure 6.8, in which the effect of increasing the number of principal components on the overall classification success is clearly demonstrated. In fact, by increasing the number of principal components, very high predication rates were reached for each of the eight players, and for five of the eight players rates of over 90% were achieved using only three principal components. Of all of the eight violinists, only the gesture data provided by player 6 achieved a success rate significantly less than 90% with four principal components as input to the classifier.

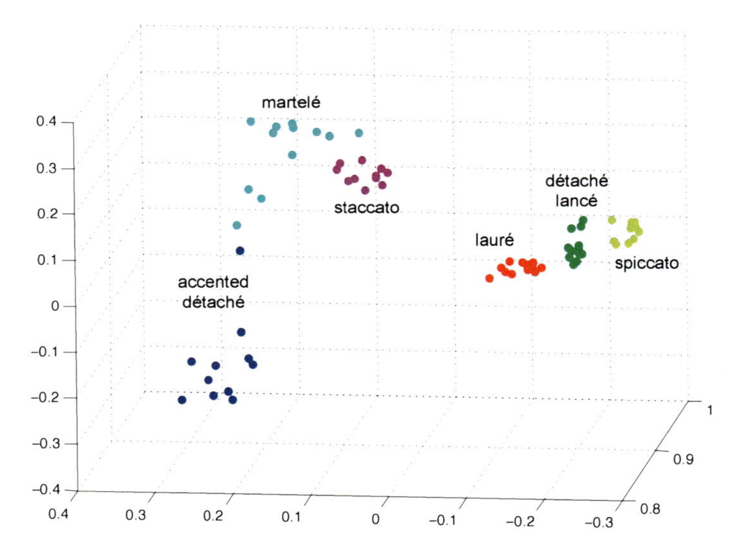

Fig. 6.7 Scatter plot of all six bowing techniques for player5: *Accented détaché*, *détaché lancé*, *louré*, *martelé*, *staccato*, *spiccato*. The axes correspond to the first three principal components.

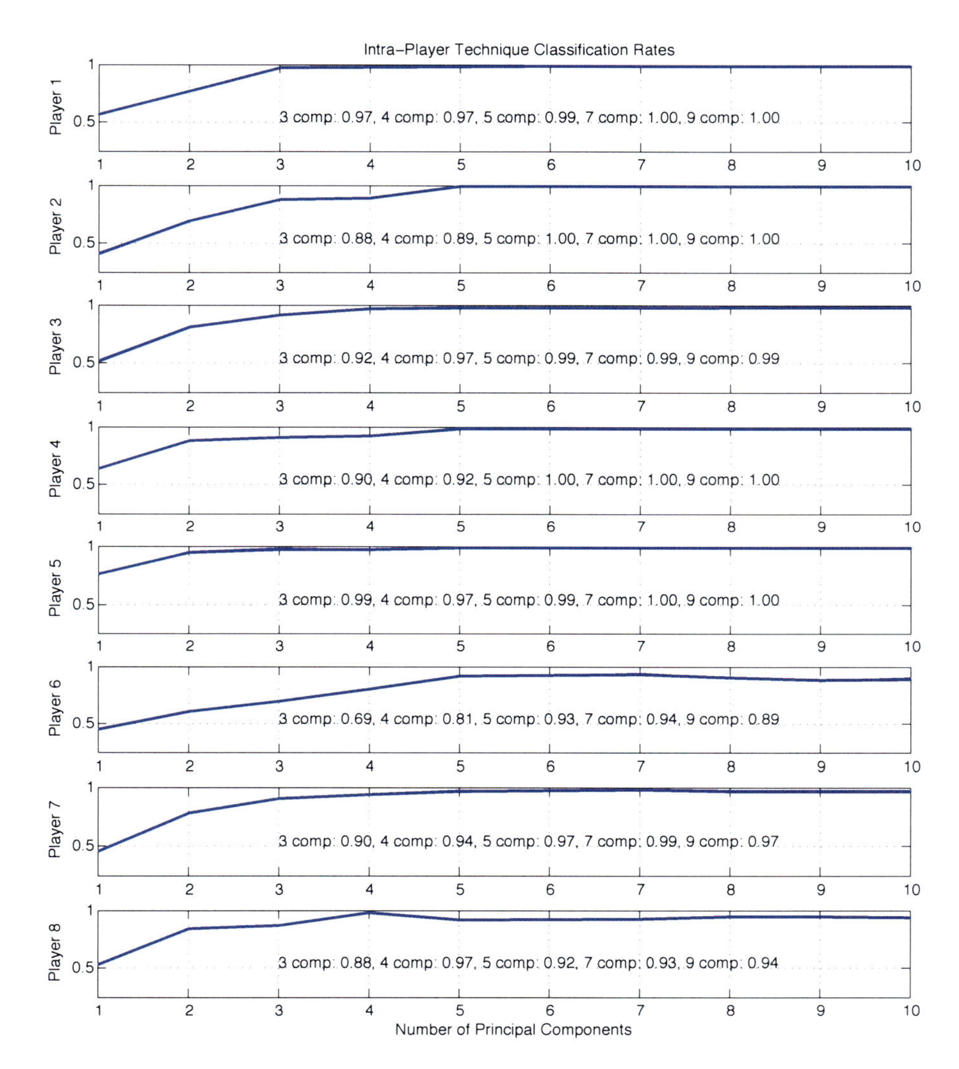

Fig. 6.8 Mean recognition rates versus the number of principal components used, produced by using the k-NN method (with one nearest neighbor) with two-thirds of each player's gesture data as training to predict the remaining third of that player's gesture data.

After performing the intra-player analysis discussed above, the k-NN classifier was trained using two-thirds of the data from *all of the players* to classify the remaining one-third of *all player data* by technique.

As in the intra-player case, the number of nearest neighbors was maintained as one, while the number of principal components was increased from one to ten. Again, three-fold cross validation was performed by rotating the training data three times and the three resultant classification rates were used to determine the mean and standard deviation of the overall classification success for this all-player case. Table 6.1 shows the confusion matrix produced by training on two-thirds of the data from each of the eight players, predicting the remaining third of each player's data (with overall prediction of 95.3±2.6%) with seven principal components. The effect on the overall success of the number of principal components is clearly illustrated by Figure 6.9. Using seven or more principal components as input to the k-NN algorithm, the mean classification rate is above 95%.

Table 6.1 Training on two-thirds of the data from each of the eight players, predicting the remaining third of each player's data by technique (with overall prediction of 95.3± 2.6%) using seven principal components.

actual \ class.	acc. detaché	det. lancé	louré	martelé	staccato	spiccato
acc. detaché	**0.938**	0.010	0.010	0.042	0.000	0.000
det. lancé	0.000	**0.917**	0.000	0.010	0.021	0.052
louré	0.000	0.000	**0.979**	0.000	0.021	0.000
martelé	0.042	0.021	0.000	**0.938**	0.000	0.000
staccato	0.000	0.010	0.010	0.000	**0.979**	0.000
spiccato	0.000	0.031	0.000	0.000	0.000	**0.969**

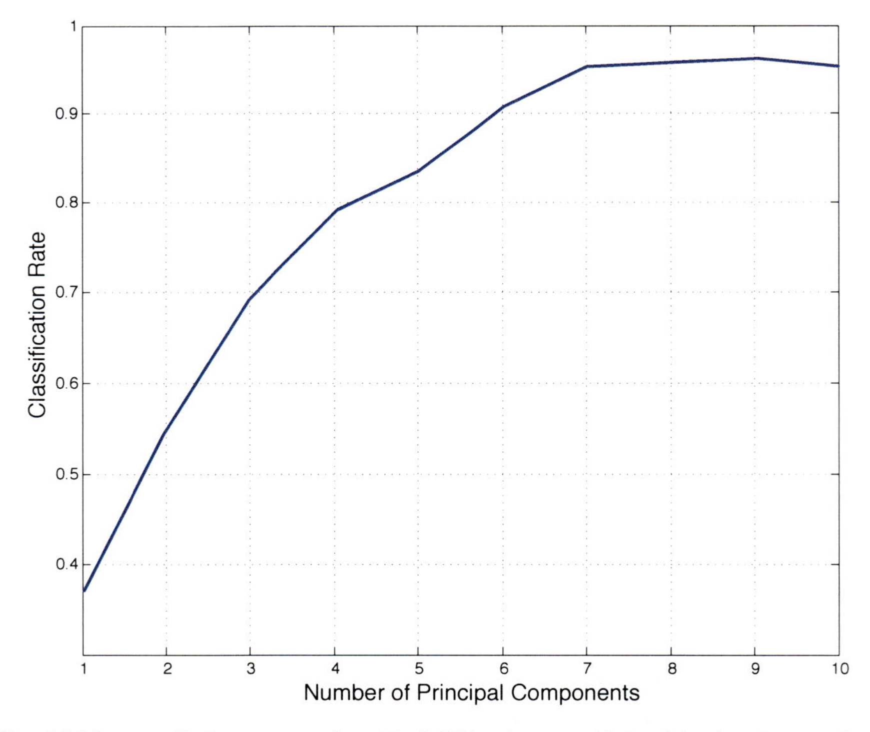

Fig. 6.9 Mean prediction rates produced by k-NN using two-thirds of the data from each of the eight players to predict the remaining one-third of all player data and increasing the principal components from one to ten.

6.5.3.2 Player Classification

After demonstrating the potential of the gesture data collected by the bowing measurement system for use in technique classification, the k-NN classification algorithm was employed to explore classification by player [9, 35].

In this exploration, only one case, the all-player case, was addressed. Just as before in performing the technique classification with the data from all of the players, the k-NN classifier was implemented using two-thirds of the data from *all of the players* to classify by player the remaining one-third of data from all of the players. But this time, the classification was done by *player*. Once again, a 3-fold cross validation procedure, in which the training data set was rotated three times, was obeyed. For each training data set, the number of principal components input to the algorithm was increased from two to twenty. The overall classification results for each number of components used were determined as the mean and standard deviation of the classification rates achieved in each of the three cross-validation trials (for each number of principal components used).

With 12 principal components, a player classification rate of nearly 90% (89.1% to be exact) was reached, and with 18 principal components, an overall classification player rate of 94.2% is obtained. Again, the effect of increasing the number of principal components was clearly demonstrated, as illustrated in the plot in Figure 6.10.

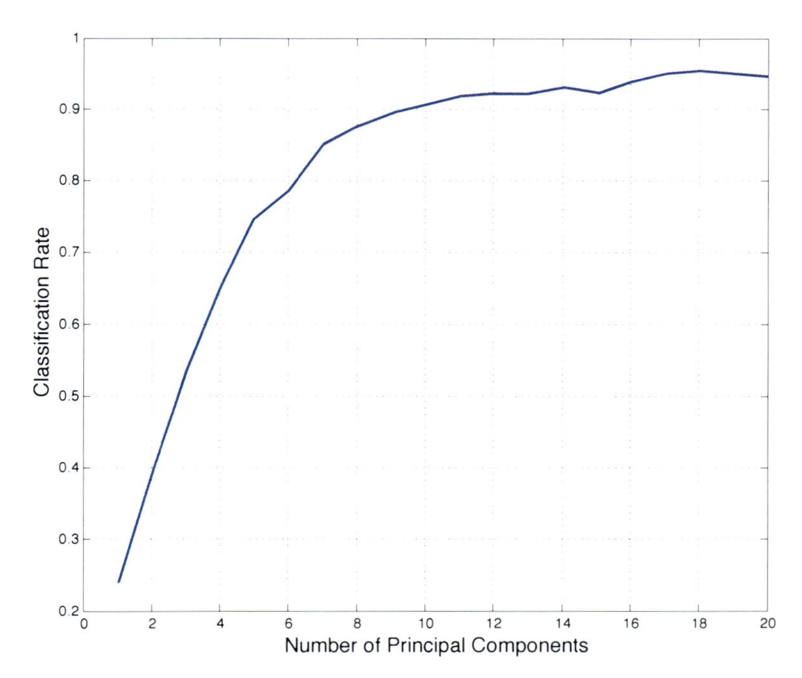

Fig. 6.10 Mean classification rates of technique produced by k-NN using two-thirds of the data from each of the eight players to classify the remaining one-third the data, classifying by *player*.

6.6 Summary

While the primary goal of the bowing technique study described above was to investigate the potential of the measurement system to provide sufficient information to enable gesture recognition of traditional violin bowing techniques, in the course of this work it was discovered that in addition to facilitating this investigation, it also enabled the identification of players as well. These early results suggest that this measurement method may be used to make quantitative comparisons between the techniques and styles of individual performers.

By capturing individual physical bowing gesture with precision and accuracy, contributions may be made in many related fields of research. These include: real-time gesture classification for use in live interactive performance; virtual instrument development using real player data to help test bowed string physical models; performance archiving that includes complete gestural recordings to complement audio and video data to preserve the techniques of our living masters; bowed string acoustics, enabling studies of bow-string interaction in realistic performance scenarios; and new music pedagogies (for more discussion please see Chapter 6 [21]).

References

1. Askenfelt, A.: Measurement of the bowing parameters in violin playing. STL-QPSR 29(1), 1–30 (1988)
2. Auer, L.: Violin Playing as I Teach It, reprint edn. Dover Publications, New York (1980)
3. Berman, J., Jackson, B.G., Sarch, K.: Dictionary of Bowing and Pizzicato Terms, 4th edn. Tichenor Publishing, Bloomington (1999)
4. Brown, R.G., Hwang, P.Y.C.: Introduction to Random Signals and Applied Kalman Filtering, 3rd edn. John Wiley & Sons, New York (1997)
5. CodaBow: Conservatory Violin Bow, http://www.codabow.com/
6. Cremer, L.: The Physics of the Violin. MIT Press, Cambridge (1984)
7. Demoucron, M.: On the control of virtual violins. Ph.D. thesis, KTH (2008)
8. Demoucron, M., Askenfelt, A., Caussé, R.: Measuring bow force in bowed string performance: Theory and implementation of a bow force sensor. Acta Acustica United with Acustica 95(4), 718–732 (2009)
9. Fjellman-Wiklund, A., Grip, H., Karlsson, J.S., Sundelin, G.: EMG trapezius muscle activity pat-tern in string players: Part I—is there variability in the playing technique? International Journal of Industrial Ergonomics 33, 347–356 (2004)
10. Flesch, C.: The Art of Violin Playing: Book One, reprint edn. Carl Fischer. Foreword by Anne-Sophie Mutter, New York (2000)
11. Fletcher, N.H.: The nonlinear physics of musical instruments. Rep. Prog. Phys. 62, 723–764 (1999)
12. Gigante, C.: Manual of Orchestral Bowing. Tichenor Publishing, Bloomington (1986)
13. Green, E.A.H.: Orchestral Bowings and Routines. American String Teachers Association, Reston (1990)
14. Guettler, K.: The bowed string: On the development of helmholtz motion and on the creation of anomalous low frequencies. Ph.D. thesis, Royal Institute of Technology - Speech, Music and Hearing (2002)

15. M-Audio: Fast Track USB, http://www.m-audio.com/
16. Micro-Measurements, V.: http://www.vishay.com/company/brands/micromeasurements/
17. Nabney, I.T.: NETLAB: Algorithms for Pattern Recognition. In: Advances in Pattern Recognition. Springer, Great Britain (2002) NETLAB toolbox, http://www.ncrg.aston.ac.uk/netlab/index.php
18. Ng, K., Larkin, O., Koerselman, T., Ong, B.: i-Maestro Gesture and Posture Support: 3D Motion Data Visualisation for Music Learning And Playing. In: Bowen, J.P., Keene, S., MacDonald, L. (eds.) Proceedings of EVA 2007 London International Conference, August 27-31. London College of Communication, p. 20.1–20.8. University of the Arts London, UK (2007)
19. Ng, K., Larin, O., Koerselman, T., Ong, B., Schwarz, D., Bevilacqua, F.: The 3D Augmented Mirror: Motion Analysis for String Practice Training. In: Proceedings of the International Computer Music Conference, ICMC 2007 – Immersed Music, Vol. II, Copen-hagen, Denmark, August 27–31, pp. 53–56 (2007)
20. Ng, K., Weyde, T., Larkin, O., Neubarth, K., Koerselman, T., Ong, B.: 3D Augmented Mirror: A Multi-modal Interface for String Instrument Learning and Teaching with Gesture Support. In: Proceedings of the 9th International Conference on Multimodal Interfaces, Nagoya, Japan, pp. 339–345. ACM SIGCHI (2007) ISBN: 978-1-59593-817-6
21. Ng, K.: Interactive Multimedia for Technology-enhanced Learning with Multimodal Feedback. In: Solis, J., Ng, K. (eds.) Musical Robots and Interactive Multimodal Systems, Tracts in Advanced Robotics, vol. 74, Springer, Heidelberg (2011)
22. Paradiso, J., Gershenfeld, N.: Musical applications of electric field sensing. Computer Music Journal 21(3), 69–89 (1997)
23. Puckette, M.: Pure Data (Pd), http://www.crca.ucsd.edu/msp/software.html
24. Rasamimanana, N.: Gesture Analysis of Bow Strokes Using an Augmented Violin, M.S. thesis, Université Piere et Marie Curie (2004)
25. Rasamimanana, N., Kaiser, F., Bevilacqua, F.: Perspectives on gesture-sound relationships in-formed from acoustic instrument studies. Organized Sound 14, 208–216 (2009)
26. Schelleng, J.C.: The Bowed string and the player. Journal of the Acoustical Society of America 53, 26–41 (1973)
27. Schoonderwaldt, E., Rasamimanana, N., Bevilacqua, F.: Combining accelerometer and video camera: Reconstruction of bow velocity profiles. In: Proceedings of the 2006 International Con-ference on New Interfaces for Musical Expression (NIME 2006), Paris (2006)
28. Schoonderwaldt, E.: Mechanics and acoustics of violin bowing: Freedom, constraints and control in performance. Ph.D. thesis, KTH (2009)
29. Schumacher, R.T., Woodhouse, J.: The transient behaviour of models of bowed-string motion. Chaos 5(3), 509–523 (1995)
30. Shan, G., Visentin, P.: A Quantitative Three-dimensional Analysis of Arm Kinematics in Violin Performance. Medical Problems of Performing Artists, 3–10 (March 2003)
31. Shan, G., Visentin, P., Schultz, A.: Multidimensional Signal Analysis as a Means of Better Un-derstanding Factors Associated with Repetitive Use in Violin Performance. Medical Problems of Performing Artists, 129–139 (September 2004)
32. Strang, G.: LinearAlgebra and Its Applications, 4th edn. Brooks Cole, Stanford (2005)

33. Turner-Stokes, L., Reid, K.: Three-dimensional motion analysis of upper limb movement in the bowing arm of string-playing musicians. Clinical Biomechanics 14, 426–433 (1999)
34. UPM: http://w3.upm-kymmene.com/
35. Winold, H., Thelen, E., Ulrich, B.D.: Coordination and Control in the Bow Arm Movements of Highly Skilled Cellists. Ecological Psychology 6(1), 1–31 (1994)
36. Woodhouse, J.: Stringed instruments: Bowed. In: Crocker, M.J. (ed.) Encyclopedia of Acoustics. Wiley-Interscience, pp. 1619–1626. Wiley, Cambridge (1997)
37. Woodhouse, J.: Bowed String Simulation Using a Thermal Friction Model. Acta Acustica United with Acustica 89, 355–368 (2003)
38. Woodhouse, J., Galluzzo, P.M.: The Bowed String As We Know It Today. Acta Acustica United with Acustica 90, 579–589 (2004)
39. Yagisan, N., Karabork, H., Goktepe, A., Karalezli, N.: Evaluation of Three-Dimensional Motion Analysis of the Upper Right Limb Movements in the Bowing Arm of Violinists Through a Digi-tal Photogrammetric Method. Medical Problems of Performing Artists, 181–184 (December 2009)
40. Yamaha: SV-200 Silent Violin, http://www.global.yamaha.com/index.html
41. Young, D.: New frontiers of expression through real-time dynamics measurement of violin bows. Master's thesis, M.I.T (2001)
42. Young, D.: The Hyperbow controller: Real-time dynamics measurement of violin performance. In: Proceedings of the 2002 Conference on New Interfaces for Musical Expression (NIME 2002), Montreal (2002)
43. Young, D.: A methodology for investigation of bowed string performance through measurement of violin bowing technique. Ph.D. thesis, M.I.T (2007)
44. Young, D., Deshmane, A.: Bowstroke Database: A Web-Accessible Archive of Violin Bowing Data. In: Proceedings of the 2007 Conference on New Interfaces for Musical Expression (NIME 2007), New York (2007)
45. Young, D., Serafin, S.: Investigating the performance of a violin physical model: Recent real player studies. In: Proceedings of the International Computer Music Conference, Copenhagen (2007)

Chapter 7
Interactive Multimedia for Technology-Enhanced Learning with Multimodal Feedback

Kia Ng

Abstract. Musical performances are generally physically demanding with high degree of control (mental and motor) and accuracy. This chapter presents the i-Maestro (www.i-maestro.org) project which explored interactive multimedia environments for technology-enhanced music education. It discusses one of the key requirements for an interactive musical robot which is to analyze and provide feedback/interaction to a "performance". This Chapter also serves as an example application of a musical robot in educational contexts. Music is not simply playing the right note at the right time. The multitude of interconnecting factors that influence and contribute to the nature of the playing is not easy to monitor nor analyze. Musical instrumentalists often use mirrors to observe themselves practicing. This Chapter briefly introduces the i-Maestro project and focuses on a gesture interface developed under the i-Maestro framework called the 3D Augmented Mirror (AMIR). AMIR captures, analyze and visualizes the performance in 3D. It offers a number of different analyses and feedback to support the learning and teaching of bowing technique and gesture.

7.1 Introduction

The i-Maestro project [19] explores novel solutions for music training in both theory and performance, building on recent innovations resulting from the development of computer and information technologies, by exploiting new pedagogical paradigms with cooperative and interactive self-learning environments, gesture interfaces, and augmented instruments. The project specifically addresses training

Kia Ng
ICSRiM – University of Leeds,
School of Computing & School of Music, Leeds LS2 9JT, UK
e-mail: k.c.ng@leeds.ac.uk, kia@kcng.org
www.kcng.org

J. Solis and K. Ng (Eds.): Musical Robots and Interactive Multimodal Systems, STAR 74, pp. 105–126.
springerlink.com

support for string instruments among the many challenging aspects of music education. Starting from an analysis of pedagogical needs, the project developed enabling technologies to support music performance and theory training, including tools based on augmented instruments, gesture analysis, audio analysis and processing, score following [10], symbolic music representation [2], cooperative support [3] and exercise generation [40].

This Chapter focuses on the i-Maestro 3D Augmented Mirror (AMIR) module which utilizes interactive multimedia technologies to offer online and offline feedback for technology-enhanced learning for strings.

7.2 Related Background

Instrumental gesture, body movement and posture are all significant elements of musical performance. The acquisition, analysis, and processing of these elements is part of an expanding area of research into new musical interfaces, which can be further grouped with research into Human-Machine Interaction (HMI). Over recent years there has been a noticeable increase in the number of conferences, workshops and research workgroups related to this area such as the International Conference on New Interfaces for Musical Expression (NIME), International Computer Music Conference (ICMC), Digital Audio Effects Conference (DAFX), COST287-ConGAS, "Gesture Workshop" (International Workshop on Gesture in Human-Computer Interaction and Simulation) and others.

Although there is a great deal of research in the area of new musical interfaces, the use of these interfaces in music pedagogy applications is at its beginning and of an experimental nature. The i-Maestro tools aim to build on innovations and technologies emerging from the fields and utilize them in a pedagogical context. A complete overview on all related areas would be beyond the scope and available space for this Chapter. Hence the following sub-sections review several selected issues and focus on a number of related pedagogical applications.

7.2.1 Musical Gesture

There is much speculation over a comprehensive definition of gesture in music [7], although a large part of current research into new musical interfaces deals in some way or the other with gesture. Sometimes confusion arises from the fact that gesture is often used in the description of musical content, for example the phrases in a lyrical interplay between two instruments might be referred to as gestures. In the context of i-Maestro we are interested in the physical gestures that are exhibited by musicians during string performance. These can be defined as body, instrument and bow movements, which may be small or large (e.g. the movements of the body as a whole).

Cadoz & Wanderley [7] describe three classifications of physical musical gesture, which were originally proposed by Delalande [12]. Two of these classifications are relevant in the development of i-Maestro Tools: (1) *Effective gestures*; (2) *Accompanist/Ancillary gestures*. Effective gestures are those that are directly involved in the sound producing mechanism of the instrument (e.g. the bowing of a violin). Ancillary gestures are those that are not directly involved with sound production, yet are related to the expression of musical features [11, 38, 39]. For example a violinist may shift his/her balance in relation to phrasing.

Another classification proposed by Delalande [12] is *figurative* gesture, which he describes as "*perceived by the audience but without a clear correspondence to a physical movement*".

For i-Maestro, it is not only expressively meaningful movements that are relevant. String instrument performance is a physical activity and performers often adopt unnecessary habitual movements that may have an adverse effect on the sound produced or even affect the performer's health. These movements have little or no connection to musical expression. For example, through discussions with several cello teachers in the i-Maestro User Group we have learnt that a common problem experienced by their students is to move the right shoulder whilst playing, where it should, in their opinion, be relaxed and held back. According to the teachers, this movement can cause tension and pain in the short term and serious problems in the long term. Ancillary gestures, although part of a performer's expressivity and individual style, may also have similarly undesirable side effects.

The different approaches proposed in i-Maestro provide access to various types of physical musical gesture, and are thus complementary. For analyzing string performance sensors allow for the direct measurement of bow dynamics while 3D motion capture allows for the measurement of body and bowing movements and the interactions between them.

7.2.2 New Musical Interfaces

Current developments in the field of new musical interfaces can generally be divided into three categories:

- Imitations of acoustic instruments. These types of controllers are the most common types of interface, which are often commercially available (e.g. MIDI wind controller). These instruments can be played using the same skills developed on acoustic instruments. Pedagogy generally follows traditional instrument education.
- Augmented Instruments: the augmentation of traditional acoustic instruments with new capabilities. Work in this field was pioneered in the 1980's at the hyper-instrument Group at the MIT Media Lab. Recent developments related to augmented string instruments include IRCAM's Augmented Violin project [4, 35], the MIT Hyperbow [41, 42, 43] and others.

- Alternative Controllers: new interfaces that are not based on traditional instrument paradigms. These may be either physical or virtual devices (in the case of motion tracking systems). These types of interfaces are original controllers which require learning new skills. Generally they are designed in a context of experimental music or new media/multimedia performance [26, 27]. Pedagogical applications are rare and of experimental nature, which can be explained partly by the community gap between music practitioners using such technologies and traditional music teachers.

As well as in composition and performance, these new interfaces are being used in the study and measurement of musical gesture in order to better understand elements such as the playability of an instrument and differences in playing style. Rasamimanana [35] describes the characterization of bow strokes using an augmented violin, Wanderley [39] provides a comprehensive study of the ancillary gestures of a clarinetist using 3D motion capture and Camurri *et al.* [8, 9] study the expressive gestures using EyesWeb computer vision software.

7.2.3 Pedagogical Applications

The majority of new musical interfaces are designed as tools for performance, studio based composition and laboratory analysis of musical gesture. However, there are several examples of the use of these technologies in pedagogical applications.

One such example is provided by Mora *et al.* [24] who discuss a system to assist piano pedagogy using 3D motion capture technology, focusing on its potential as a tool for self observation. They present evidence to support the use of video recording and playback of instrumental performance, stressing its positive effects on a learner's skill acquisition and cognitive processes. The basic premise of the system is to capture and reconstruct the posture of a professional piano player so that it may be compared against the posture of a student. An OpenGL 3D visualization of the "ideal" posture is overlaid on top of 2D video recordings of the student's recital. Video recordings are made from multiple angles and the 3D visualization can be rotated to match the camera angle. The motion capture of the professional pianist is performed using a large number of reflective markers placed all over the body. This provides enough information to make a detailed anatomical model of the performer's skeleton.

When the 3D skeleton is overlaid onto the video of the student, the proportions of each bone may be adjusted to match the student's body, whilst still displaying the relevant information for posture adjustment. The new individualized skeleton can be saved. Once the comparison has been made, the authors suggest the enhancement and correction of the student's posture by adjusting the height of the piano stool, adding a foot stool etc. in order to make the posture match that of the professional.

Sturm & Rabbath [37] used 3D motion capture technology in a commercial instructional DVD for the double-bass, based on motion studies done at Ball State

University, Indiana, USA. It features several clips of 3D motion capture and many video recordings from different angles in order to illustrate various playing techniques. Reflective markers are placed on the arms, hands and fingers of the performer, on the bow and on the instrument strings. 3D motion is shown in parallel to video footage, but is made out of small looped segments and is non-interactive.

Baillie *et al.* [1] used a haptic model of violin and viola bowing to provide real-time haptic feedback to help a student maintain the bow strokes midway between the bridge and the fingerboard. Using a force feedback device the authors simulate the sensation of bowing a string instrument but apply constraints so that the movement of the bow is limited.

Ferguson [14] discusses several audio-analysis based "sonification studies" which use auditory feedback to help teach musical instrument skills. The examples presented include tools to aid the performer to study their intonation, vibrato, rhythmic accuracy, and control of dynamics. The sound of the instrument is analyzed using a microphone so it is necessary for the performer to wear headphones in order to avoid confusing the audio analysis algorithms with the sound of the sonification. The author suggests that with careful consideration it is possible to choose a sonification that may accompany the sound of the instrument. The studies presented deliver feedback in several different ways. One study is designed to help inform a student about their rhythmic accuracy. Rather than sonifying all data, the system only makes a sound when the user makes a mistake. It uses an interesting approach which indicates the degree of deviation from the "correct" rhythm. The performer must play a rhythm in time with a fixed pulse. If they play a note ahead of the beat, the system plays a sound from the note-onset until the pulse. If they play behind the beat, the sound is emitted from the beat until the note-onset.

The sonification of gesture data has been the focus of several systems which aim to help people to learn certain body movements through auditory feedback. Kapur *et al.* [20] present a technical framework for the sonification of 3D motion data. They discuss three preliminary case studies in which they use the motion capture data, one of which links continuous walking gestures to sound synthesis in order to aid proprioception in people with body movement related disabilities. They note that sonification is very effective at representing the speed of movement and suggest that using the auditory feedback from the system may help people with the motor co-ordination required in walking. The authors used several different sonifications, mapping the raw position data and extracted features such as velocity and acceleration to the parameters of additive and FM synthesis algorithms. A similar application of sonification is discussed in Ghez *et al.* [16] who sonify spatial location and joint motion to see if auditory signals can provide proprioceptive information normally obtained through muscle and joint receptors. Their approach to sonification is to control the timing of a melody using the subject's arm movements. The subject develops proprioceptive control through auditory feedback.

Effenberg *et al.* [13] developed "MotionLab Sonify" which is a system for the sonification of human body movement based on 3D motion capture data. The system uses standard kinematic data (marker position, velocity, acceleration) but also uses inverse dynamics algorithms to extract pseudo realistic kinetic data (the forces/torques in operation at certain joints). MotionLab is a framework for the playback and visualization of 3D motion capture data. MotionLab Sonify is a plug-in for the framework which processes 3D Motion data from a skeletal model and allows the processed data to be linked to MIDI pitch messages such as Note-On/Note-Off and pitch bend. MIDI was chosen because the authors believe that extended periods of listening to simple synthesized tones such as sine waves is fatiguing to the ear. They suggest that due to the wide range of well known sounds offered by MIDI, humans can adapt easily to these sounds. The MotionLab Sonify system includes a wizard for rapidly choosing which data to monitor and adjust the mapping of this data to midi parameters. The sonification may either be delivered using MIDI pitch bend on sustained sounds, or by adjusting the pitch using different midi note values. To address the problem of continuous sonification during periods of little movement, the user may set a threshold underneath which values will not be sonified.

Ho [17] developed a Violin Monitoring System designed to monitor four aspects of bowing technique (bow position, velocity, pressure and sounding point) to help violinists to solve the problem of "how to make a good sound". The system used an violin augmented with strain gauge sensors on the tip and frog of the bow for measuring bow pressure, a resistive wire attached to the bow hairs for measuring the bow position (the part of the bow that is used) bow velocity, and a "bow hook" and slide resistor for measuring the bow sounding point (point of contact in relation to bridge and base of fingerboard). The system also comprised a computer and A/D converter which displayed graphs of the sensor voltage readings in real time. Ho explains that the sensors gave a "strange and uncomfortable feel to the player" requiring extra effort in order to move the bow. She describes the process of using the system to iteratively improve performance by recording the output of the sensors and adjusting the technique based on what is shown.

7.3 Augmented Mirror

Musicians often use mirrors to study themselves practicing. More recently we have seen the increased use of video recording in music education as a practice and teaching tool. For the purpose of analyzing a performance these tools are not ideal since they offer a limited 2D perspective view of the performance.

Using 3D Motion capture technology it is possible to overcome this limitation by visualizing the instrument and the performer in a 3D environment. Visualization such as this can serve as a "3D Augmented Mirror" (AMIR) [28, 29, 30] to help students improve their technique and develop self awareness. It assists teachers to identify and explain problems to students (see Fig. 7.1).

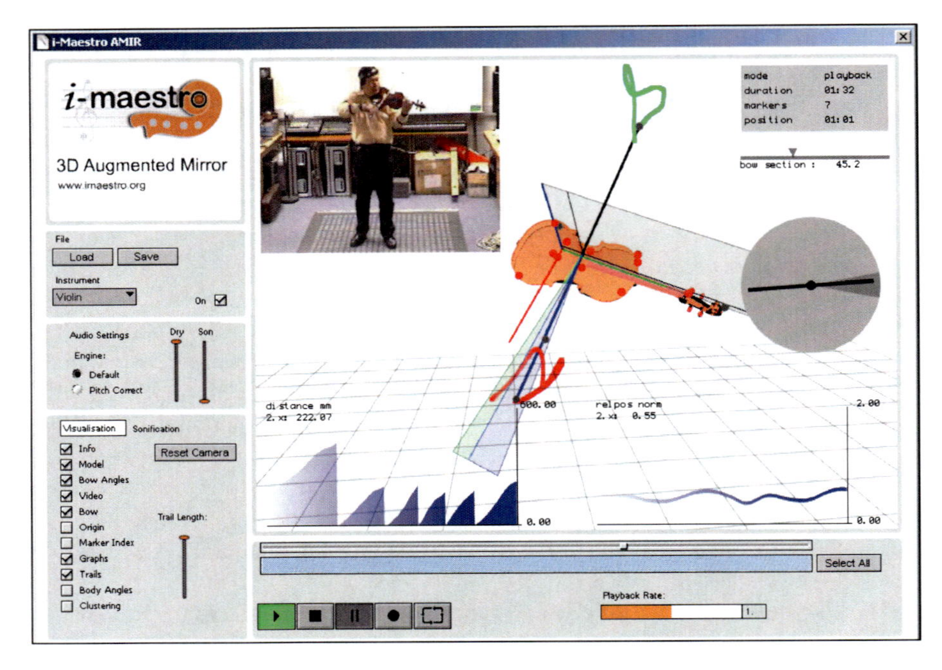

Fig. 7.1 i-Maestro 3D Augmented Mirror interface.

By analyzing and processing 3D motion capture data it is possible to obtain comprehensive information about a performance. For example, in the case of a string player, we can analyze the different characteristics of bow movements and the relationships between the bow, the instrument and the body. This information can be extremely useful for both student and teacher. It can be used in real-time to provide instantaneous feedback about the performance and may also be used to make recordings for in-depth study after the performance has taken place [34].

Based on the pedagogical analysis, the i-Maestro AMIR tool has been developed to provide real-time gesture analysis and feedback with visualization and sonification using 3D motion capture and sensor technologies. 3D motion capture allows us to study the gesture and the posture of a performer in detail and is particularly appropriate for studying the movements involved in playing a string instrument. The system provides a tool for teachers and students to better understand and illustrate techniques/elements of the performance and to achieve learning objectives faster.

At a basic level, the system allows real-time visualization of 3D data as well as the synchronized recording of motion capture data, video and audio data and manipulation on playback (looping, adjusted speed etc). At a more advanced level, the 3D motion capture data is processed to provide different analyses of the performance, the results of which are displayed using visual and auditory feedback.

7.3.1 Pedagogical Context, Considerations and Requirements

Through our experience, discussion with teachers and existing string pedagogy literature we identified areas of interest for analyzing string performance in relation to music practice training using 3D motion capture. In addition there are many properties of string performance that are not fully understood (and are debated by different methods and teachers) which may be clarified through motion analysis performed in the context of i-Maestro.

We want to allow teachers and students to be able to study the performance in three dimensions both in real-time (online) and offline contexts (i.e. after having recorded a performance). Providing this means allowing various manipulations of the 3D environment including magnification, rotation, changing the (virtual) camera location and viewpoint. The users should be able to slow down the playback and maintain synchronization with the audio and video recordings. There should also be the option to correct the pitch of the audio playback when the recording is slowed down, in order to always make clear reference to the musical piece.

7.3.2 Motion Trajectories/Shapes of Bowing Movement

Pedagogic literature on string/bowing technique often includes 2D illustrations of bowing movements. An early example of this can be seen in Percival Hodgson's 1934 book "Motion Study and Violin Bowing" [18], which uses an industrial engineering process known as a "*cyclegraph*" to study the movements of a violinist's bowing arm. This process involved placing a small light on the player's hand and capturing the trajectories drawn by various bow movements in 2D using long exposure photography. Using this primitive form of motion capture, Hodgson was able to prove that bowing motions trace curved rather than linear paths. The book presents many different musical excerpts and their corresponding *cyclegraphs* which illustrate a variety of shapes that are drawn by the bowing hand. The shapes range from simple ellipses and figures of eight, to complex, spiralling patterns. In the latter chapters of the book Hodgson uses this information to provide detailed instruction on various bowing techniques and issues. Other significant string pedagogy literature such as Paul Rolland's "The teaching of action in string playing" [36] and Robert Gerle's "The Art of Bowing Practice: The Expressive Bow Technique" [15] also use 2D images to illustrate 3D bowing shapes and trajectories. The importance of bowing shape has been noted by several teachers. For example, they have informed us that they instruct their pupils to "make a figure of eight" shape. The passage in Fig. 7.2 displays the "figure of eight" shown in Fig. 7.3.

Fig. 7.2 Excerpt from Bach's Partita No. 3 in E major. BMW 1006 (Preludio).

Fig. 7.3 "Figure of eight" motion trails of the above cross-string passage.

Using 3D motion capture it is possible to trace the 3D trajectories drawn by bowing movements which can then be used by the teacher to or student to assess the quality of the bow stroke. This information is clearly useful in the context of a string lesson although the implications of different shapes are not certain at this stage, and may largely depend on the teacher's preference. If the student is not playing the shape smoothly it may mean, for example, that they are making the performance difficult for themselves, reducing their freedom of movement. 3D motion capture provides a unique advantage here over 2D methods such as video tracking or Hodgson's *cyclegraph* technique since movements of the bow can be isolated from movements of the body and visualized in their true 3D form.

The i-Maestro system provides the user with the functionality to draw 3D motion trails from each marker, the length of which can be set in seconds. Trails may be drawn for one or several markers. It is also possible to freeze the trails, zoom in and study them for detailed analysis.

Areas of interest lie in the analysis of the consistency of the trajectories/shapes that are drawn by the same bowing movements both by the same performer and by different performers, and in the automatic segmentation of shapes. Also we are interested in extracting and studying features such as the shape's orientation, smoothness and bounding volume and storing gesture shapes in a motion data repository/database so that multiple gestures can be compared.

7.3.3 Common Practice Support Units and Feature Measurements

In collaborations with pedagogical and requirement analysis with the user groups, several common key practice support units have been developed based on the AMIR framework. These include:

7.3.3.1 Angle of Bow in Relation to the Bridge

This is a common topic for string instrument study also known as "parallel bowing". It is the student's ability to play with the bow parallel to the bridge or, more accurately, perpendicular to the strings. This issue is mentioned in many bowing technique instruction books and has also been identified as a recurring problem that is observed by teachers. The natural tendency is to play without straight, parallel bow movement (following the natural movements of the joints) which is why this is something that students must work to correct for a long time.

The bowing angles are calculated using the markers on the bow in relation to the local coordinate system on the bridge.

Bowing parallel does not necessarily mean bowing linearly back and forth. At the beginning or end of the bow stroke, the bow may change angle and this is also likely if the piece requires the performer to play across several strings with one stroke. This must be accounted for when analyzing this data. Also, it should be noted that bowing parallel is not always necessary, and when watching an advanced performer they will often bow at a slight angle, especially when using different bowing techniques, depending on the bow section in used. Nevertheless, when learning to play a string instrument, parallel bowing motion is one of the important trainings.

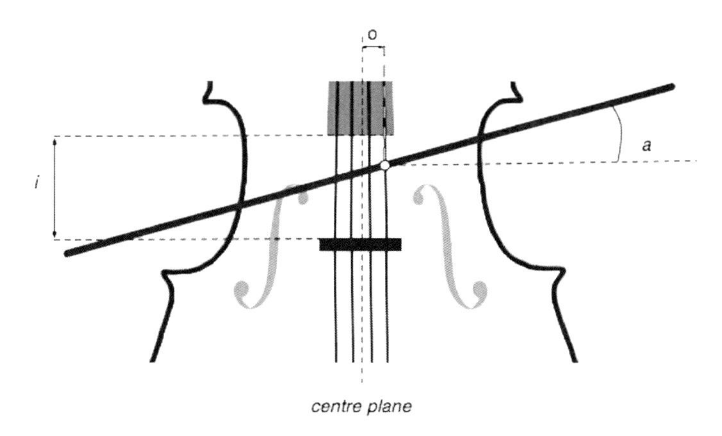

Fig. 7.4 Diagram to show various measurements.

7.3.3.2 Bow Height

The position of bow in relation to the bridge and the base of the fingerboard is identified as *bow height*. Like the bowing angle (above), the point at which the bow meets the strings has a large effect on the tone production, and hence a monitoring and feedback of this feature will be useful. For example, to analyze the playing of "*ponticello*" (to bow close to the bridge) and "*tasto*" (to bow close to the fingerboard) techniques.

This is monitored by a value to indicate the point on the *instrument-y axis* at which the bow crosses the *centre plane* (offset depending on the string being played), where the bridge is equal to zero and the base of the fingerboard is equal to one.

7.3.3.3 Bow Section

This is to monitor which part of the bow that is used. Bowing technique literature often features exercises where the student must use a particular part of the bow, and several teachers have agreed on this approach which helps the student to learn to be economical and accurate in their bow movements.

First, we track the *contact point* where the bow meets the string. Next, we measure the *bowing magnitude*, which is the distance in millimeters from the marker at the tip of the bow to the contact point. The *bow section* indicates the part of the bow that is being used as a factor of the bow length. By dividing the *bowing magnitude* by the length of the bow (or more accurately the length of the bow hair) we obtain a factor in the range 0 to 1.

7.3.3.4 Bow Stroke Segmentation

The segmentation of bow strokes is based on the *bowing magnitude* value:

- The bow is moving up (up bow) if the current bowing magnitude is larger than the previous bowing magnitude
- The bow is moving down (down bow) if the current bowing magnitude is smaller than the previous bowing magnitude

In order to tell if the bow is not on the strings we infer:

- The bow is not on the strings if both the markers at the tip and the tail of the bow are on one side of the centre plane. This would indicate that the whole bow is on one side of the centre plane.
- The bow is not on the strings if the distance between the *contact point* to the finger board is greater than the *finger board height* constant, because that would indicate that the bow is above the strings.

In some cases, the performer may play with the very bottom or very end of the bow hair. This may result in both markers being detected on one side of the centre plane. To get around this problem, we apply an offset the marker at the tip of the bow.

7.3.3.5 Motion Properties

By analyzing and understanding the distance travelled, velocity and acceleration of bow movements, we can obtain information about the regularity and consistency of the gestures, which may be relevant in the production of good tone quality. This can be studied on a per stroke basis and/or over the duration of a performance. Additionally, the distance travelled by markers on the bow can be used to analyze the consistency of bowing gestures. By segmenting the distance travelled by each marker the users can study the different stages of the bow stroke to look at the consistency of the bow gesture.

Bow velocity and acceleration are calculated by taking the first and second derivatives of the *bow magnitude*. In contrast to taking these measurements from the position values of the bow markers, this method gives us information relevant to the part of the bow that is actually producing sound. Using our current setup, we have found that these calculations often result in noisy data, which needs to be smoothed in order to be useful. We have found that taking an average of the values over ten frames provides a suitable tradeoff between smoothness of the data and timing resolution.

The *distance travelled* is measured by accumulating the *bow velocity* value over successive frames. This analysis can be segmented by the bowing direction to compare the curve of distance travelled between bow strokes.

7.3.3.6 Other Features and Measurements

Other features and measurements include:

- Joint angles – to monitor the angles between different parts of the performer's body
- Sensor channel – This is an addition to the motion capture based analyses. AMIR has also been designed to be ready to support analysis, visualization and sonification of sensor data. This could be used to record accelerometer data with AMIR or other kind of sensor data. One such example would be to measure the pressure with which a violinist grips the instrument, by placing a pressure sensor beneath their chin rest. The data from this sensor could be sonified or visualized, to provide feedback that can help the student to avoid pain caused by "clamping" the instrument between the chin and the shoulder.

7.3.3.7 Feature Clustering

Using data from the extracted features and analyses above we developed an approach to visualize the similarity of bowing features in order to facilitate their study in relation to bowing technique and musical expression descriptors. Fig. 7.5 shows an example visualization of bow strokes based on analysis of the *bow-velocity*, *bow-height*, and *bow-length* features from a performance by linking each feature to one of the dimensions in the graph.

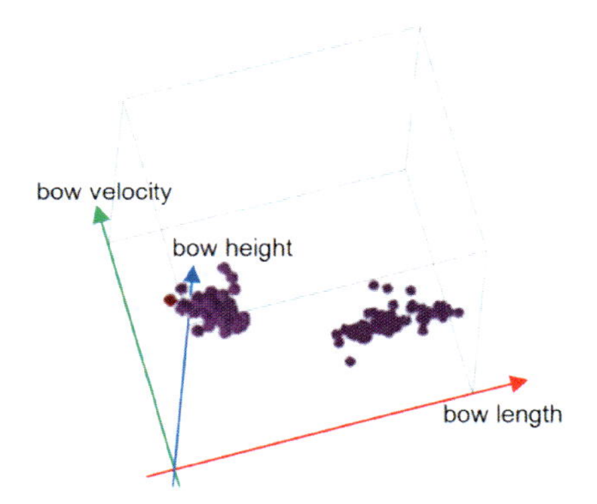

Fig. 7.5 A 3D clustering visualization.

The visual feedback is a valuable way of looking at complex data sets, since it is both informative and easy to understand at the same time. Bow strokes played with a similar expression appear in the same area of the graph as clusters, and as a new stroke is added the matching cluster is highlighted to show that the stroke was similar. The more similar bow strokes are; the more compact the clusters will appear in the graph. We also indicate the distance from the centroid of the cluster. We implement this technique in such a way to provide an easy to understand, high-level interface to non-technical users like musicians and teachers to study the similarity of different bow strokes. As well as using this system to compare strokes to previous strokes in the performance, it is also possible to load in analysis data sets from other recordings to facilitate the comparison between two performances.

The k-Means clustering method is used to compute the mean of a particular dataset in order to study the differences in captured data for a range of these qualities and to identification which bow stroke technique is being used in order to provide assessment support such as consistency measure.

7.3.3.8 Further Calculations and Processing

The different analysis objects can be used in an offline processing context to extract certain features from the overall performance data or any selected region of the capture data. For example we can calculate the mean of the distance travelled for each bow stroke in the capture or the range of a particular angle (e.g. the angle of the performer's head with respect to the shoulders). The data for up and down bow strokes can be separated and a correlation of the data can be performed.

The different types of offline statistics calculations we can perform on the various types of analysis data include mean/average and standard deviation.

In an offline context we create line graphs of a number of features to show the analysis on a per stroke basis. These line graphs are visualised and sonified after a performance has taken place. Using a signed line graph display it is possible to show a deviation from a base value. This can be used for example with the *bow-bridge angle* to show the deviation from a parallel bow.

One line graph we call a distance "displacement curve" which is a graph of the difference in distance travelled by the bow in comparison to a linear velocity stroke [21]. To calculate the displacement curve, we first accumulate the velocity for each frame in the bow stroke. This gives us $g(t) = |d_{t+1} - d_t|,\ t \geq 0$ which represents the distance travelled over the stroke. Next we subtract a normalised linear function $f(t) = t,\ t \geq 0$, which leaves us $s(t) = \int_0^t |f(u) - g(u)|\, du$ representing the displacement curve in relation to a linear-velocity bow stroke (see Fig. 7.6).

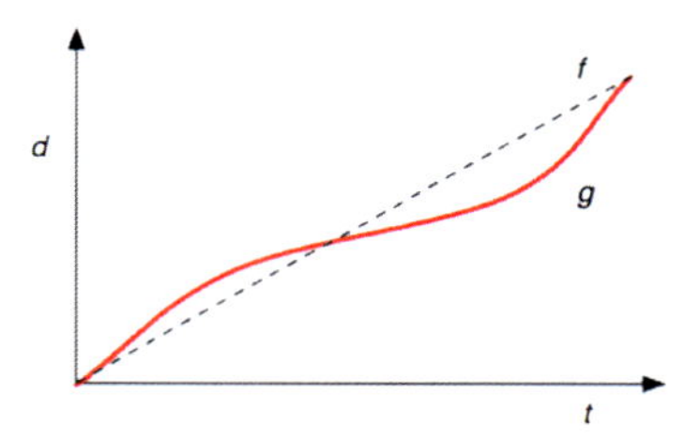

Fig. 7.6 Graph of the displacement curve for a bow stroke segment.

The line graphs are useful to show how the velocity varies over the bow stroke. However, in some cases this could be too much information. By taking the standard deviation of the values from the bow stroke array it is possible to give a more general assessment of the feature for that stroke. For instance, in the case of the bowing angle, the standard deviation can provide an indication of the overall degree of difference from a parallel bow. This is particularly relevant in order to simplify the sonification.

7.3.4 Sonification in AMIR

Features we sonify within the AMIR interface include: bow change (up/down), bow-bridge angle, bow height, which string, Bow section, bow velocity and acceleration, distance "displacement curve".

In our initial experiments with sonification of bow gesture analysis data, we attempted to sonify continuously changing parameters in real-time, synchronous to the performance. We linked the *bow-bridge angle* to different parameters of a synthesis patch and also investigated processing the sound of the instrument by applying distortions and harmonization to reflect changes in the extracted features. We found that when performing, continuous real-time auditory feedback was too distracting to the performer and difficult for them to interpret simultaneously with the instrument sound. For this reason we decided not to work further on continuous auditory feedback. We found that a much more useful and appropriate modality for real-time feedback is to inform the player using the simplest of auditory displays: an auditory alert/notification.

7.3.4.1 Alerts

We allow the user to set thresholds for certain parameters and when these thresholds are crossed an alert is sounded. This informs the performer that certain criteria have been met with minimum distraction. For these alert notification sounds a simple electronic "beep" or an audio sample is used. AMIR consists of a selection of different notification samples appropriate for this purpose.

The character of the notification can be changed depending on how frequently the threshold has been crossed. For example, if a player continuously repeats an error, the sound will increase in volume each time. If they start to improve and correct their error, the volume of the alert will decrease. We have used alert sounds to monitor the bow angle, which string is played, the bowing height and the bow section. In each case the user defines an acceptable range for the parameter using a visual interface.

One significant issue with real-time sonification as described here is the latency of the alert. This is particularly important when the system is meant to be providing feedback on time-critical events such as the bowing rhythm or string crossings. This is a situation when the 200 fps capture frequency of our motion capture system is not ideal, since it could introduce ~5ms latency in some cases. Also there is the latency of the network and the sound card used on the computer. Network latency should be negligible on a cabled gigabit network.

7.3.4.2 Sonified Line Graphs

For offline analysis of the performance we explore the use of sonified line graphs [23] to display information about analysis features for each bow stroke. Sonified line graphs have been shown to be useful in a number of different situations and can help the visually impaired to interpret trends in data sets that would otherwise be difficult to access [6, 23].

To create the line graph, the continuous analysis parameters listed above are split into variable length 1D arrays based on the bowing segmentation, with each element in the array representing the analysis for one frame of motion data. For instance, a bow stroke that lasts one second will be represented by an array of two hundred elements when captured at 200fps. The line graph is visualised, and may be aligned with a waveform representation of the part of the audio recording that corresponds to the bow stroke (see Fig. 7.7). The graph may be sonified synchronous to playback or alternatively, the user can step through the piece stroke by stroke.

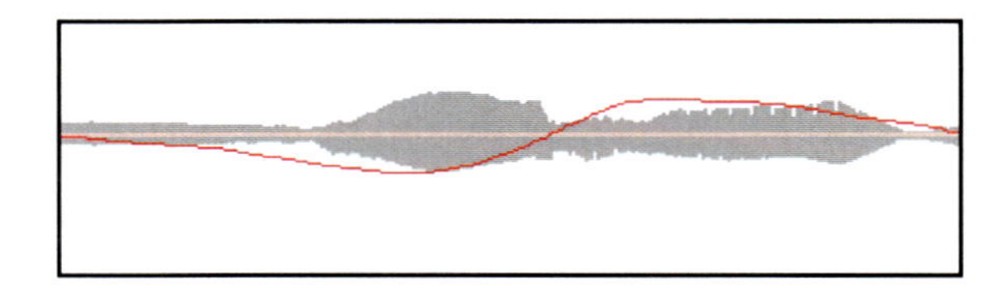

Fig. 7.7 Displacement curve visualization aligned to corresponding audio waveform.

We have chosen to use pitch modulation of a synthesised waveform to sonify the line graph. For each bow stroke played in the performance, a fixed pitch tone is sounded from the left channel and a modulated pitch tone is sounded from the right. Spatial separation such as this should aid cognition, see [5]. In this way it is possible to hear the degree of deviation from the base pitch, which can be perceptually linked to the deviation of the analysis feature. We use a table-lookup oscillator containing the waveform described by Neuwirth [25] which has been found to be particularly suitable for distinguishing pitch differences and easy to listen to in comparison to pure sine tones. An additive synthesis waveform was chosen over MIDI instrument sounds (as recommended in [6]) for a number of reasons.

Synthesising the tone in AMIR guarantees a stable fundamental frequency and can be more precisely controllable in terms of note onset and duration. This is important for sonification that is synchronised with the playback of a recording.

The tone can be adjusted so that it either equals the duration of the bow stroke, or is a fixed duration for each stroke. The first option means that the feature in question can be temporally matched to the music being played, where as the second option will give a more uniform representation of the stroke in comparison to other strokes. It should be noted that the playback of the recording may be slowed down and the duration of the sonified line graph is adjusted accordingly. This makes it possible to assess very fast strokes.

The user may set the base pitch of the tone, for example they might set it to the tonic of the key of the passage that they are playing. The mapping of the modulation can be changed, depending on the resolution to which the user wishes to study the performance, for example a very determined player could set the mapping to an extreme level in order to help him/her develop precise control of bowing movements.

We have found that the features that are best suited to the line graph sonification are the displacement curve, bowing height and the angle of the bow, since these can be represented as deviation from a base value. These features can naturally be compared to a zero base value which would represent a) a linear-velocity stroke, b) a bowing with an equal distance between the fingerboard and bridge or c) bowing parallel to the bridge. Other features such as bow section, bow velocity and acceleration are not referenced against a base value. For these features only one tone is sounded since there is no base value.

7.4 AMIR Validation

To investigate quantitative measurements of the effectiveness of a new practice training tool and pedagogical scenario is a challenging and arguably musically dubious pursuit due to the vast number of factors involved and the highly subjective nature of music performance. We cannot measure *a performance* quantitatively as it involves many different aspects, interrelations and contexts. However we can measure *a specific aspect of a performance* quantitatively, for example, for bowing technique we can look at the bow-bridge angle, usage of the correct part of the bow, amongst other things.

Each instrumentalist is different in their playing style and in their pedagogical requirements. Likewise each piece of music demands different abilities from a player. For these reasons we designed the validation procedure so as to only compare a student's performance against their own performance of the same piece. The validation is designed with particular pieces that the subjects are working on and we use the measurements session by session to see the progress and improvement focusing on specific features.

By using different performers, and comparing their progress (improving specific aspects) achieved with and without using the i-Maestro technology, we believe we have created a fair test.

The validation sessions took place at the ICSRiM motion capture lab at the School of Music, University of Leeds. Test subjects were found through adverts placed in the University. Players were chosen mainly based on their availability (for the number of sessions and regular intervals). Ideally we would have liked to have been able to select a range of different abilities and ages, with the aim to collect data for a selection of different players. Unfortunately we were limited by the availability of the test subjects.

In total fifteen students were able to fully complete the six, hour long sessions that we required and commit to keeping up their practice. The participants ranged in ability and although all had achieved the equivalent to Trinity/ABSRM Grade 8, not all had kept up their studies and level of standards. This is potentially a weakness in our test, but one that we had to make do with.

7.4.1 Procedure

To validate the Augmented Mirror scenario, we focus on two bowing features pertinent to each participant that can be analysed using AMIR. The player performs two passages from a piece (or two pieces) that are selected for suitability (by the subjects, with guidance from the validators). A specific section is chosen from each piece for the analysis and the player performs it three times in a row in order to increase the number of bow strokes used for the analysis, and therefore improve the reliability of the statistic.

The player records the piece/s in a total of six sessions at roughly weekly intervals, the first three are used to see how their performance changes when practicing without the help of the i-Maestro AMIR system. The improvement of playing technique may be measured by the changes of the percentages listed below between the recording sessions, which are calculated by manually observing the recordings with the player and counting the errors in each of their two selected features. The total number of bow strokes in the passage is counted and a percentage of bow strokes where there was an error related to the feature in question is calculated. The subjects were encouraged to be quite strict with their judgments. The fact that one player may judge the same playing differently to another does not affect the validity of the results, since we only compare the player's performances with their own performances – their judgments should hopefully stay the same between sessions.

Players are instructed to practice at home as well as using the system, and to continue to focus on improving the two features. They are asked to maintain a consistent amount of practice between sessions, as best as possible, so as not to influence the results.

The procedure is designed in this way in order to minimise the effects of the different levels of the participants and the different requirements and difficulties of the music pieces, including Debussy's Cello concerto, Glasunow's "Chant Du Menestrel, Bruch's Violin Concert No. 1, Bach's Partita No.3 and Mendelssohn's Violin Concerto No. 2 in E minor, Op. 64.

7.4.2 Analysis

The results of the summative validation have provided some interesting insights into the application of AMIR in a pedagogical scenario. In some cases the student continues to improve whilst they use the interactive feedback from the system. In other cases, the use of AMIR appears to bring a sudden awareness of the problem and they can react appropriately and resolve the technical issue quickly. After that, no further significant improvement can be easily achieved. In fact, in some cases the initial introduction of the tool disturbs the playing which can be seen by an increased percentage of errors (e.g. Student D). In the subsequent session the student can adjust to the interactive feedback and take advantage of the support provided by AMIR.

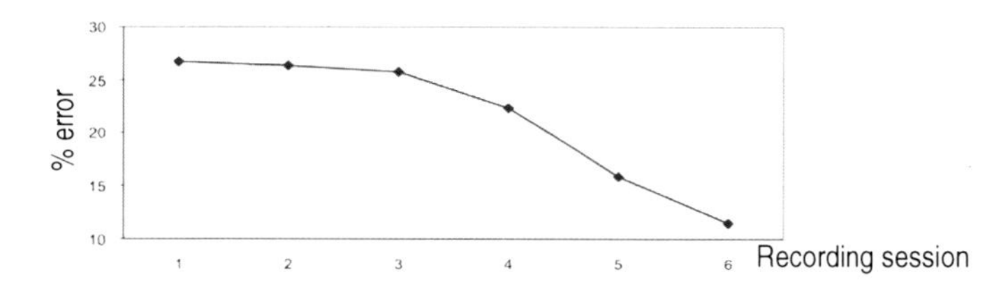

Fig. 7.8 Average % error for features for all students.

From the comments received, it seems that the students are very enthusiastic about the system. It seemed that all the students involved got on well with the system and quickly learnt how to use it. It was also clear to us that their performance improved as the sessions went on. Although we were controlling the set up and recording procedure, we have no doubt that the students would have been able to do this, since they did not have any problems operating and understanding the other aspects, which are no more difficult really.

7.5 Conclusions

Learning to play an instrument is a physical activity. If a student develops a bad posture early on this can be potentially damaging later in his/her musical career. The technology discussed here may be used to develop and enhance awareness of body gesture and posture and to avoid these problems. This technology can be used to capture and analyze a performance in detail and has the potential to assist both teachers and students in numerous ways.

AMIR has been successfully validated with a group of music student at the University of Leeds as well as several professional virtuosos. It has been found that 3D motion tracking, visualization and sonification can enhance the practicing of movement control by displaying/sonifying movement parameters that are not directly or objectively accessible to the student. For example, the perception of bow movement relative to the bridge may interfere with the student's viewing perspective. 3D motion tracking can provide him/her with an objective recording of movement. It adds to the visual feedback, which can be used in adapting the motor program to achieve a particular goal.

The system can be used in different modalities, but has mainly been designed for use in a typical one to one instrumental lesson situation. The interaction model is therefore written to deal with this situation so the teacher is responsible for operating the system. The system could also be used by the student alone in a self-learning environment, or even in a group lesson.

This technology can also be used to capture current playing style (particularly body gesture and movements) beyond the typical audio visual recording. It is being used to capture and analyze multimodal data of musical performance for the contemporary performing arts test-bed of the CASPAR project [32, 33] which develops ontology modals to describe and represent the multimedia data their inter-relationship which can be used for long term preservation [22] for future re-performance, stylistic analysis and more generally a more detailed representation for the preservation of the performing arts.

Acknowledgments

The research is supported in part by the European Commission under Contract IST-026883 I-MAESTRO. The authors would like to acknowledge the EC IST FP6 for the partial funding of the I-MAESTRO project (www.i-maestro.org), and to express gratitude to all I-MAESTRO project partners and participants, for their interests, contributions and collaborations.

References

1. Baillie, S., Brewster, S., Hall, C., O'Donnell, J.: Motion space reduction in a haptic model of violin and viola bowing. In: 1st Joint Eurohaptics Conference and Symposium on Haptic Interfaces for Virtual Environment and Teleoperator Systems, pp. 525–526. IEEE CS (2005)
2. Bellini, P., Nesi, P., Zoia, G.: MPEG symbolic music representation: a solution for multimedia music applications. In: Ng, Nesi (eds.) Interactive Multimedia Music Technologies. IGI Global (2008), doi:10.4018/978-1-59904-150-6.
3. Bellini, P., Frosini, F., Mitolo, N., Nesi, P., Paolucci, M.: Collaborative working for music education. In: Ng (ed.) Proceedings of the 4th i-Maestro Workshop on Technology-Enhanced Music Education, Co-Located with the 8th International Conference on New Interfaces for Musical Expression (NIME 2008), Genova, Italy, pp. 5–10 (2008) ISBN: 9780853162698
4. Bevilacqua, F., Rasamimanana, N., Fléty, E., Lemouton, S., Baschet, F.: The augmented violin project: research, composition and performance report. In: Proceedings of the Int. Conference on New Interfaces for Musical Expression (NIME), Paris, France (2006)
5. Brewster, S.A., Wright, P.C., Edwards, A.D.N.: Parallel Earcons: reducing the length of audio messages. International Journal of Human-Computer Studies 43(22), 153–175 (1995)

6. Brown, L.M., Brewster, S.A.: Drawing by ear: interpreting sonified line graphs. In: Proceedings of Int. Conference on Auditory Display (ICAD), Boston, Massachusetts (2003)
7. Cadoz, C., Wanderley, M.M.: Gesture-Music. In: Trends in Gestural Control of Music. IRCAM-Centre Pompidou (2000)
8. Camurri, A., Mazzarino, B., Ricchetti, M., Timmers, R., Volpe, G.: Multimodal analysis of expressive gesture in music and dance performances. In: Camurri, A., Volpe, G. (eds.) GW 2003. LNCS (LNAI), vol. 2915, pp. 20–39. Springer, Heidelberg (2004a)
9. Camurri, A., Mazzarino, B., Volpe, G.: Analysis of Expressive Gesture: The EyesWeb Expressive Gesture Processing Library. In: Camurri, A., Volpe, G. (eds.) GW 2003. LNCS (LNAI), vol. 2915, pp. 460–467. Springer, Heidelberg (2004b)
10. Cont, A., Schwarz, D.: Score following at IRCAM. In: Cont, A., Schwarz, D. (eds.) MIREX 2006, Second Annual Music Information Retrieval Evaluation eXchange, Abstract Collection, The International Music Information Retrieval Systems Evaluation Laboratory (IMIRSEL), p. 94. University of Illinois at Urbana-Champaign, Victoria, Canada (2006), http://www.music-r.org/evaluation/MIREX/2006abstracts/MIREX2006Abstracts.pdf (last accessed December 8, 2010)
11. Dahl, S., Friberg, A.: Visual perception of expressiveness in musicians' body movements. Music Perception 24 (2004)
12. Delalande, F.: Le Geste, outil d'analyse: quelques enseignements d'une recherche sur la gestique de Glenn Gould. Analyse Musicale, 1er trimestre (1988)
13. Effenberg, A., Melzer, J., Weber, A., Zinke, A.: MotionLab Sonify: a framework for the sonification of human motion data. In: Proceedings of Ninth International Conference on Information Visualisation, IV 2005 (2005)
14. Ferguson, S.: Learning musical instrument skills through interactive sonification. In: Proceedings of the International Conference on New Interfaces for Musical Expression (NIME 2006), Paris, France (2006)
15. Gerle, R.: The art of bowing practice: the expressive bow technique. Stainer & Bell (1991)
16. Ghez, C., Rikakis, T., DuBois, L., Cook, P.: An auditory display system for aiding interjoint coordination. In: Proceedings of the International Conference on Auditory Display (2000)
17. Ho, T.K.-L.: A computer-assisted approach to the teaching of violin tone production. ACM SIGCUE Outlook 21(2), 73–83 (1991), doi:10.1145/122463.122472
18. Hodgson, P.: Motion study and violin bowing. The Strad, Lonon (1934)
19. i-Maestro project website: http://www.i-maestro.org (last accessed December 8, 2010)
20. Kapur, A., Tzanetakis, G., Virji-Babul, N., Wang, G., Cook, P.R.: A framework for sonification of vicon motion capture data. In: Proceedings of the International Conference on Digital Audio Effects DAFX 2005, Madrid, Spain (2005)
21. Larkin, O., Koerselman, T., Ong, B., Ng, K.: Sonification of bowing features for string instrument training. In: Proceedings of the 14th International Conference on Auditory Display, Paris, France (2008)
22. Manson, P.: Digital preservation research: An evolving landscape. In: The European Research Consortium for Informatics and Mathematics (ERCIM) News 80, Special Theme: Digital Preservation (2010), http://ercim-news.ercim.eu/en80

23. Mansur, D.L.: Graphs in sound: a numerical data analysis method for the blind, M.Sc. Thesis, Department of Computing Science, University of California (1975)
24. Mora, J., Won-Sook, L., Comeau, G., Shirmohammadi, S., El Saddik, A.: Assisted piano pedagogy through 3D visualization of piano playing. In: Proceedings of HAVE 2006, Ottawa, Canada (2006)
25. Neuwirth, E.: Designing a pleasing sound mathematically. MATHMAG: Mathematics Magazine 74 (2001)
26. Ng, K.: Sensing and mapping for interactive performance. Organised Sound 7(2), 191–200 (2002)
27. Ng, K.: Music via Motion (MvM): trans-domain mapping of motion and sound for interactive performances. Proceedings of the IEEE, Special Issue on: Engineering and Music – Supervisory Control & Auditory Communication 92(4), 645–655 (2004)
28. Ng, K., Larkin, O., Koerselman, T., Ong, B.: i-Maestro gesture and posture support: 3d motion data visualization for music learning and playing. In: Bowen, J.P., Keene, S., MacDonald, L. (eds.) Proceedings of EVA 2007 London International Conference, July 11-13, vol. 8, pp. pp. 20.1-20.8. London College of Communication, University of the Arts London, UK (2007a)
29. Ng, K., Larkin, O., Koerselman, T., Ong, B., Schwarz, D., Bevilacqua, F.: The 3D augmented mirror: motion analysis for string practice training. In: Proceedings of the International Computer Music Conference, ICMC 2007 – Immersed Music, Copenhagen, Denmark, Vol.II, pp. 53–56 (August 2007b), ISBN: 0-9713192-5-1
30. Ng, K., Weyde, T., Larkin, O., Neubarth, K., Koerselman, T., Ong, B.: 3D augmented mirror: a multimodal interface for string instrument learning and teaching with gesture support. In: Proceedings of the 9th International Conference on Multimodal Interfaces, Nagoya, Japan, pp. 339–345. ACM SIGCHI (2007), `http://doi.acm.org/10.1145/1322192.1322252`, ISBN: 978-1-59593-817-6
31. Ng, K., Nesi, P. (eds.): Interactive multimedia music technologies, p. 394 pages. IGI Global, Information Science Reference, Library of Congress 2007023452 (2008) ISBN: 978-1-59904-150-6 (hardcover) 978-1-59904-152-0 (ebook)
32. Ng, K., Pham, T.V., Ong, B., Mikroyannidis, A., Giaretta, D.: Preservation of Interactive Multimedia Performances. International Journal of Metadata, Semantics and Ontologies (IJMSO) 2008 3(3), 183–196 (2009), doi:10.1504/IJMSO.2008.023567
33. Ng, K.: Digital preservation of interactive multimedia performances. In: The European Research Consortium for Informatics and Mathematics (ERCIM) News 80, Special Theme: Digital Preservation (2010), `http://ercim-news.ercim.eu/en80`
34. Ong, B., Ng, K., Mitolo, N., Nesi, P.: i-Maestro: interactive multimedia environments for music education. In: Proceedings of the 2nd Int. Conference on Automated Production of Cross Media Content for Multi-channel Distribution (AXMEDIS 2006), Workshops, Tutorials, Applications and Industrial, December 13-15, pp. 87–91. Firenze University Press (FUP), Leeds (2006)
35. Rasamimanana, N.: Gesture analysis of bow strokes using an augmented violin - mémoire de stage de DEA ATIAM année 2003-2004. IRCAM-Centre Pompidou (2004)
36. Rolland, P.: The teaching of action in string playing. Illinois String Research Associates (1974)
37. Sturm, H., Rabbath, F.: The art of the bow DVD (2006), `http://www.artofthebow.com` (last accessed December 18, 2009)

38. Wanderley, M.M.: Non-obvious Performer Gestures in Instrumental Music. In: Braffort, A., Gibet, S., Teil, D., Gherbi, R., Richardson, J. (eds.) GW 1999. LNCS (LNAI), vol. 1739, p. 37. Springer, Heidelberg (2000)
39. Wanderley, M.M.: Quantitative analysis of non-obvious performer gestures. In: Wachsmuth, I., Sowa, T. (eds.) Gesture and Sign Language in Human-Computer Interaction, pp. 241–253. Springer, Heidelberg (2002)
40. Weyde, T., Ng, K., Neubarth, K., Larkin, O., Koerselman, T., Ong, B.: A systemic approach to music performance learning with multimodal technology support. In: Bastiaens, T., Carliner, S. (eds.) Proceedings of E-Learn 2007, World Conference on E-Learning in Corporate, Government, Healthcare, & Higher Education (AACE), October 15-19, Québec City (2007)
41. Young, D.: Wireless sensor system for measurement of violin bowing parameters. In: Proceedings of the Stockholm Music Acoustics Conference, SMAC 2003, Stockholm (2003)
42. Young, D.: A methodology for investigation of bowed string performance through measurement of violin bowing technique, PhD Thesis, M.I.T (2007)
43. Young, D.: Capturing Bowing Gesture: Interpreting Individual Technique. In: Solis, J., Ng, K. (eds.) Musical robots and Interactive Multimodal Systems. Tracts in Advanced Robotics, vol. 74, Springer, Heidelberg (2011) (hardcover), ISBN: 978-3-642-22290-0

Chapter 8
Online Gesture Analysis and Control of Audio Processing

Frédéric Bevilacqua, Norbert Schnell, Nicolas Rasamimanana,
Bruno Zamborlin, and Fabrice Guédy

Abstract. This chapter presents a general framework for gesture-controlled audio processing. The gesture parameters are assumed to be multi-dimensional temporal profiles obtained from movement or sound capture systems. The analysis is based on machine learning techniques, comparing the incoming dataflow with stored templates. The mapping procedures between the gesture and the audio processing include a specific method we called temporal mapping. In this case, the temporal evolution of the gesture input is taken into account in the mapping process. We describe an example of a possible use of the framework that we experimented with in various contexts, including music and dance performances, music pedagogy and installations.

8.1 Introduction

The role of gesture control in musical interactive systems has constantly increased over the last ten years as a direct consequence of both new conceptual and new technological advances. First, the fundamental role of physical gesture in human-machine interaction has been fully recognized, influenced by theories such as enaction or embodied cognition [24]. For example, Leman laid out an embodied cognition approach to music [38], and, like several other authors, insisted on the role of action in music [27, 28 and 35]. These concepts resonate with the increased availability of cost-effective sensors and interfaces, which also drive the development of new musical digital instruments where the role of physical gesture is central. From a research perspective, important works have been reported in the

Frederic Bevilacqua · Norbert Schnell · Nicolas Rasamimanana ·
Bruno Zamborlin · Fabrice Guedy
IRCAM - Real Time Musical Interactions Team, STMS IRCAM-CNRS-UPMC,
Place Igor Stravinsky, Paris, 75004, France
e-mail: {Frederic.Bevilacqua,Norbert.Schnell,Nicolas.Rasamimanana,
Bruno.Zamborlin,Fabrice.Guedy}@ircam.fr

J. Solis and K. Ng (Eds.): Musical Robots and Interactive Multimodal Systems, STAR 74, pp. 127–142.
springerlink.com

community at the NIME conferences (New Interfaces for Musical Expressions) since 2001.

Using gesture-controlled audio processing naturally raises the question about the relationship between the gesture input data and the sound control parameters. This relationship has been conceptualized as a "mapping" procedure between different input-output parameters. Several approaches of gesture-sound mapping have been proposed [1, 33, 34, 39, 63, 66 and 67]. The mapping can be described for low-level parameters, or for high-level descriptors that can be comprehended from a cognitive standpoint (perceived or semantic levels) [15, 32, 33 and 65]. Some authors also proposed to use emotional characteristics that could be applied to both gesture and sound [16 and 19].

Most often in mapping procedures, the relationship is established as instantaneous, i.e. the input values at any given time are linked to output parameters [1, 23, 34, 62 and 60]. The user must dynamically modify the control parameters in order to imprint any corresponding time behavior of the sound evolution. Similarly to traditional instruments, this might involve important practice to effectively master such a control. Nevertheless, mapping strategies that directly address the evolution of gesture and sound parameters, e.g. including advanced techniques of temporal modeling, are still rarely used. Modeling generally includes statistical measures over buffered data. While such methods, inspired by the "bag of words" approach using gesture features or "bag of frames" using sound descriptors [3], can be powerful for a classification task, they might be unfit for real-time audio control where the continuous time sequence of descriptors is crucial.

We present here a framework that we developed for gesture-controlled interactive audio processing. By refining several prototypes in different contexts, music pedagogy, music and dance performances, we developed a generic approach for gesture analysis and mapping [6, 7, 10, 29, 30 and 52]. From a technical point of view, this approach is based on mapping strategies using gesture recognition techniques, as also proposed by others [4, 5, 25 and 22]. Our approach is based on a general principle: the gestures are assumed to be temporal processes, characterized by temporal profiles. The gesture analysis is thus based on a tool that we specifically developed for the analysis of temporal data in real-time, called the *gesture follower*. Furthermore, we introduced the notion of *temporal mapping*, as opposed to *spatial mapping*, to insist on the temporal aspects of the relationship between gesture, sound and musical structures [10 and 54]. This distinction between spatial and temporal views of control was also pointed out by Van Nort [63]. *Temporal mapping* brings to focus the time evolution of the data rather than their absolute values for the design of musical interaction systems. This approach relies on the modeling of specific temporal profiles or behavior obtained through either a training process or set manually [11, 20, 23, 40, 51 and 53].

This chapter is structured as follows: first, we present the general architecture, followed by a description of the gesture recognition system and temporal mapping procedure; finally, we describe one possible interaction paradigm with this framework that was used in several different applications.

8.2 Gesture Capture and Processing

The general architecture of our system is illustrated in Fig. 8.1. Similar to the description of digital musical instruments given by [66], we considered four distinct parts: gesture capture, gesture processing, mapping, and audio processing. The specific elements of the gesture analysis and mapping are described in detail in the next sections. Note that this architecture is supported by a set of software libraries that greatly facilitates its design and rapid prototyping [9, 56, 57 and 58], in providing tools for the storage and processing of time-based data, such as gesture and sound data.

8.2.1 Gesture Capture System

The system was designed for a broad range of input data, obtained from various heterogeneous capture systems (see [42] for a review). We cite below the different types of systems that we have involved in our projects, clearly illustrating the variety of data types that can be taken into account.

- 3D spatial data obtained using motion capture data
- Image analysis parameters obtained from video capture
- Gesture data obtained from inertial measurement units, including any combination of accelerometers, gyroscopes and magnetometers (obtained through prototypes or commercial devices including game and mobile phone interface)
- Gesture data obtained from sensors such as FSR (Force Sensing Resistor), bend, strain gauge, piezoelectric, Hall, optical, ultrasound sensors
- Tablet and multitouch interfaces
- Sliders and potentiometers
- MIDI interfaces
- Sound descriptors derived from sound capture
- Any combination of the above

Taking into account the heterogeneity of possible input data, the term "gesture" can refer to completely different types of physical quantities. Nevertheless, any of these inputs can still be considered as multidimensional temporal data. In order to work efficiently with such different input, we separate the data stream analysis into two procedures: 1) the *preprocessing* that is specific to each input system and 2) the *online temporal profile analysis* that is generic (as illustrated in Figure 8.1).

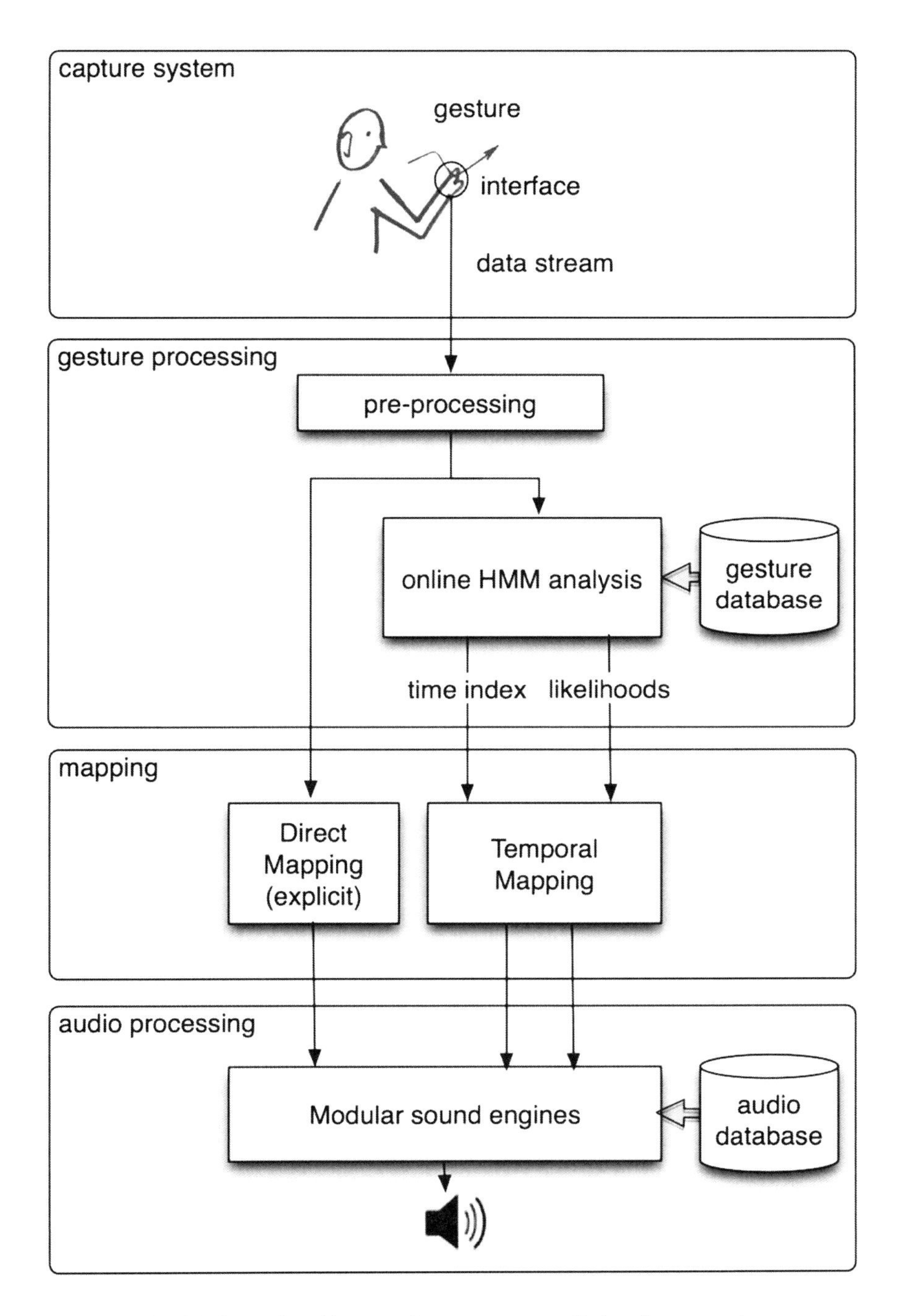

Fig. 8.1 General architecture for gesture-controlled audio processing

8.2.2 Pre-processing

The pre-processing procedure includes:

- Filtering (e.g. low/high/band pass)
- Re-sampling to ensure a constant sampling rate
- Data fusion and dimension reduction (e.g. using Principal Component Analysis)
- Normalization
- Segmentation (optional)

The pre-processing is designed specifically for a given gesture capture system, taking into account its particularities. The preprocessing should ensure that the data is formatted as a temporal stream of vectors x(t) of dimension M, regularly sampled over a time interval Δt. Thus, the preprocessed data can be seen as series $\boldsymbol{x}_1, \boldsymbol{x}_2, ..., \boldsymbol{x}_n$. A recorded gesture from $t = 0$ to $t = (N-1)\Delta t$ is stored in a matrix G of dimension $N \times M$.

8.2.3 Temporal Profiles Processing

The processing is based on machine learning techniques, and thus requires proceeding in two steps. The first step corresponds to the training of the system using a database: this is called learning procedure. During this step, the system computes model parameters based on the database (described in the next section). The second step is the online processing, which corresponds to the actual use of the system during performance. During this step, the system outputs parameters used for the audio control.

8.2.3.1 Modeling and Learning Procedure

Let us first summarize our requirements. First, we need a very fine-grained time modeling system so that we can consider "gesture data" profiles at several time scales. This is desirable when dealing with musical gestures [39]. Second, for practical reasons, we wish to be able to use a single example in the learning process. This is necessary to ensure that the gesture vocabulary can be set very efficiently or easily adaptable to idiosyncrasies of a particular performer. We found that this requirement is of crucial importance when working in pedagogical or artistic contexts.

These requirements led us to develop a hybrid approach between methods such as Hidden Markov Models (HMM), Dynamic Time Warping (DTW) and Linear Dynamic Systems (LDS) [41, 43, 46, 47, 61 and 68]. Interestingly, Rijko and coworkers also working in a performing arts context, proposed similar approaches [48, 49 and 50].

DTW is a common method for gesture recognition consisting in temporally aligning a reference and a test gesture, using dynamic programming techniques.

This allows for the computation of a similarity measure between different gestures that is invariant to speed variation. HMM have also been widely used for gesture recognition. They can also account for speed variations, while benefiting for a more general formalism. HMM allows for data modeling with a reduced number of hidden states, which can be characterized through a training phase with a large number of examples. While more effective, the standard HMM implementation might suffer from coarse time modeling. This might be improved using Semi-Markov Model [21] or Segmental HMM [2, 12 and 13].

We proposed a method that is close to DTW, in the sense that we keep the usage simplicity of comparing the test gesture with a single reference example, as well as the advantage of applying fine-grained time warping procedure. Nevertheless, we present our method using an HMM formalism [47], for the convenience provided by a probabilistic approach. As such, our method is based on the forward procedure in order to satisfy our constraint of real-time computation, while standard DTW techniques require operating on completed gesture [5]. Nevertheless, to guarantee our requirements, i.e. a fine-grained time modeling and a simplified training procedure, we adopted a non-standard HMM implementation, previously described in [7, 8 and 10].

We recall here only the main features of our implementations. Similar to example-based methods, we associate each gesture template to a state-based structure: each data sample represents a "state" [14]. Furthermore, a probability density function is associated to each state, setting the observation probability of the data. This structure can then be associated to an HMM (Figure 8.2).

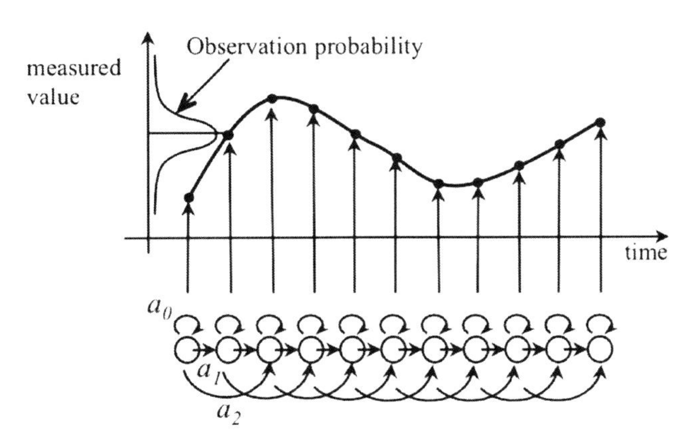

Fig. 8.2 HMM structure associated to a gesture template

The fact that the sampling rate is regular simplifies the learning procedure. In this case, all transition coefficients of the same type between states (a_0=stay, a_1=next, a_2=skip, etc) must share identical values, which can be manually set using prior knowledge. See [10] for choice examples in setting these transition probability values.

The learning procedure consists simply of recording at least one gesture, stored in a matrix G $(N \times M)$. Each matrix element is associated to the mean μ_i of a normal probability function b_i, corresponding to the observation probability function. Using several gesture templates corresponds to recording and storing of an array of G_k matrices.

The value of the variance can also be set using prior knowledge. For example, prior experiments can establish typical variance values for a given type of gestures and capture systems. A global scaling factor, which operates on all the variance values, can be manually adjusted by the user.

8.2.3.2 Online Processing: Time Warping and Likelihood Estimation

The gesture follower operates in real-time on the input dataflow. The live input is continuously compared to the templates stored in the system using an HMM online decoding algorithm. The gesture follower provides in real-time two types of parameters that are used in the mapping procedure.

First, during the performance, it reports the time progression index of the live gesture given a pre-recorded template. This corresponds to computing the time warping during the performance as shown in Fig. 8.3. We call this procedure "following", since it is similar to the paradigm of score following [21 and 59]. Note that as explained above, in the case of the gesture follower, the Markov chains are built based on recorded templates, while in the case of score following the Markov chains are built using a symbolic score.

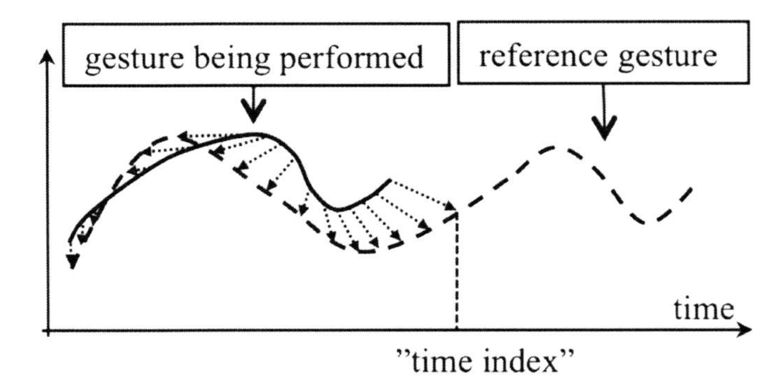

Fig. 8.3 Online time warping of the gesture profile

Second it returns the *likelihood* value that can be interpreted as the probability of the observed gesture being generated by the template. This can be used as a similarity measure between the gesture being performed and the templates [18].

The decoding is computed using the forward procedure [47]. The state probability distribution $\alpha_t(i)$ of a series of observations $O_1 \ldots O_t$ is computed as follows, considering the observation probability $b_i(O_t)$:

1. Initialization

$$\alpha_1(i) = \pi_i b_i(O_1) \qquad 1 \le i \le N$$

where π_i is the initial state distribution.

2. Induction

$$b_i(O_t) = \frac{1}{\sigma_i \sqrt{2\pi}} \exp\left[-(O_t - \mu_i)^2 / 2\sigma_i^2\right]$$

$$\alpha_{t+1}(i) = \left(\sum_{j=1}^{N} \alpha_t(j) a_{ij} \right) b_i(O_t) \qquad 1 \le t \le T-1,\ 1 \le i \le N$$

where a_{ij} are the state transition probabilities (note that $b_i(O_t)$ is expressed here for the case M=1, the generalization to M>1 is straightforward)

The time progression index and the likelihood associated to the observation series is updated at each new observation from the $\alpha_i(t)$ distribution:

$$\textit{time progression index}(t) = \arg\max\left[\alpha_t(i)\right]$$

$$\textit{likelihood}(t) = \sum_{i=1}^{N} \alpha_t(i)$$

These parameters are output by the system continuously, from the beginning of the gesture. The likelihood parameters are thus available before the end of the gestures, which could allow for the determination of early recognition, as also proposed by Mori [45].

Generally, the computation is run in parallel for k templates, returning k values of the time progression index and the likelihood. The *argmax*[*likelihood(t)*] returns the likeliest template for the current gesture at time t.

8.3 Temporal Mapping

Different types of mappings have been identified, such as *explicit* or *implicit* mapping [63]. In explicit mapping, the mathematical relationships between input and output are directly set by the user. On the contrary, indirect mapping generally refers to the use of machine learning techniques, implying a training phase to set parameters that are not directly accessed by the user [17, 22 and 25].

As shown in Fig. 8.1,, our architecture includes both explicit and implicit types of mapping that are operated simultaneously. Explicit mapping has been largely discussed and is commonly used. We discuss below more specifically the implicit mapping used in conjunction with the *gesture follower*. Precisely, as discussed previously, we introduce a *temporal mapping* procedure by establishing relationships between the gesture and audio temporal profiles. The temporal mapping can be seen as a synchronization procedure between the input gesture parameters and the sound process parameters. Technically, this is made possible by the use of the time progression index that the gesture follower provides continuously: the pacing of the gesture can therefore be synchronized with specific time processes.

Fig. 8.4 illustrates a simple example of temporal mapping: two temporal profiles are mapped together, namely hand acceleration and audio loudness. Please note that hand acceleration and audio loudness were chosen here for the sake of clarity, and that any gesture data or audio processing parameters could be used. In this example the gesture data is composed of three phases illustrated by colored regions: one first phase (dark grey), an exact repetition of this first phase (dark grey), and a third different phase (light grey). The audio loudness data is structured differently: it is constantly increasing up to a maximum value, and then constantly decreasing.

The temporal mapping consists here in explicitly setting a relationship between these two temporal profiles: values of acceleration and loudness are linked together according to their history. The mapping is therefore dependant on a sequence of values rather than on independent single values. For example, vertical lines in Fig. 8.4 all indicate similar acceleration values. In particular, lines 1 and 3 (resp. 2 and 4) have strictly identical values (all derivatives equal) as they correspond to a repeated gesture. One can note that interestingly these identical gesture values are mapped to different audio values, as shown by the values pointed by lines 1 and 3 (resp. 2 and 4). In our case, the mapping clearly depends on the time sequencing of the different gesture values and must be considered as a time process. While gesture reference profiles are generally directly recorded from capture systems, audio input profiles can be created by the user manually, or derived from other modeling procedures. For example, recorded bowing velocity profiles could be used as parameters to control physical models. Using the temporal mapping, these velocity profiles could be synchronized to any other gesture profile. This is especially interesting when the target velocity profile might be difficult to achieve, due to biomechanical constraints, or particularities in the response of the capture system (e.g., non linearities).

This formalization of temporal mapping includes also the possibility to use particular temporal markers to trigger processes at specific times (generally associated with the use of cue-lists). For example, in Fig. 8.4, line 5 (as well as any other) could be used as such a marker. Therefore, temporal mapping can be extended to the control of a combination of continuous time profiles and discrete time events, synchronized to the input gesture data.

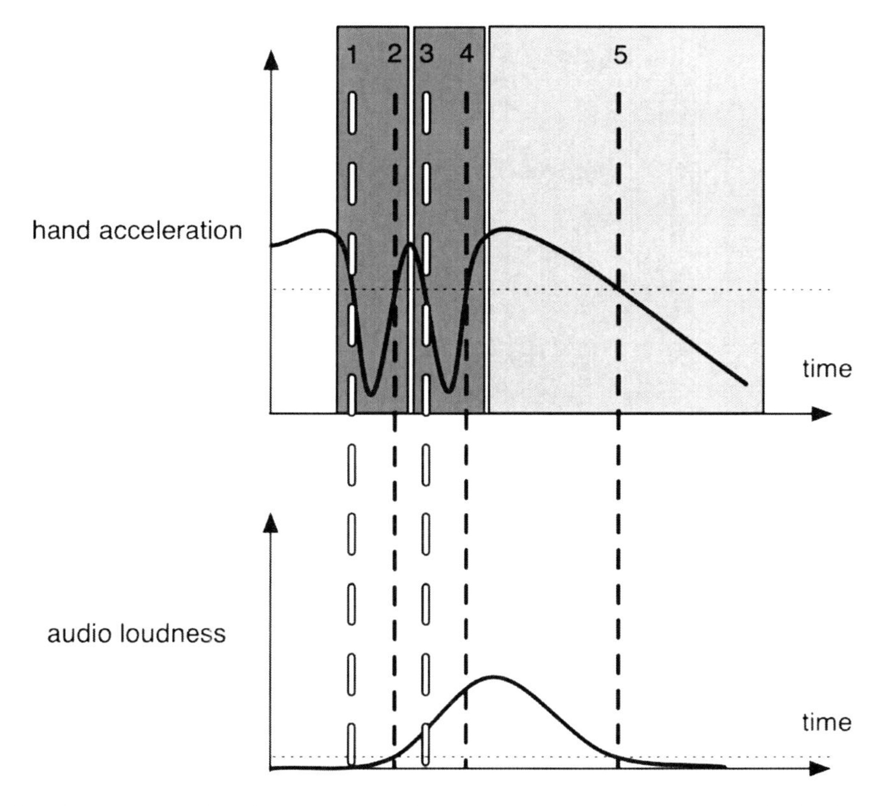

Fig. 8.4 Toy example of temporal mapping between hand acceleration audio loudness. Mapped gesture data is composed of three phases: one first phase (dark grey), a repetition of this first phase (dark grey) and a third different phase (light grey). Vertical lines indicate similar gesture values.

Momeni and Henry previously described an approach that also intrinsically takes into account temporal processes in the mapping procedure. Precisely, they used physical models to produce dynamic layers of audio control [31 and 44]. In our approach, we rather leave users free to define any shape for the audio control, whether based on physical dynamics or designed by other means.

8.4 Example Use with Phase-Vocoder Techniques

The direct combination of the gesture follower with an advanced phase-vocoder system [55] allows for the implementation of a set of applications where a gesture can control continuously the playing speed of an audio file [7 and 10]. In practice, this type of application can be efficiently built using the method illustrated in Fig. 8.5.

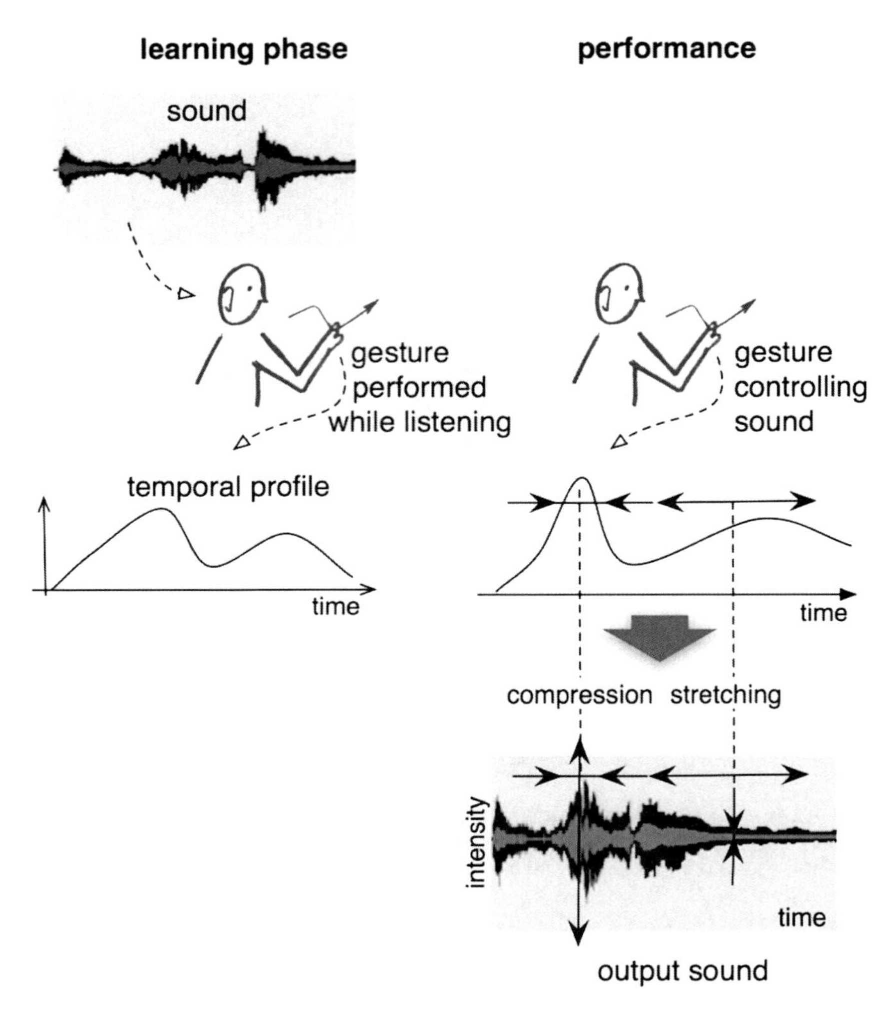

Fig. 8.5 Possible use of the general framework: the user first proposes a gesture while listening to the sound, and then play the sound with temporal and spatial variations.

First, the user must record a gesture while listening to an audio file. This step is necessary for the system to learn a gesture template that is actually synchronous with the original audio recording. The audio recording and the gesture can be several minutes long.

The second step corresponds for the user to re-perform the "same gesture", but introducing speed and intensity variations compared to the reference template. The gesture follower is used to temporally synchronize the gesture with the rendering of the audio file. In other words, the temporal mapping procedure allows for setting a direct correspondence between the gesture time progression and the audio time progression. The audio rendering is performed using a phase-vocoder, ensuring that only the audio playback speed is changed while preserving the pitch and timbre of the original audio recording.

This application allows for the design of a "conducting scenario", where the gesture chosen by the user can be used to control the playing speed of a recording. From an application perspective, this could be applied to "virtual conducting" applications (see for example [36 and 37]. The advantage of our application resides in the fact that the gesture can be freely chosen by the user (e.g. dance movements) by simply recording it once. Precisely, this application can accommodate directly both standard conducting gestures and original gestures invented by the user, which can lead to novel interaction design strategies.

8.5 Conclusion

We described a general framework for gesture-controlled audio processing which has been experimented with in various artistic and pedagogical contexts. It is based on an online gesture processing system, which takes into account temporal behavior of gesture data, and a temporal mapping procedure. The gesture processing makes use of a machine learning technique, and requires pre-recorded gesture templates. Nevertheless, the learning procedure operates using a single recording, which makes the learning procedure simple and easily adaptable to various gesture capture systems. Moreover, our approach could be complemented using Segmental HMM [2, 12 and 13] or hierarchical HMM [26] in modeling transitions between gesture templates, which is missing in the current approach. This would allow for a higher structural level of sound control.

Acknowledgements

We acknowledge partial support of the following projects: the EU-ICT projects SAME and i-Maestro, the ANR (French National Research Agency) projects EarToy ANR-06-RIAM-004 02)and Interlude (ANR-08-CORD-010). We would like to thank Riccardo Borghesi for his crucial role in software development and the students of the "Atelier des Feuillantines" for contributing to experiments.

References

1. Arfib, D., Couturier, J., Kessous, L., Verfaille, V.: Strategies of mapping between gesture data and synthesis model parameters using perceptual spaces. Organized Sound 7(2), 127–144 (2002)
2. Artieres, T., Marukatat, S., Gallinari, P.: Online handwritten shape recognition using segmental hidden markov models. IEEE Transactions on Pattern Analysis and Machine Intelligence 29(2), 205–217 (2007)
3. Aucouturier, J., Daudet, L.: Editorial: Pattern recognition of non-speech audio. Pattern Recognition Letters 31(12), 1487–1488 (2010)

4. Bell, B., Kleban, J., Overholt, D., Putnam, L., Thompson, J., Kuchera-Morin, J.: The multimodal music stand. In: NIME 2007: Proceedings of the 7th International Conference on New Interfaces for Musical Expression, pp. 62–65 (2007)
5. Bettens, F., Todoroff, T.: Real-time DTW-based Gesture Recognition External Object for Max/MSP and PureData. In: SMC: Proceedings of the 6th Sound and Music Computing Conference, pp. 30–35 (2009)
6. Bevilacqua, F.: Momentary notes on capturing gestures. In (capturing intentions). Emio Greco/PC and the Amsterdam School for the Arts (2007)
7. Bevilacqua, F., Guédy, F., Schnell, N., Fléty, E., Leroy, N.: Wireless sensor interface and gesture-follower for music pedagogy. In: NIME 2007: Proceedings of the 7th International Conference on New Interfaces for Musical Expression, pp. 124–129 (2007)
8. Bevilacqua, F., Muller, R.: A gesture follower for performing arts. In: Proceedings of the International Gesture Workshop (2005)
9. Bevilacqua, F., Muller, R., Schnell, N.: MnM: a max/msp mapping toolbox. In: NIME 2005: Proceedings of the 5th International Conference on New Interfaces for Musical Expression, pp. 85–88 (2005)
10. Bevilacqua, F., Zamborlin, B., Sypniewski, A., Schnell, N., Guédy, F., Rasamimanana, N.: Continuous Realtime Gesture Following and Recognition. In: Kopp, S., Wachsmuth, I. (eds.) GW 2009. LNCS, vol. 5934, pp. 73–84. Springer, Heidelberg (2010)
11. Bianco, T., Freour, V., Rasamimanana, N., Bevilaqua, F., Caussé, R.: On Gestural Variation and Coarticulation Effects in Sound Control. In: Kopp, S., Wachsmuth, I. (eds.) GW 2009. LNCS, vol. 5934, pp. 134–145. Springer, Heidelberg (2010)
12. Bloit, J., Rasamimanana, N., Bevilacqua, F.: Modeling and segmentation of audio descriptor profiles with segmental models. Pattern Recognition Letters 31, 1507–1513 (2010)
13. Bloit, J., Rasamimanana, N., Bevilacqua, F.: Towards morphological sound description using segmental models. In: Proceedings of DAFx (2009)
14. Bobick, A.F., Wilson, A.D.: A state-based approach to the representation and recognition of gesture. IEEE Transactions on Pattern Analysis and Machine Intelligence 19(12), 1325–1337 (1997)
15. Camurri, A., De Poli, G., Friberg, A., Leman, M., Volpe, G.: The mega project: analysis and synthesis of multisensory expressive gesture in performing art applications. The Journal of New Music Research 34(1), 5–21 (2005)
16. Camurri, A., Volpe, G., Poli, G.D., Leman, M.: Communicating expressiveness and affect in multimodal interactive systems. IEEE MultiMedia 12(1), 43–53 (2005)
17. Caramiaux, B., Bevilacqua, F., Schnell, N.: Towards a Gesture-Sound Cross-Modal Analysis. In: Kopp, S., Wachsmuth, I. (eds.) GW 2009. LNCS, vol. 5934, pp. 158–170. Springer, Heidelberg (2010)
18. Caramiaux, B., Bevilacqua, F., Schnell, N.: Analysing Gesture and Sound Similarities with a HMM-based Divergence Measure. In: Proceedings of the Sound and Music Computing Conference, SMC (2010)
19. Castellano, G., Bresin, R., Camurri, A., Volpe, G.: Expressive control of music and visual media by full-body movement. In: NIME 2007: Proceedings of the 7th International Conference on New Interfaces for Musical Expression, pp. 390–391 (2007)
20. Chafe, C.: Simulating performance on a bowed instrument. In: Current Directions in Computer Music Research, pp. 185–198. MIT Press, Cambridge (1989)

21. Cont, A.: Antescofo: Anticipatory synchronization and control of interactive parameters in computer music. In: Proceedings of the International Computer Music Conference, ICMC (2008)
22. Cont, A., Coduys, T., Henry, C.: Real-time gesture mapping in pd environment using neural networks. In: NIME 2004: Proceedings of the 2004 Conference on New Interfaces for Musical Expression, pp. 39–42 (2004)
23. Demoucron, M., Rasamimanana, N.: Score-based real-time performance with a virtual violin. In: Proceedings of DAFx (2009)
24. Dourish, P.: Where the action is: the foundations of embodied interaction. MIT Press, Cambridge (2001)
25. Fels, S., Hinton, G.: Glove-talk: a neural network interface between a data-glove and a speech synthesizer. IEEE Transactions on Neural Networks 3(6) (1992)
26. Fine, S., Singer, Y.: The hierarchical hidden markov model: Analysis and applications. Machine Learning 32(1), 41–62 (1998)
27. Godøy, R., Haga, E., Jensenius, A.R.: Exploring music-related gestures by sound-tracing - a preliminary study. In: 2nd ConGAS International Symposium on Gesture Interfaces for Multimedia Systems, Leeds, UK (2006)
28. Godøy, R., Leman, M. (eds.): Musical Gestures: Sound, Movement and Meaning. Routledge, New Yark (2009)
29. Guédy, F., Bevilacqua, F., Schnell, N.: Prospective et expérimentation pédagogique dans le cadre du projet I-Maestro. In: JIM 2007-Lyon
30. Guédy, F.: Le traitement du son en pédagogie musicale, vol. 2. Ircam – Editions Léo, L'Inouï (2006)
31. Henry, C.: Physical modeling for pure data (pmpd) and real time interaction with an audio synthesis. In: Proceedings of the Sound and Music Computing Conference, SMC (2004)
32. Hoffman, M., Cook, P.R.: Feature-based synthesis: Mapping from acoustic and perceptual features to synthesis parameters. In: Proceedings of International Computer Music Conference, ICMC (2006)
33. Hunt, A., Wanderley, M., Paradis, M.: The importance of parameter mapping in electronic instrument design. The Journal of New Music Research 32(4) (2003)
34. Hunt, A., Wanderley, M.M.: Mapping performer parameters to synthesis engines. Organised Sound 7(2), 97–108 (2002)
35. Jensenius, A.R.: Action-sound: Developing methods and tools to study music-related body movement. PhD thesis, University of Oslo, Department of Musicology, Oslo, Norway (2007)
36. Lee, E., Grüll, I., Kiel, H., Borchers, J.: Conga: a framework for adaptive conducting gesture analysis. In: NIME 2006: Proceedings of the 2006 Conference on New Interfaces for Musical Expression, pp. 260–265 (2006)
37. Lee, E., Wolf, M., Borchers, J.: Improving orchestral conducting systems in public spaces: examining the temporal characteristics and conceptual models of conducting gestures. In: CHI 2005: Proceedings of the SIGCHI Conference on Human Factors in Computing Systems, pp. 731–740 (2005)
38. Leman, M.: Embodied Music Cognition and Mediation Technology. Massachusetts Institute of Technology Press, Cambridge (2008)
39. Levitin, D., McAdams, S., Adams, R.: Control parameters for musical instruments: a foundation for new mappings of gesture to sound. Organised Sound 7(2), 171–189 (2002)

40. Maestre, E.: Modeling Instrumental Gestures: An Analysis/Synthesis Framework for Violin Bowing. PhD thesis, Universitat Pompeu Fabra (2009)
41. Minka, T.P.: From hidden markov models to linear dynamical systems. Technical report, Tech. Rep. 531, Vision and Modeling Group of Media Lab, MIT (1999)
42. Miranda, E., Wanderley, M.: New Digital Musical Instruments: Control And Interaction Beyond the Keyboard. A-R Editions, Inc (2006)
43. Mitra, S., Acharya, T., Member, S., Member, S.: Gesture recognition: A survey. IEEE Transactions on Systems, Man and Cybernetics - Part C 37, 311–324 (2007)
44. Momeni, A., Henry, C.: Dynamic independent mapping layers for concurrent control of audio and video synthesis. Computer Music Journal 30(1), 49–66 (2006)
45. Mori, A., Uchida, S., Kurazume, R., Ichiro Taniguchi, R., Hasegawa, T., Sakoe, H.: Early recognition and prediction of gestures. In: Proceedings of the International Conference on Pattern Recognition, vol. 3, pp. 560–563 (2006)
46. Myers, C.S., Rabiner, L.R.: A comparative study of several dynamic time-warping algorithms for connected word recognition. The Bell System Technical Journal 60(7), 1389–1409 (1981)
47. Rabiner, L.R.: A tutorial on hidden markov models and selected applications in speech recognition. Proceedings of the IEEE, 257–286 (1989)
48. Rajko, S., Qian, G.: A hybrid hmm/dpa adaptive gesture recognition method. In: International Symposium on Visual Computing (ISVC), pp. 227–234 (2005)
49. Rajko, S., Qian, G.: Hmm parameter reduction for practical gesture recognition. In: 8th IEEE International Conference on Automatic Face and Gesture Recognition (FG 2008), pp. 1–6 (2008)
50. Rajko, S., Qian, G., Ingalls, T., James, J.: Real-time gesture recognition with minimal training requirements and on-line learning. In: CVPR 2007: IEEE Conference on Computer Vision and Pattern Recognition, pp. 1–8 (2007)
51. Rank, E.: A player model for midi control of synthetic bowed strings. In: Diderot Forum on Mathematics and Music (1999)
52. Rasamimanana, N., Guedy, F., Schnell, N., Lambert, J.-P., Bevilacqua, F.: Three pedagogical scenarios using the sound and gesture lab. In: Proceedings of the 4th i-Maestro Workshop on Technology Enhanced Music Education (2008)
53. Rasamimanana, N.H., Bevilacqua, F.: Effort-based analysis of bowing movements: evidence of anticipation effects. The Journal of New Music Research 37(4), 339–351 (2008)
54. Rasamimanana, N.H., Kaiser, F., Bevilacqua, F.: Perspectives on gesture-sound relationships informed from acoustic instrument studies. Organised Sound 14(2), 208–216 (2009)
55. Roebel, A.: A new approach to transient processing in the phase vocoder. In: Proceedings of DAFx (September 2003)
56. Schnell, N., Borghesi, R., Schwarz, D., Bevilacqua, F., Müller, R.: Ftm - complex data structures for max. In: Proceedings of the International Computer Music Conference, ICMC (2005)
57. Schnell, N., Röbel, A., Schwarz, D., Peeters, G., Borghesi, R.: Mubu and friends: Assembling tools for content based real-time interactive audio processing in max/msp. In: Proceedings of the International Computer Music Conference, ICMC (2009)
58. Schnell, N., et al.: Gabor, Multi-Representation Real-Time Analysis/Synthesis. In: Proceedings of DAFx (September 2005)

59. Schwarz, D., Orio, N., Schnell, N.: Robust polyphonic midi score following with hidden markov models. In: Proceedings of the International Computer Music Conference, ICMC (2004)
60. Serafin, S., Burtner, M., Nichols, C., O'Modhrain, S.: Expressive controllers for bowed string physical models. In: DAFX Conference, pp. 6–9 (2001)
61. Turaga, P., Chellappa, R., Subrahmanian, V., Udrea, O.: Machine recognition of human activities: A survey. IEEE Transactions on Circuits and Systems for Video Technology 18(11), 1473–1488 (2008)
62. Van Nort, D., Wanderley, M., Depalle, P.: On the choice of mappings based on geometric properties. In: Proceedings of the International Conference on New Interfaces for Musical Expression, NIME (2004)
63. Van Nort, D.: Modular and Adaptive Control of Sonic Processes. PhD thesis, McGill University (2010)
64. Verfaille, V., Wanderley, M., Depalle, P.: Mapping strategies for gestural and adaptive control of digital audio effects. The Journal of New Music Research 35(1), 71–93 (2006)
65. Volpe, G.: Expressive gesture in performing arts and new media. Journal of New Music Research 34(1) (2005)
66. Wanderley, M., Depalle, P.: Gestural control of sound synthesis. Proceedings of the IEEE 92, 632–644 (2004)
67. Wanderley, M.: (guest ed.) Mapping strategies in real-time computer music. Organised Sound, vol. 7(02) (2002)
68. Wilson, A.D., Bobick, A.F.: Realtime online adaptive gesture recognition. In: Proceedings of the International Conference on Pattern Recognition (1999)

Section II: Musical Robots and Automated Instruments

Chapter 9
Automated Piano: Techniques for Accurate Expression of Piano Playing

Eiji Hayashi

Abstract. The challenge of developing an automated piano that accurately produce the soft tones of a desired performance which is a problem encountered by pianists themselves, led to a reconsideration of the touch involved in producing soft piano tones. For this purpose, the behavior of the piano's action mechanism was measured and observed based on a weight effect which is one of the pianist's performance techniques, and the accurate expression of soft tones was realized. Furthermore, although double-strikes by the hammer and non-musical sounds occur during the performance of soft tones, such sound effects were cancelable. This Chapter describes the development of a superior automated piano which is capable of reproducing a desired performance with expressive soft tones. The piano's hardware and software have been developed, and the piano's action mechanism has been analyzed with reference to the "touch." This fundamental input waveform can be used to accurately and automatically reproduce a key touch based on performance information for a piece of classical music. The user can accurately create an emotionally expressive performance according to an idea without moving the body of the pianist.

9.1 Introduction

Automated pianos; including barrel-operated, roll-operated and reproducing pianos, were commonly manufactured in the late 19th and early 20th centuries [21, 23]. Called the "Pianista"; a pneumatically-operated piano, was made and patented by Fourneauz (a Frenchman) in 1863. The Pianola, manufactured in 1902 by Aeolian Company, and the same company later made the Duo-Art reproducing piano in 1913. In America the Welte Mignon (M.Welte & Sons), the Duo-Art (Aeolian Company) and the Ampico (American Company) were famous reproducing piano called the big three. In Europe, the "Hupfield-Dea" and the "Philipps-Duca" were well-known. The earlier models played back the notes by simply following the length of slits on a roll, which represented the actual length of the

J. Solis and K. Ng (Eds.): Musical Robots and Interactive Multimodal Systems, STAR 74, pp. 143–163.
springerlink.com

notes to be played. They could not, however, vary the volume of the note being played. The reproducing pianos that followed in the early 1910's; however, featured volume control and those had considerably improved the ability to record and play back music compared to the earlier roll-operated pianos.

In recent years, Yamaha has marketed a piano player, while Bösendorfer marketed a computerized grand piano. Such pianos use optical sensors attached to the keys, hammers, and pedals, allowing performance information to be created from actual key, hammer, and pedal functions when the pianist performs. An actuator using a solenoid for the drive system is attached to the bottom of the rear edge of the key, enabling an automatic performance. The performances are characterized by a narrow dynamic range, due to the unstable reproduction of soft tones as well as difficulties playing the same key with fast repetitions. These automated pianos, often used in musical education or in the lounge of a hotel, are also used for the evaluation of automatic computer-generated performance in music-related research fields.

Currently available commercial automated pianos have some limitations in reproducing soft tones, resulting in difficulties to accurately reproduce the full range of expression of a complete performance. In addition, little experimental research has been done on the motions of the action of a grand piano [1, 2, 3 and 4], which has many individual parts, including the keys and hammers. A device that can more accurately reproduce a grand piano performance is desired. Furthermore, if the performance data of a gifted, deceased player can be obtained, it will also become possible to reproduce that player's live performances on an automated piano using the performance data.

Fig. 9.1 Automated piano: FMT-I.

The author has developed an enhanced automated piano with which a user can reproduce a desired performance. See Figure 9.1 [5, 6]. The piano's hardware and software have been developed, and the piano's action mechanism has been analyzed [7, 8]. The proposed automated piano employs feedback control to follow up an input waveform for a touch actuator which uses the position sensor of an eddy current for striking a key. This fundamental input waveform is used to accurately and automatically reproduce a key touch based on the performance information of a piece of classical music.

In order to reproduce an accurate musical performance by an automated piano, the user may needs to edit thousands of notes in a score of music [9, 10, and 11]. However, since the automated piano can accurately reproduce music, the user can accurately create an emotionally expressive performance according to an idea without having dexterity skills like a pianist. Since a user can create a desired expression with the automated piano, he or she can also make variations in the performance following repeated listening, allowing the user to make changes in their expressive performance.

The automated piano was exhibited at EXPO 2005 held in Aichi (Japan) and The Great Robot Exhibition in National Museum of Nature and Science in 2007 held in Tokyo (Japan), where the functions of the proposed system were demonstrated.

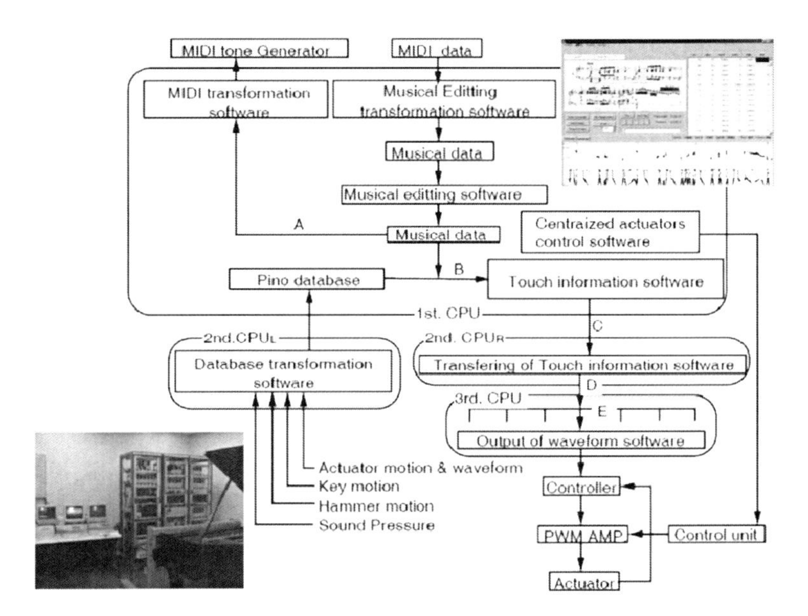

Fig. 9.2 Automated piano system.

In addition, the author has developed an interactive musical editing system that utilizes a database to edit music more efficiently and user-friendly [24]. The proposed system has been conceived so that it can search for similar phrases throughout a musical score and evaluates the style of the performance. The method of searching for similar phrases uses Dynamic Programming (DP) matching.

This chapter describes the automated piano, the behavior of the piano's action, and the "touch" system, which allows the piano to achieve the stable reproduction and repetition of a soft tone on the same key.

9.2 System Architecture of the Automated Piano

The system is composed of a generalization/management control computer system, a control system, and an actuation system, as shown in Figure 9.2. The touch actuator [12]; as shown in Figure 9.3 uses a non-contact displacement sensor (AEC: Gap sensor) which applies an eddy-current to ensure that a key is touched correctly. A PWM (Pulse Width Module) amplifier (Servo land corporation) which can output a maximum of 9 A was used for the touch actuator. To place the actuator unit for each of the 88 keys, the width was set to 23 mm less than that of a white key, which is about 23.5mm. To obtain rapid response, we use a lightweight rotor in the actuator, made by manufacturing an epoxy resin laminated plate, and 2 coils of aluminum ribbon wire (1.4 x 0.25 mm) were embedded in the rotor, as shown in Figure 9.3. The developed touch actuator could realize high-speed response to a band width of 80 Hz. The pedaling actuator, as shown in Figure 9.4, was designed as a combination of a dc motor and a self-locking mechanism utilizing a worm gear and wheel.

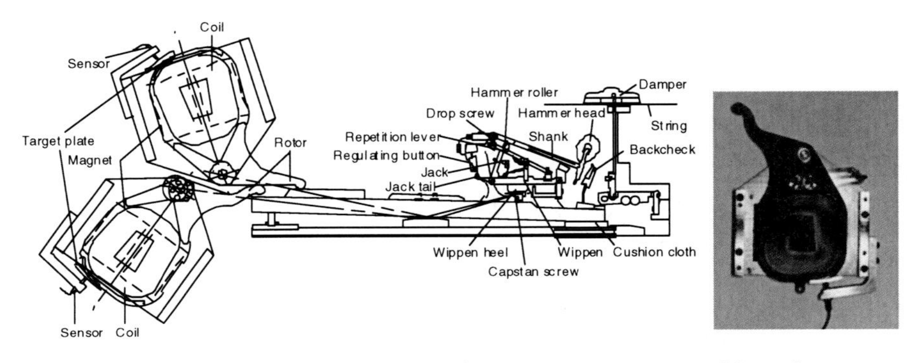

(a) Arrangement on a grand piano (b) touch actuator

Fig. 9.3 Touch actuator.

In the generalization/management control computer system, 90 sets of onboard computers supply an input waveform to several touches and pedal actuators simultaneously during musical reproduction. A dual-ported RAM was used for the interface, allowing a high-speed transfer of data for creating a waveform.

The proposed system and the user's interface of the system are shown in Figures 9.5 and 1.6 [13, 14, 15, 22, 24], respectively. A user edits the performance information data using the editing software shown in Figure 9.6. The performance information data consists of the five parameters: tone, volume, the beginning time of the tone, the interval time of the tone and bar information. Those parameters are referred as: *key*, *velo*, *step*, *gate*, and *bar*, respectively. The user edits the music by changing these values in the performance information.

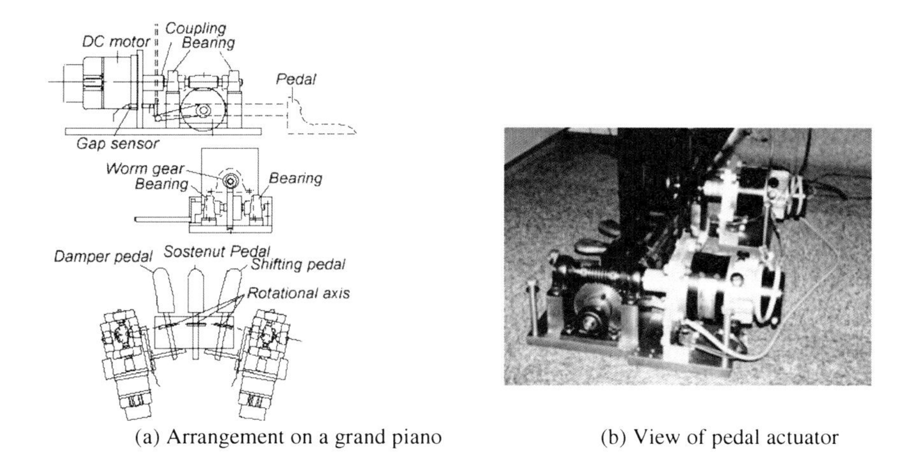

(a) Arrangement on a grand piano (b) View of pedal actuator

Fig. 9.4 Pedal actuator.

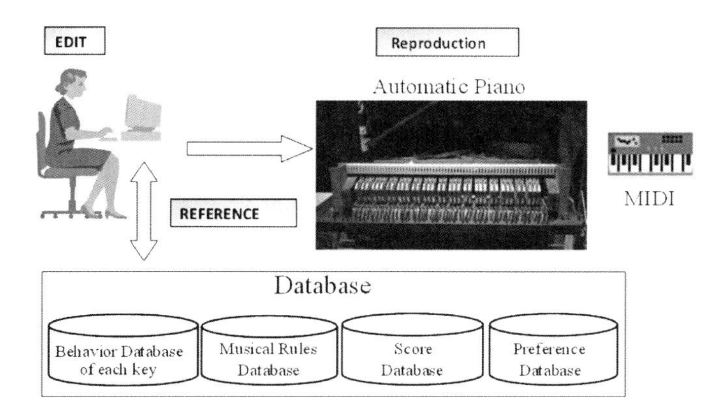

Fig. 9.5 Sketch of software for editing and reproducing.

- ***Key***: The key number corresponds to note number of MIDI from 21 to 108.
- ***Velo***: The velo shows key-depressing strength. The value of *velo* corresponds to the velocity of MIDI from 0 to 127.
- ***Gate***: The gate time shows duration of sound. The unit of gate time is a millisecond, and it corresponds to the interval of time from note-on to note-off for a certain key in MIDI.
- ***Step***: The step time shows an interval between a given note and the next note. The unit of step time is a millisecond.
- ***Bar***: The bar information is convenient for the user to edit a piece of classical music.

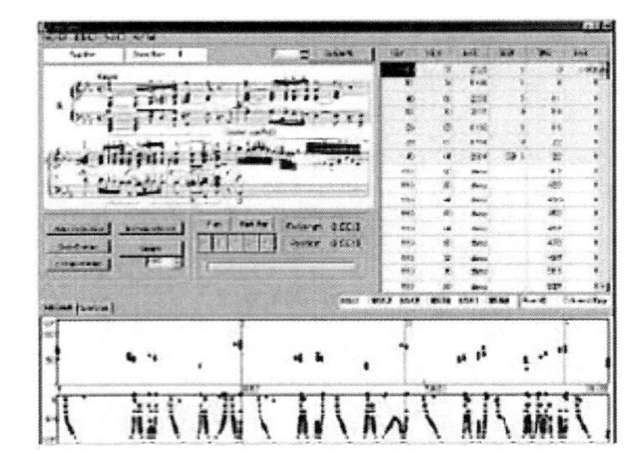

Fig. 9.6 Application display of the editing support system.

In order to assist the user in arranging music according to their intentions, a series of databases were implemented with information about the behavior of the piano's action, music theory, e.g., information about musical grammar and basic musical reinterpretation, and the user's preferences regarding musical reinterpretation, as follows:

- **Behavior database:** This database, which is used to build the dynamic characteristics of each key, has parameters describing the volume of the sound based on the analysis of the piano-action mechanism's behavior (see Figure 9.7). When the user reproduces the performance information data, the software transforms it into the drive waveform. The author has created drive waveforms for the touch actuator for every key corresponding to the notes in the performance information data using this database.
- **Rules Database:** This database has the architecture of the musical grammar that is necessary to interpret symbols in the musical notation.
- **Score Database:** This database has symbols including notes, and time signatures, rests, and so on in musical notation. Symbols were pulled together in order bar by bar, and also the bar symbols were arranged a time series.
- **Preference Database:** This database stores the user's preferences regarding performance characteristics. The expressions are defined by the relationships between tempo and dynamic, and their basic data structures are described in the following section.

9.3 Method of Transformation Process for Reproduction

In order to create a drive waveform for the touch actuators, the author proposed to divide a key strike and its return to its original position into four regions, A-D as shown in Figure 9.7, and the result of the analysis was included in the behavior database for each key. The key strike was divided into the four sections as follows:

- **Section A:** The duration of time between the moment when the key is struck and the hammer strikes the strings.
- **Section B:** The contact time of the hammer against the strings.

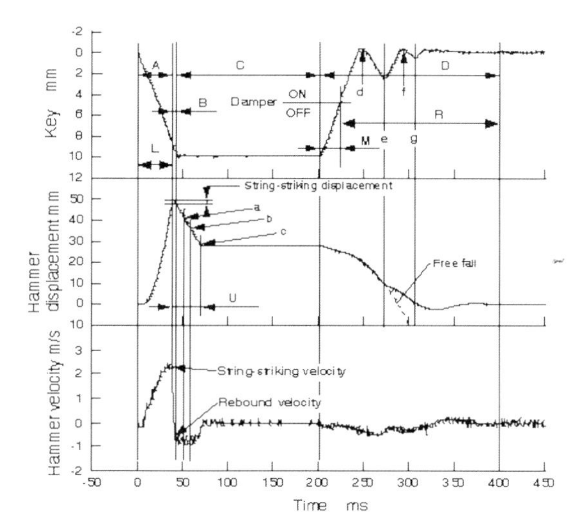

Fig. 9.7 General histories of key and hammer in the general touch case, since the hammer has an initial velocity at the escapement point, it independently breaks contact from other parts in the piano's-action and strikes a string. The hammer keeps in contact with the string from first contact of the string until it breaks contact again. After that the hammer rebounding from the string drops to the repetition lever, and the back check holds the hammer if the key is being held at the key's bottom-most point.

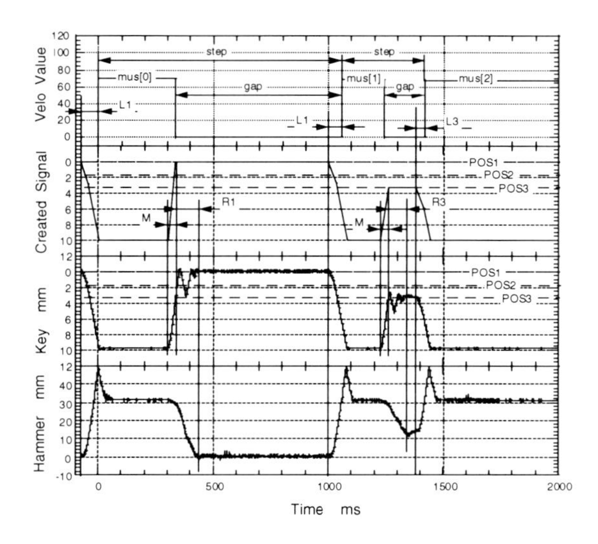

Fig. 9.8 Translating method from performance information

- **Section C:** The time when the hammer and strings separate until the hammer returns to its original position, or the time when the key is fully depressed (section U) until the hammer returns to its back check, or until the hammer comes to rest on a repetition lever, depending on how the hammer is set up. At point "a" the hammer roller makes contact with the repetition lever, and the hammer pushes down the repetition lever. At point "b" the bottom of hammer head makes contact with the back check. Finally, at point "c" the hammer goes into a stall.
- **Section D:** The time when the key is released until the hammer comes to rest at its original position by the influence of its gravity. If a key for the next touch is depressed again before "g" point in time, it is difficult to obtain the desired sound, because the key separates from other parts during the term of d-e and f-g.

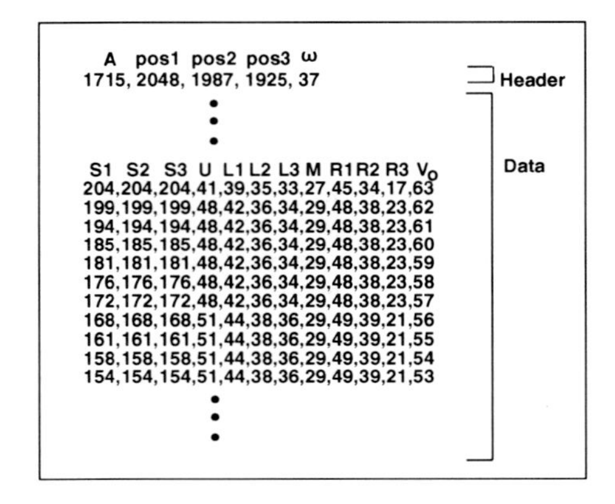

Fig. 9.9 Sample of behavior database data.

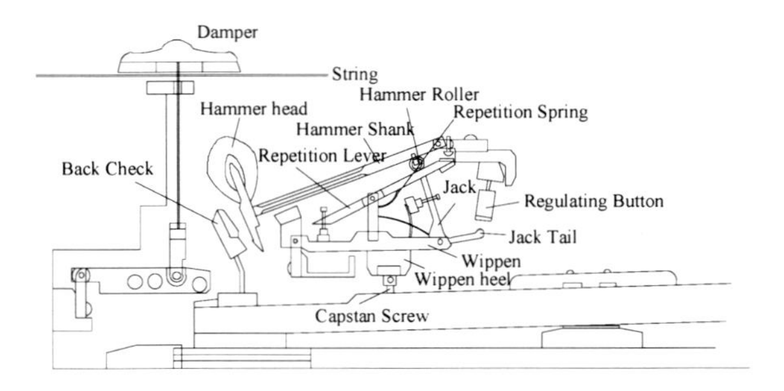

Fig. 9.10 Piano's- action mechanism

When the input waveform for the touch cannot be created, the input waveform is created using the data indicating an increased volume. It has been found that even if the music has a slow tempo, the desired performance cannot be achieved at the part of the performance that involves the repetition of the same key with a soft tone.

The transformation processing method [5] creates an input waveform for the driving actuator from performance information, as shown in Figure 9.8. The input waveform for each touch is created for each key, referring to the data in the behavior database describing the volume of the sound (*Vo*), as shown in Figure 9.9. The letters in Figure 9.8 correspond to the letters in Figure 9.9.

9.4 Performance of Soft Tones

With sufficient training, pianists can learn to understand the behavior of a piano's action mechanism, which is a complex device made of wood, leather, and felt (see Figure 9.10). This mechanism allows the pianist to accurately control the dynamics of his or her performance, even if small mistakes are made.

The sound of many currently available automated pianos is inaccurate regarding musical dynamics because the solenoid does not allow a key to be struck in a way that accurately mimics what occurs during a real performance.

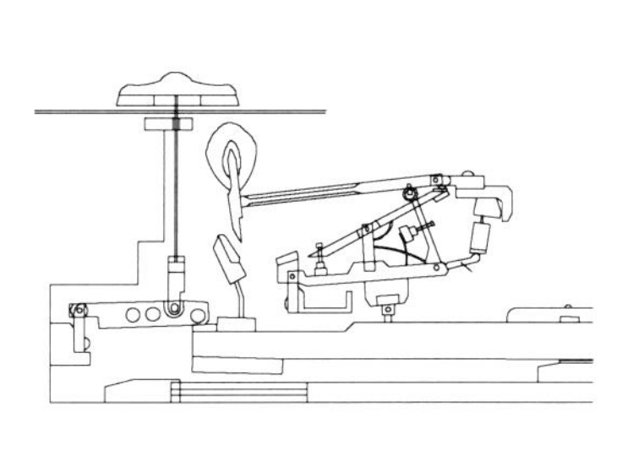

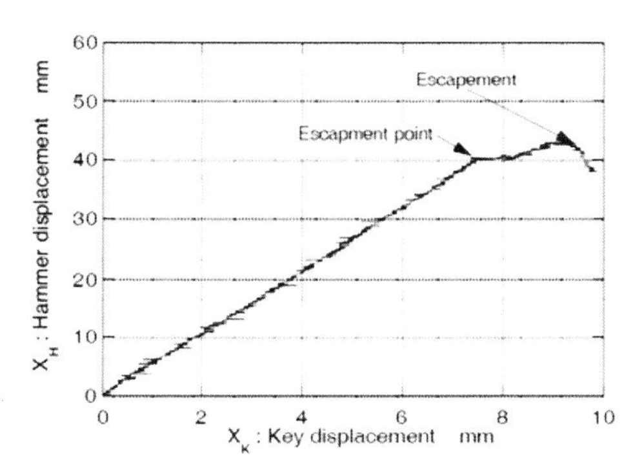

Fig. 9.11 Escapment in piano's- action.

9.4.1 The Piano's - Action Mechanism

When a key is depressed slowly, the wippen moves upward. In the initial movement, the repetition lever pushes up the hammer roller (attached near the rotational center of the shank), and then the jack pushes up the hammer roller, and the hammer head moves toward the string. As the key is depressed further, the jack tail reaches the regulating button (referred to as the "escapement point" in the following discussion) as the hammer head comes within several millimeters of the string. After that, the jack is forced by the regulating button to rotate and disconnect itself from the hammer roller. This is referred to as the "escapement" in the following discussion (see Figure 9.11), and it becomes the end spot for the rising motion of the hammer. If the key is depressed extremely slowly, the hammer falls on the repetition lever without striking the string. In cases where the key is struck with the usual force, the hammer will have acquired sufficient velocity at the escapement point to continue and strike the string.

The suggested dynamic model of the piano's action is shown in Figure 9.12. The relationship between the hammer string-striking speed and various touches, the dynamic response at constant speed, at constant acceleration, and a modified constant speed of key, are shown in Figures 9.13 – 1.15. Problems occur when there is a constant speed and a constant acceleration in soft tone less than about 1 m/s of hammer string-striking speed, i.e., along a large gradient. This problem was solved by developing a touch of a modified constant speed.

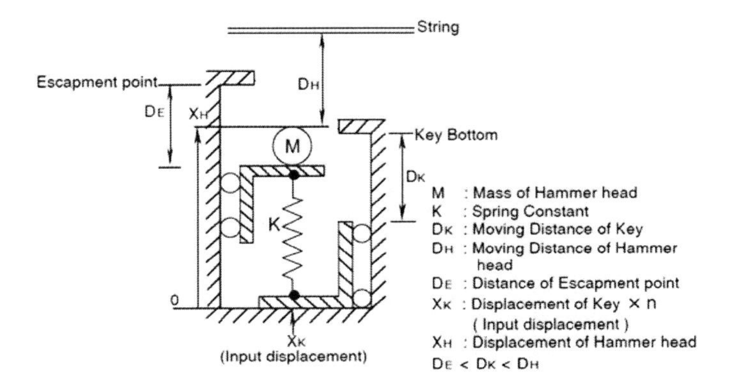

Fig. 9.12 Suggested piano's action model. M is the mass of the hammer head, and K is the spring constant, which translates the elasticity of all the piano-action parts, including the elasticities of the key, the wippen, the hammer roller, and the shank, to the position of the hammer head.

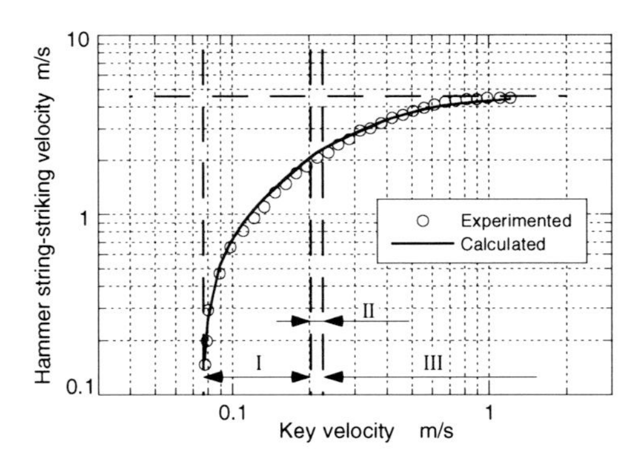

Fig. 9.13 Relationship between the key speed and the hammer string-striking velocity.

In the piano's action described in this section, the ω_1 and ω_2 natural frequencies used in this analysis were 81.7 rad/s and 113 rad/s, respectively. When the hammer shank was clamped as shown in Figure 9.16, its natural frequency (*f*) was found to be approximately 440 rad/s. Based on this value, it can be predicted that the natural frequencies of the individual wooden parts that make up the action, except for the felt found on the wippen heel and the hammer roller, are greater than the natural frequencies ω_1 and ω_2. Furthermore, the natural frequency of the piano's action (except for the felt), estimated by a series-type synthetic method, will be greater than the natural frequency of ω_1 and ω_2. This indicates that the stiffness of the felt has a significant effect on the physical values of the piano's action mechanism.

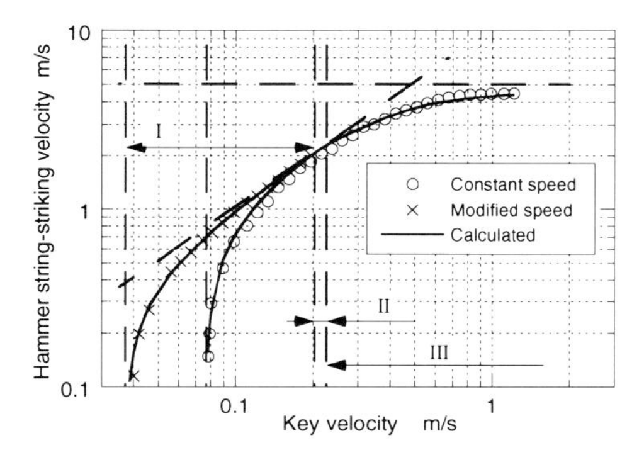

Fig. 9.14 Relationship between the key speed and the hammer string striking velocity.

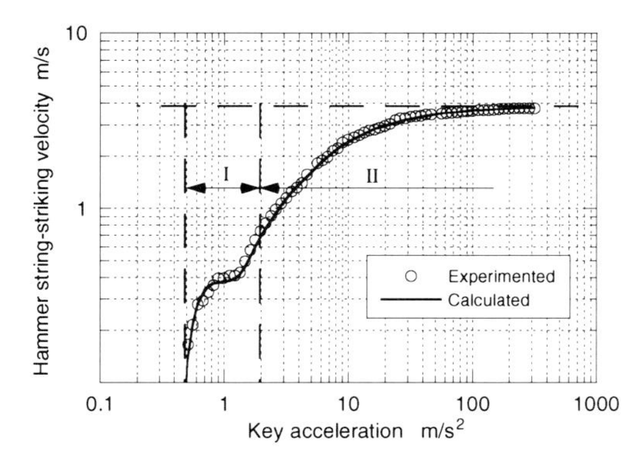

Fig. 9.15 Relationship between the key acceleration and the hammer string striking velocity.

9.4.2 The Problem of Reproducing Soft Tones

The creation of a soft tone appears to depend largely on the hammer's behavior in section C after the hammer has struck the strings. In an actual performance, the portion marked with an asterisks in Figure 9.17, is increased greatly and the hammer may approach the strings again, and a double- strike may be generated instantly as the hammer "bounces" against the string.

The sound generated by such a movement differs from the one produced by the usual hammer strike. In this chapter, the double-strike sound during a soft sound will be called a "non-musical sound". Although such a non-musical sound is faint, during a soft section of a performance such a sound will be clear. In an actual performance, dissonance is caused by a combination of various elements, such as tempo, the length of the sound between the notes, the strength with which notes are played, and the pedal operation. Therefore, when pianists hear dissonance, they will deal with it in terms of these factors. In the same way, the user of our automated piano can change performance data and they can erase the sound of the dissonance.

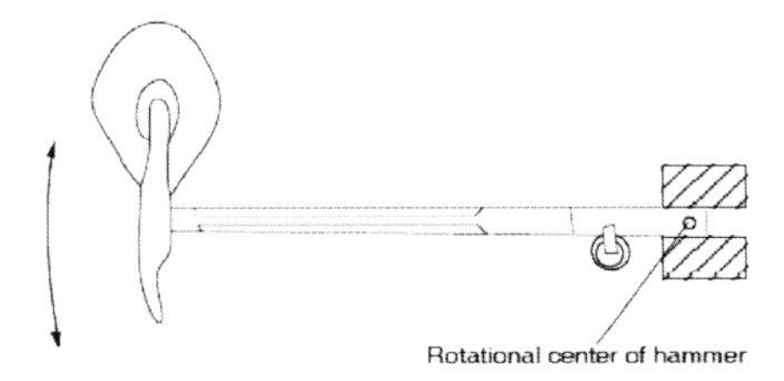

Fig. 9.16 Free vibration of hammer clamped at center of rotation.

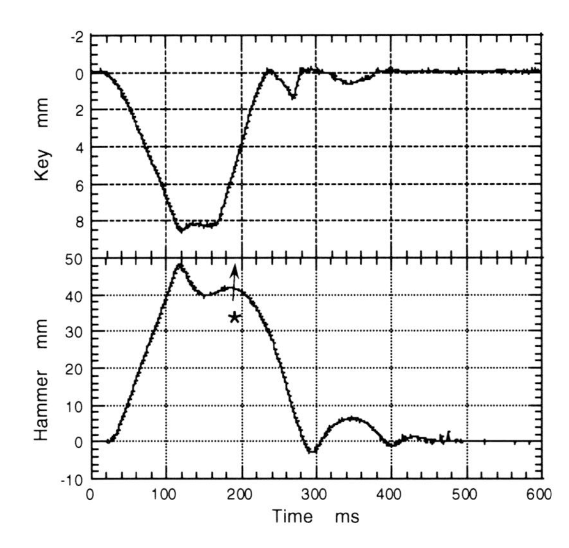

Fig. 9.17 Behaviors of key and hammer during a soft tone.

Although we thought until now that such a non-musical sound was peculiar to the developed automated piano, such sounds can be heard on the recordings of great pianists if listened to closely.

An acoustic specialist told us that old master tapes that have not been processed after a performance allow the listener to really appreciate the performances of the great pianists of the past. Although recognizing non-musical sounds in a performance requires experience, such sounds can be discovered in the performances of Kempff, Gould, Backhaus, and Michelangeli. The parts of the performances, in which they generated such kind of sounds, were the same parts of the performances where we were taking pains to remove sounds.

As an example, bars 5 to 7 of the score of Beethoven's piano sonata No. 14, The Moonlight Sonata, are shown in Figure 9.18, and the sound waveforms of bars 5 and 7 are shown in Figure 9.19 (a) and (b), respectively. The difference between (a) and (b) is that (a) does not contain a non-musical sound, while (b) does. The place where the non-musical sound is generated is in the note (dotted half note: G#) marked by the arrows in (b).

Fig. 9.18 Bars 5-7 from Beethoven's Piano Sonata No.14, "Moonlight".

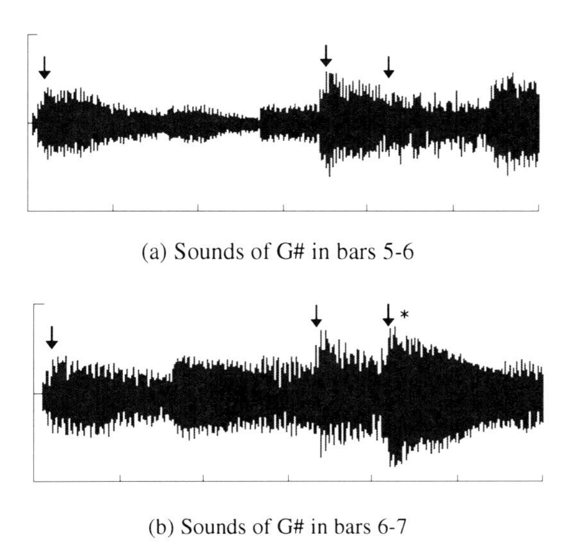

(a) Sounds of G# in bars 5-6

(b) Sounds of G# in bars 6-7

Fig. 9.19 Sound waveforms of (a) G# in bars 5-6 and (b) G# in bars 6-7.

Actually, several people were asked to listen to the performance without first being informed of what to listen for on the first audition. We found that the difference in sound between (a) and (b) could not be recognized on the first trial. After being told what to listen for, after several auditions the non-musical sound could be recognized and the difference between the performances was noted.

Though such sounds can be identified aurally, the difference between their presence and absence, as shown in Figure 9.19 (a) and (b), is not easily measured or analyzed electronically. Thus, we tried to clarify this sound based on the measurement of the behaviors of the key and hammer.

9.5 Behavior of the Piano's Action during a Soft Tone

In order to understand the non-musical sound, the author considered a behavior of the piano's action again on a soft tone. The behavior of the piano's action during the creation of a soft tone is shown in Figure 9.20. The behavior of the hammer describes a key velocity of 0.075 m/s and a hammer velocity of 0.6 m/s, at a volume of about *mp* - *pp* (mezzo piano to pianissimo).

In the general touch case, since the hammer has an initial velocity at the escapement point, it independently breaks contact from other parts in the piano's action and strikes a string. The hammer keeps in contact with the string from first contact of the string until it breaks contact again. After that the hammer rebounding from the string drops to the repetition lever, and the back check holds the hammer if the key is being held at the key's bottom-most point.

Along the hammer's movement from the hammer breaks contact again, the hammer-roller touches the repetition lever, and then the hammer pushes down the repetition spring with the repetition lever by a power of the reaction when it rebounded from the string. When the hammer roller pushes the repetition lever further down, the lower side of the hammer comes into contact with the back check. The force of the hammer decreases by degrees as a consequence of the contact between the back check and the lower hammer head. Finally, the back check holds the hammer. As the key is retained at its bottom-most point, the sound is sustained because the damper is released.

The author opinions regarding the soft tone [17, 18 and 19] are as follows:

- **Section A:** The velocities of the key and the hammer are slow, and the time between the key strike and the striking of the hammer is long, making the hammer strike the strings at a low velocity.
- **Section B and C:** As shown in Figure 9.20, the hammer isn't fixed to the back check when the key is kept fully depressed, because the rebound velocity of the hammer is slow after it strikes the string. Consequently, the hammer comes to rest at a high position on the repetition lever (Figure 9.21).

  The rebounded hammer depresses the spring in the repetition lever because its force is less than the force to fix it by the back check. The hammer is then flipped by the back check. After that, according to the slow restoration of the spring, the hammer goes up with the repetition lever.

Since the hammer is not fixed to the back check after the hammer strikes the strings, the duration of section U until the hammer comes to rest stably on the repetition lever increases sharply.

- **Section D:** The potential energy becomes large due to the high position, and the duration until the key and the hammer come to rest in their original position also increases.

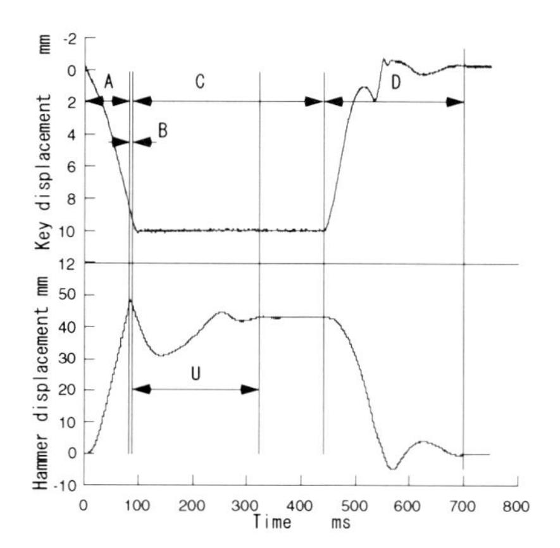

Fig. 9.20 General behavior of a soft tone.

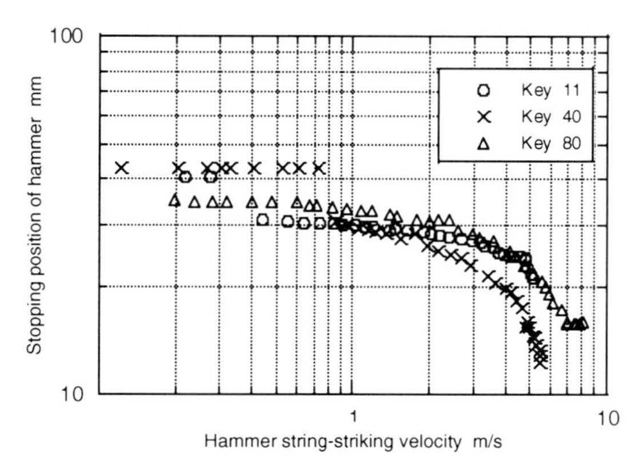

Fig. 9.21 Relationship between the string-striking and the stopping position when the hammer is held by the back check. The rebound velocity of the hammer increases in velocity, and the stopping position of the hammer becomes low with increasing string-striking velocity of the hammer.

The height at which the hammer is fixed on the back check and the interval time of section U change according to the volume of sound. In the case of Figure 9.21 for example, the interval time of section U reaches about 250 ms. Moreover, when the key is struck to produce a volume at which the hammer is fixed to the back check, the interval time of section U is about 20 to 80 ms; accordingly, the behavior of the soft tone becomes unstable suddenly. However, if the time during which the key is depressed fully is shorter than the time of section U, a subsequent movement of the hammer will become unstable. Moreover, if the time of section U is considered, touches of the same key in which a soft tone is repeated will be affected greatly and it will become difficult during the next touch for the hammer to remain stable. This results in the next touch becoming louder than the desired sound, or generating a double-strike so that the hammer after striking the strings may float too much, as mentioned above. Furthermore, although a double-strike may not happen, a non-musical sound is produced since the hammer floats too near the string. In order to solve this problem, the author adopted a performance technique involving an after-touch based on the weight effect.

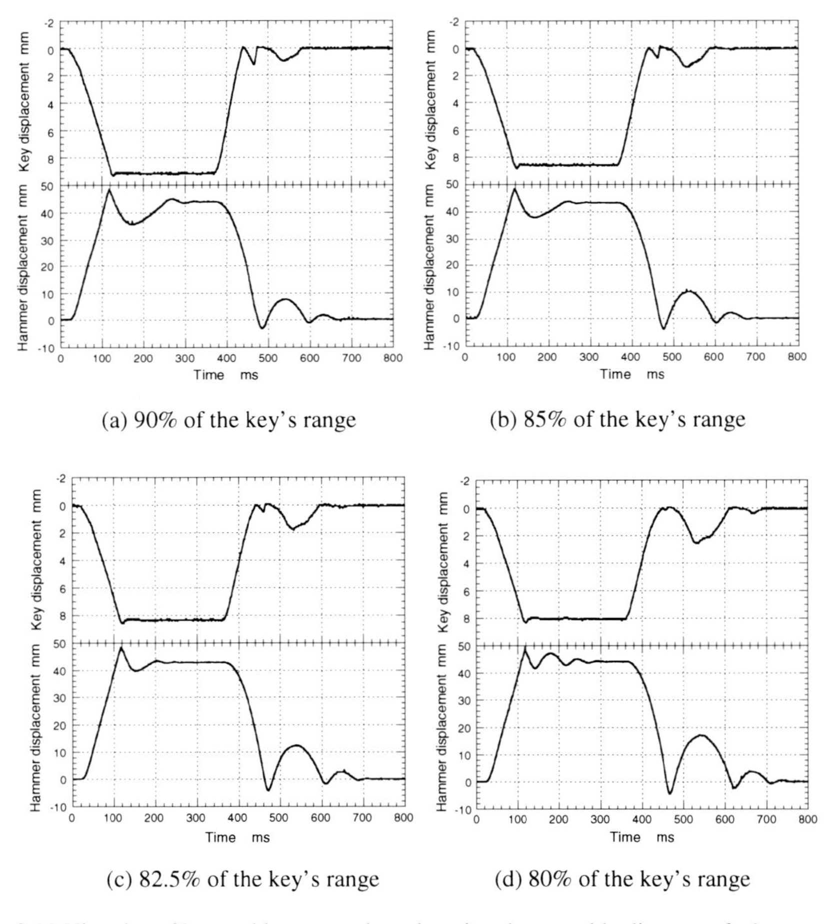

Fig. 9.22 Histories of key and hammer when changing the movable distance of a key.

9.6 Touch after the Hammer Strikes the Strings

A touch can be performed with either a direct stroke or an indirect stroke [20]. In the developed automated piano, the swing-stroke is performed in a condition in which the fingertip of the actuator remains constantly in contact with the surface of a key; this is equivalent to a direct swing-stroke. In case of a human player, it is necessary to touch the keys while balancing and feeling suitable elasticity to the force of the rebound of a key and a hammer; this is called the "weight effect." If the touch after the hammer strikes the strings based on the weight effect is considered, a pianist carries out a suitable counteracting motion for the force of the hammer immediately after completing a swing-stroke. Such performance motions derive from a combination of the feeling and the experience of the pianist's arm, hand, and fingers and are an important part of the technique of a piano performance. Naturally they are difficult to mechanically emulate. In the developed automated piano, before the key is fully depressed and after touch is completed, the effect was measured using the touch signal when held in this position. The key's behaviors when held before being fully depressed are shown in Figure 9.22.

As shown in the figures above, the hammer behavior of 82.5 - 85 % of a key range of movement comes to rest on the repetition lever quicker than the hammer behavior of other key range, and the hammer's movement also comes to rest at its original position early in the subsequent section D.

9.6.1 Discussion

The author has inferred that in some case the hammer moves without involving the back check because the lower end of the hammer head is flipped by the back check after the hammer strikes the strings.

If 82.5 - 85 % of the key's range of movement is considered according to the piano action's mechanism, the other end of the jack reaches the regulating bottom and a repetition lever reaches a drop screw. Since the back end of the key was not fully raised, the function in which the hammer is held to a back check was found to be unnecessary. Additionally, the author considers that the energy of the hammer which caused it to rebound after it struck the strings was absorbable with the repetition lever and its spring at this position. The key must stay at 82.5 - 85 % of the key's range within about 30 ms after the moments when the hammer rebounds from the string.

A non-musical sound can be easily recognized during the soft tone sections of famous pianists' performances. Therefore, in the performance of such a soft sound on an automated piano, it is necessary to reproduce the strength of the note and the interval between the notes appropriately, as a little stronger sound or a little slower tempo.

9.7 Experiment

Based on the above results, the database of the behavior of each key was rewritten and the algorithm of the transformation process was changed. The experimental results regarding the repetition of a soft key strike were then evaluated.

Figure 9.23a shows a reproduction in which a 90% range of movement was applied. The strength of the desired sound was reproducible in several repetitions, as were double-strikes and non-musical sounds. Figure 9.23b shows a reproduction applying an 85% range of movement. As the history of the hammer's velocity when striking the strings shows, the volume of a desired sound could be reproduced without the generation of double-strikes or non-musical sounds.

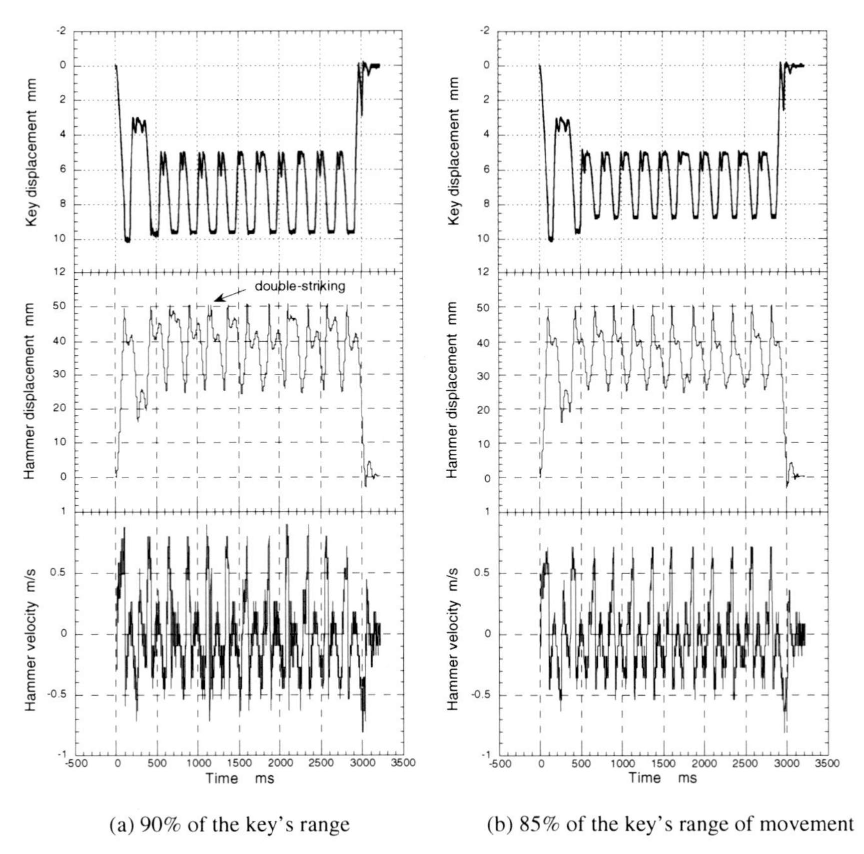

(a) 90% of the key's range (b) 85% of the key's range of movement

Fig. 9.23 Behavior of repetition of the same key.

9.7.1 Discussion

In order to express thoughts and emotions in a performance, it is necessary to extend the dynamic range, especially including soft tones. Using the method described here, the author has been able to make remarkable progress in the performance of an extremely stable soft tone. When one listen the soft tones, we felt they were clear.

A grand piano has a dynamic range of about 40 dB. Our automated piano has achieved a desired performance including representation of dynamics within a dynamic range of about 32 dB from fortissississmo (*fff*) up to pianississimo (*ppp*).

9.8 Conclusion

The author has developed an actuators control system, four databases and a translation process that support the editing of a piece of computer music on our automated piano. The system could offer users performance data containing musical expression, making it easier for a user to edit performance data. In the performance of soft tones, an expression that a gentle touch is done with exquisite senses is suitable. Furthermore, the author has had the wonderful opportunity for a famous pianist, Ms. Ryoko Fukazawa, to give a lesson on the piano. Performances of the automated piano have very much improved.

Pianists continue to train hard to play ideal performances due to the pitilessness that sounds cannot be muffled once they are produced. The moving distance of a key is approximately 10 millimeters. It is extremely difficult for pianists to create the ideal touch for each key within the range. Additionally, an interval between the initial input as a touch and producing sound will be different according to the produced sound strength, and will take approximately 0.02 to 0.2 seconds. The interval is just a fraction of time considering a blink of the eyes takes 0.1 seconds.

The concept of how to "touch" uses various symbolic descriptions of what to do, and we have encountered countless numbers of literal expressions such as lightly, heavily, airily, and other expressions that may be beyond understanding of engineer. The author believes that these words guided us to control the necessary motions during performance at the beginning stage of how to produce "touch" motions effectively for each key. Additionally, when the proposed automated piano is programmed to reproduce a musical performance, one can experience disappointment and joy due to other sounds besides the sounds created by touches, caused by touching the hammer to the strings, bouncing off sounds of the jack against the hammer roller, and interface sounds between parts. These sound effects may all be part of the "touch" activity.

Although a user can certainly create a desired expression with the automated piano, he or she can also vary the performance after listening repeatedly and make changes in their expressive performance. A change brings interest to humans, and humans will never lose interest as long as something is changing. Our findings suggest that a future robot will also need to have slight variations in their behavior to make interactions with them more pleasing.

Acknowledgments. The author would like to acknowledge the advice, assistance, and financial support of Mr. Hajime Mori, ex-president of Apex Corporation, who died in 2003.

References

1. White, W.B.: The human element in piano tone production. Journal of Acoustic Society in America, 357–367 (1930)
2. Hart, H.C., Fuller, M.W., Lusby, W.S.: A Precision Study of Piano Touch and Tone. Journal of Acoustic Society in America 6, 80–94 (1934)
3. Askenfelt, A., Jansson, E.V.: From Touch to String Vibration. I: Timing in the grand piano action. Journal of Acoustic Society in America 88(1), 52–63 (1990)
4. Askenfelt, A., Jansson, E.V.: From Touch to String Vibration. II: The motion of the key and hammer. Journal of Acoustic Society in America 90(5), 2383–2393 (1991)
5. Hayashi, E., Yamane, M., Ishikawa, T., Yamamoto, K., Mori, H.: Development of a Piano Player. In: Proceedings of the 1993 International Computer Music Conference, Tokyo, Japan, pp. 426–427 (September 1993)
6. Hayashi, E., Yamane, M., Mori, H.: Development of Moving Coil Actuator for an Automatic Piano. International Journal of Japan Society for Precision Engineering 28(2), 164–169 (2000)
7. Hayashi, E., Yamane, M., Mori, H.: Behavior of Piano-Action in a Grand Piano. I: alysis of the Motion of the Hammer Prior to String Contact. Journal of Acoustical Society of America 105, 3534–3544 (2000)
8. Hayashi, E.: Development of an Automatic Piano that Produce Appropriate Touch for the Accurate Expression of a Soft Tone. In: International Symposium on Advanced Robotics and Machine Intelligence (IROS 2006), Workshop CD-ROM, Beijing, China (2006)
9. Asami, K., Hayashi, E., Yamane, M., Mori, H., Kitamura, T.: Intelligent Edit of Support for an Automatic Piano. In: Proceedings of the 3rd International Conference on Advanced Mechatronics, KAIST, Taejon, Korea, pp. 342–347 (August1998)
10. Asami, K., Hayashi, E., Yamane, M., Mori, H., Kitamura, T.: An Intelligent Supporting System for Editing Music Based on Grouping Analysis in Automatic Piano. In: IEEE Proceedings of RO-MAN 1998, kagawa, Japan, pp. 672–677 (September 1998)
11. Hikisaka, Y., Hayashi, E.: Interactive musical editing system to support human errors and offer personal preferences for an automatic piano. Method of searching for similar phrases with DP matching and inferring performance expression. In: Artificial Life and Robotics (AROB 12th 2007), CD-ROM, Oita Japan (January 2007)
12. Hayashi, E., Yamane, M., Mori, H.: Development of Moving Coil Actuator for an Automatic Piano. International Journal of The Japan Society Precision Engineering 28(2), 164–169 (1994)
13. Hayashi, E., Takamatu, Y., Mori, H.: Interactive musical editing system for supporting human errors and offering personal preferences for an automatic piano. In: Artificial Life and Ro-botics (AROB 9th 2004), CD-ROM, Oita Japan (January 2004)
14. Hayashi, E., Takamatsu, Y.: Iinteractive musical editing system for supporting human errors and offering personal preferences for an automatic piano-Preference database for crescendo and decrescendo. In: Artificial Life and Robotics (AROB 10th 2005), CD-ROM (Febuary 2005)
15. Hikisaka, Y., Takamatsu, Y., Hayashi, E.: Interactive musical editing system for supporting human errors and offering personal preferences for an automatic piano - A system of infer-ring phrase expression. In: Artificial Life and Robotics AROB 11th 2006), CD-ROM, Oita Japan (January 2006)

16. Hayashi, E., Yamane, M., Mori, H.: Development of Piano Player 2nd Report, Study on Repetition of Same key Based on Analysis of Behavior of Piano-Action. Journal of Japan Society of Mechanical Engineering 61(587), 339–345 (1995) (in Japanese)
17. Hayashi, E., Yamane, M., Mori, H.: Behavior of Piano Action in a Grand Piano. I: Analysis of the Motion of the Hammer Prior to String. Journal of Acoustical Society of America 105(6), 3534–3544 (1999)
18. Hayashi, E., Yamane, M., Mori, H.: Development of Piano Player 2nd Report, Study on Repetition of Same key Based on Analysis of Behavior of Piano-Action. Journal of Japan Society of Mechanical Engineering 61(587), 339–345 (1995) (in Japanese)
19. Hayashi, E., Yamane, M., Mori, H.: Development of Piano Player 1st Report, Analysis of Behavior of Piano Action until String Striking of Hamme. Journal of Japan Society of Mechanical Engineering 60(579), 325–331 (1994) (in Japanese)
20. Gat, J.: The Technique Of Piano playing, London and Wellingborough,1974.
21. Daivid Bowers, Q.: Encyclopedia of Automatic Musical Instruments. The Vestal Press, New York (1994)
22. Rowe, R.: Machine Musicianship. The MIT Press, Cambridge (2001)
23. McElhone, K.: Mechanical Music. AMICA, USA
24. Temperiety, D.: The Cognition of Basic Musical Structure. The MIT Press, Cambridge (2001)

Chapter 10
McBlare: A Robotic Bagpipe Player

Roger B. Dannenberg, H. Ben Brown, and Ron Lupish

Abstract. McBlare is a robotic bagpipe player developed by the Robotics Institute and Computer Science Department at Carnegie Mellon University. This project has taught us some lessons about bagpipe playing and control that are not obvious from subjective human experience with bagpipes. From the artistic perspective, McBlare offers an interesting platform for virtuosic playing and interactive control. McBlare plays a standard set of bagpipes, using a custom air compressor to supply air and electromechanical "fingers" to control the chanter. McBlare is MIDI controlled, allowing for simple interfacing to a keyboard, computer, or hardware sequencer. The control mechanism exceeds the measured speed of expert human performers. McBlare can perform traditional bagpipe music as well as experimental computer-generated music. One characteristic of traditional bagpipe performance is the use of ornaments, or very rapid sequences of up to several notes inserted between longer melody notes. Using a collection of traditional bagpipe pieces as source material, McBlare can automatically discover typical ornaments from examples and insert ornaments into new melodic sequences. Recently, McBlare has been interfaced to control devices to allow non-traditional bagpipe music to be generated with real-time, continuous gestural control.

10.1 Introduction

In 2004, Carnegie Mellon University's Robotics Institute celebrated its twenty-fifth anniversary. In preparations for the event, it was suggested that the festivities

Roger B. Dannenberg
Carnegie Mellon University, Computer Science Department,
5000 Forbes Avenue, Pittsburgh, PA 15213
e-mail: rbd@cs.cmu.edu

H. Ben Brown
Carnegie Mellon University, Robotics Institute,
5000 Forbes Avenue, Pittsburgh, PA 15213
e-mail: hbb@cs.cmu.edu

Ron Lupish
SMS Siemag, 100 Sandusky St., Pittsburgh, PA 15212
e-mail: ron.lupish@sms-siemag.us

J. Solis and K. Ng (Eds.): Musical Robots and Interactive Multimodal Systems, STAR 74, pp. 165–178.
springerlink.com

should include a robotic bagpiper to acknowledge Carnegie Mellon's Scottish connection[1] using a Robotics theme. Members of the Robotics Institute set out to build a system that could play an ordinary, off-the-shelf traditional set of Highland Bagpipes with computer control. The system is now known as "McBlare". Mechanized instruments and musical robots have been around for centuries. [8] Although early mechanical instruments were usually keyboard-oriented, many other electro-mechanical instruments have been constructed, including guitars and percussion instruments [9, 10, 13]. Robot players have also been constructed for wind instruments including the flute [11, 12] and trumpet [1, 14].

There have been at least two other robotic bagpipe projects. Ohta, Akita, and Ohtani [7] developed a bagpipe player and presented it at the 1993 International Computer Music Conference. In this player, conventional pipes are fitted to a specially constructed chamber rather than using the traditional bag. Their paper describes the belt-driven "finger" mechanism and suggests some basic parameters as a starting point for the design:

- 4 mm finger travel;
- 20 ms total time to open and close tone hole;
- 100 gf minimum closing force for tone holes.

Sergi Jorda also describes bagpipes used in his work, consisting of single pitched pipes that can only be turned on and off [4]. In a separate email communication, Jorda indicated that "Pressure is very tricky" and may depend on humidity, temperature and other factors. In contrast to previous efforts, the Carnegie Mellon project decided to use off-the-shelf bagpipes to retain the traditional bagpipe look and playing characteristics.

Additional basic information was obtained by meeting with Alasdair Gillies, CMU Director of Piping, and Patrick Regan, a professional piper. These experts were observed and videotaped to learn about the instrument and playing techniques. From slow-motion video (25% speed) the fastest fingering appeared to be about 15 Hz. Required finger pressure on the chanter appeared to be very light. We noted breathing cycle periods of about 4 seconds, and measured the time to exhaust the air from the bag playing a low A: 12 seconds; and a high A: 8 seconds. (However, we now know that the lower pitches actually use a higher air flow at a given air pressure.) The numbers give a rough indication of the air flow requirement: between 0.045 and 0.07 cubic meters per minute (1.6 and 2.5 cubic feet per minute), based on a measured bag volume of 0.0093 cubic meters (0.33 cubic feet). Alasdair said he maintains a pressure of 32" water column (7.9 kPa or 1.15 PSI) in the bag. Soshi Iba, experienced piper and then PhD candidate in Robotics, also provided substantial input and served as a primary test subject, and the third author who joined the project later is also an accomplished piper.

The next section presents an overview of McBlare, beginning with a brief description of bagpipes and how they work. There are two major robotic components

[1] Andrew Carnegie, who founded Carnegie Mellon (originally the Carnegie Institute of Technology), was born in Scotland. The University has an official tartan, the School of Music offers a degree in bagpipe performance, and one of the student ensembles is the pipe band.

of McBlare: the air supply, and the chanter control, which are described in following sections. One of the major difficulties we encountered has been properly setting up the bagpipes and coaxing them into playing the full melodic range reliably. The final two sections report on our findings, current status, and some recent developments in interactive music control of McBlare.

10.2 Bagpipes

Bagpipes are some of the most ancient instruments, and they exist in almost all cultures. There are many variations, but the most famous type is the Highland Bagpipes (see Figure 10.1), and this is the type played by McBlare. There are three long, fixed pipes called drones. Two tenor drones are tuned to the same pitch, which is traditionally called A, but which is closer to B♭4. The third drone (bass drone) sounds an octave lower. Drones each use a single reed, traditionally a tongue cut into a tube of cane, more recently a cane or artificial tongue attached to a hollow body of plastic or composite material. The fourth pipe is the chanter, or melody pipe. The chanter is louder than the drones and uses a double reed, similar in size to a bassoon reed, but shorter in length and substantially stiffer. Unlike a bassoon reed, however, it is constructed around a small copper tube, or "staple".

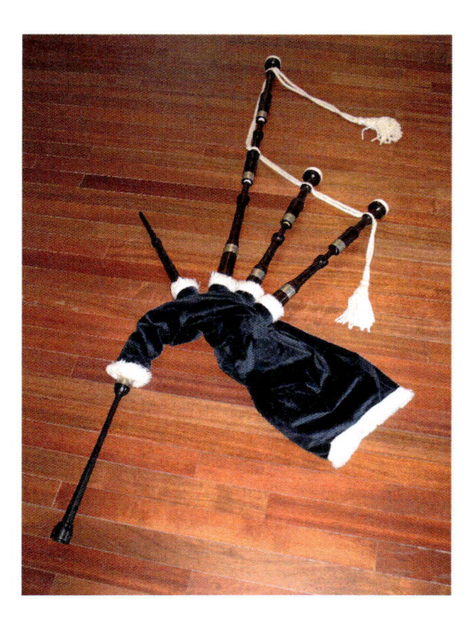

Fig. 10.1 Traditional Highland Bagpipes.

The chanter (lower left of Fig. 10.1) has sound holes that are opened and closed with the fingers, giving it a range from G_4 to A_5 (as written). All four pipes are inserted into the bag, a leather or synthetic air chamber that is inflated by the player's lung power through a fifth pipe, the blowstick or blowpipe (pointing to the upper left of Fig. 10.1). This tube has a one-way check-valve, so the player can take a breath while continuing to supply air to the reeds by squeezing the bag under his or her arm to regulate pressure.

Reeds at rest are slightly open, allowing air to pass through them. As pressure increases and air flow through the open reed increases in response, the Bernoulli effect decreases the pressure inside the reed, eventually causing the reed to close. The resulting loss of airflow reduces the pressure drop inside the reed, and the reed reopens. When things are working properly, the pressure fluctuations that drive the reed are reinforced by pressure waves reflected from the open end of the pipe, thus the oscillation frequency is controlled by the pipe length. The acoustic length of the chanter is mainly determined by the first open sound hole (i.e., the open sound hole nearest to the reed), allowing the player to control the pitch. For more technical details, see Guillemain's article on models of double-reed wind instruments [3].

It should be noted that the bagpipe player's lips are nowhere near the reeds of the bagpipe, unlike the oboe, bassoon, or clarinet. The bagpipe player's lips merely make a seal around the blowstick when inflating the bag. The reeds are at the ends of the four pipes where they enter the bag (see Fig. 10.1).

Pressure regulation is critical. It usually takes a bit more pressure to start the chanter oscillating (and more flow, since initially, the reeds are continuously open). This initial pressure tends to be around 8.3 kPa (1.2 pounds per square inch). Once started, the chanter operates from around 5.5 to 8.3 kPa (0.8 to 1.2 psi). The drone reeds take considerably less pressure to sound than does the chanter reed, and drones operate over a wider pressure range, so it is the chanter reed that determines the pressure required for the overall instrument. Unfortunately, the chanter tends to require lower pressure at lower pitches and higher pressure at higher pitches. At the low pitches, too high a pressure can cause the pitch to jump to the next octave or produce a warbling multiphonic effect (sometimes called "gurgling"). If insufficient pressure is maintained on the chanter reed for the higher pitches, it will cease vibrating (referred to as "choking"). Thus, there is a very narrow pressure range in which the full pitch range of the chanter is playable at a fixed pressure. Furthermore, pressure changes affect the chanter tuning (much more than the drones), so the chanter intonation can be fine-tuned with pressure changes. Typically, this is not done; rather, experienced pipers carefully attempt to adjust the stiffness and position of the reed in the chanter so as to be able to play the full 9-note range of the chanter with little or no pressure variation.

In some informal experiments, we monitored air pressure using an analog pressure gauge while an experienced player performed. We observed that air pressure fluctuated over a range from about 6.2 to 7.6 kPa (0.9 to 1.1 psi), with a tendency to use higher pressure in the upper register. Because of grace notes and some fast passages, it is impossible to change pressure with every note, and we speculate that players anticipate the range of notes and grace notes to be played in the near future and adjust pressure to optimize their sound and intonation.

Whether pressure should be constant or not is not well understood, although constant pressure is generally considered the ideal. For example, Andrew Lenz's "bagpipejourney" web site described how to construct and use a water manometer. He says "Theoretically you should be playing all the notes at the same pressure, but it's not uncommon for people to blow harder on High-A [5]."

10.3 The McBlare Robot

From a scientific and engineering perspective, the main challenge of building a robot bagpipe player was lack of information. How critical is pressure regulation? How fast do "fingers" need to operate? Is constant pressure good enough, or does pressure need to change from low notes to high notes? Is a humidifier necessary? Building and operating McBlare has provided at least partial answers to these and other questions.

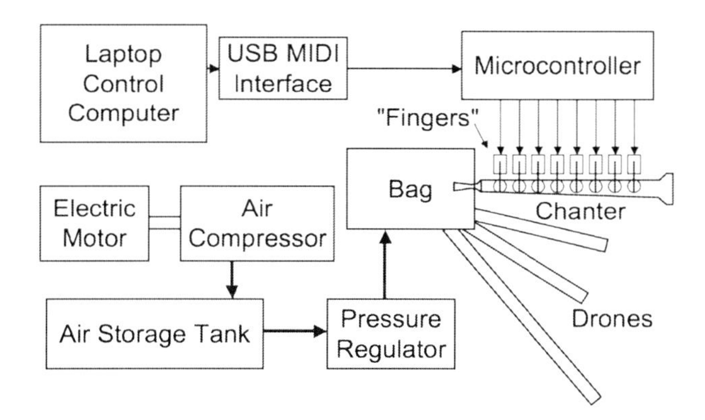

Fig. 10.2 System diagram of McBlare.

Our bagpipe-playing robot, McBlare, uses a computer system to control electro-mechanical "fingers" that operate the chanter, and an air compressor and regulator to provide steady air pressure and flow to the bag. The system is diagramed in Figure 10.2. High-level control is provided via MIDI from a laptop computer (a MIDI keyboard may be substituted for direct control). MIDI is decoded by a microcontroller to drive 8 fingers (thus, McBlare has 8 degrees of freedom). The air supply uses a standard mechanical diaphragm-based pressure regulator and sends air to the bag via the blowstick. The pump is about 700mm wide, 300mm deep, and 400mm high. The chanter (a standard chanter) is about 330mm long (not counting the reed), and the minimum "finger" and tone hole spacing is about 19mm. The air supply and chanter control are describe in more detail below.

10.3.1 The Air Supply

McBlare uses a custom-built air compressor. A 1/16 HP, 115VAC electric motor drives a gearbox that reduces the speed to about 250 rpm. Two 76 mm (3") diameter air pump cylinders, salvaged from compressors for inflatable rafts, are driven in opposition so that they deliver about 500 pump strokes per minute (see Figure 10.3). The radius of the crank arm driving the cylinders is adjustable from 15 mm to 51 mm (0.6" to 2.0"); we found that the smallest radius provides adequate air flow, calculated to be 0.034 cubic meters per minute (1.2 cubic feet per minute[2]). The air flow exhibits considerable fluctuation because of the pumping action of the cylinders. A small air storage tank sits between the pump and the pipes and helps to smooth the air pressure. Moreover, a high flow-rate, low pressure regulator drops the tank pressure of about 35 kPa (5 psi) down to a suitable bagpipe pressure. The pressure ripple on the bagpipe side of the regulator is a few percent with a frequency of about 8 Hz. This gives McBlare a barely audible "vibrato" that can be detected by listening carefully to sustained notes. The wavering pitch and amplitude might be eliminated with a rotary pump or a large storage tank, but the effect is so slight that even professional players rarely notice it.

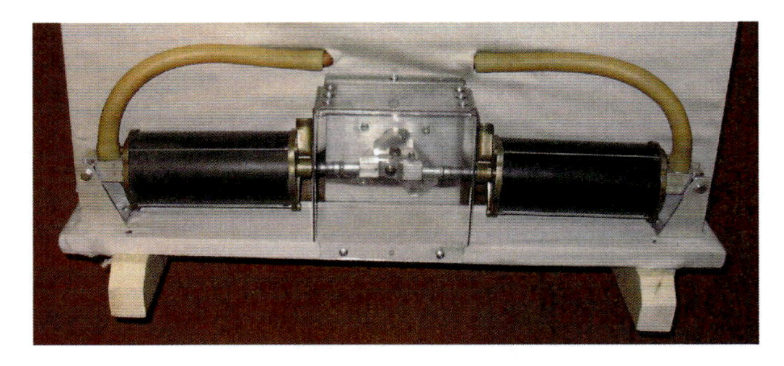

Fig. 10.3 The McBlare air compressor. Electric motor (not visible) drives eccentric (center) through a gearbox. Eccentric drives two air pump cylinders (right and left) in opposition.

The bagpipes are connected with a rubber hose that slips over the same tube that a human performer would blow into (the blowstick). By blowing in air at a constant, regulated pressure, we can maintain pressure without squeezing the bag. (Earlier designs called for a mechanical "squeezer" but at 7 kPa (1 psi), a squeezer in contact with many square inches would have to be very powerful, adding significantly to McBlare's weight and complexity.)

Pressure regulation is adjusted manually using pump crank arm radius to control the rough flow rate, a bleed valve on the tank to relieve tank pressure that could stall the motor, and the pressure regulator. Fine adjustments are typically required using the pressure regulator to find the "sweet spot" where the lowest note sounds without gurgling and the highest note does not cut out.

[2] This is less than the 0.045-0.07 cubic meters per minute based on bag deflation measurements above. This may be due to differences in instruments and/or measurement errors.

The original reason to construct the pump was that such a powerful, low-pressure, high-volume pump is not readily available. The pistons were used rather than a rotary pump simply because they were available as salvage parts. After constructing the air compressor, we did locate an off-the-shelf rotary compressor that also works well, but is certainly not as fun to watch as the crank-and-cylinder pump.

10.3.2 The Chanter Control

The chanter requires "fingers" to open and close sound holes. Analysis of video indicates that bagpipers can play sequences of notes at rates up to around 25 notes per second. Human players can also uncover sound holes slowly or partially, using either an up-down motion or a sideways motion. The design for McBlare restricts "fingers" to up-and-down motion normal to the chanter surface. Fortunately, this is appropriate for traditional playing. The actuators operate faster than human muscles, allowing McBlare to exceed the speed of human pipers.

McBlare's "fingers" are modified electro-mechanical relays (see Figure 10.4). Small coils pull down a metal plate, which is spring loaded to return. Lightweight plastic tubes extend the metal plate about 3 cm, ending in small rubber circles designed to seal the sound hole. The length of travel at the sound hole is about 2.5 mm, and the actuators can switch to open or closed position in about 8 ms. The magnet coils consume about 1W each, enough to keep the mechanism warm, but not enough to require any special cooling. The magnet mechanism has the beneficial characteristic that the finger force is maximum (around 100 gf) with the magnet closed, the point at which finger force is needed for sealing the tone hole.

The whole "hand" assembly is designed to fit a standard chanter, but the individual finger units can be adjusted laterally (along the length of the chanter) and vertically. The lateral adjustment accommodates variations in hole spacing. The vertical adjustment is critical so that the magnet closure point corresponds to the point of finger closure.

The actuator current is controlled by power transistors, which in turn are controlled by a microcontroller. The microcontroller receives MIDI, decodes MIDI note-on messages to obtain pitch, and then uses a table lookup to determine the correct traditional fingering for that pitch. The full chromatic scale is decoded, although non-standard pitches are not in tune. MIDI notes outside of the bagpipe range are transposed up or down in octaves to fall inside the bagpipe range. Additional MIDI commands are decoded to allow individual finger control for non-standard fingerings. For example, an E will sound if the highest 3 tone holes (high A, high G, and F#) are closed and the E tone hole is open. The standard fingering also closes the D, C#, and B and opens the low A tone holes, but in fact, any of the 16 combinations of these 4 lowest tone holes can be used to play an E. Each combination has a subtle effect on the exact pitch and tone quality.

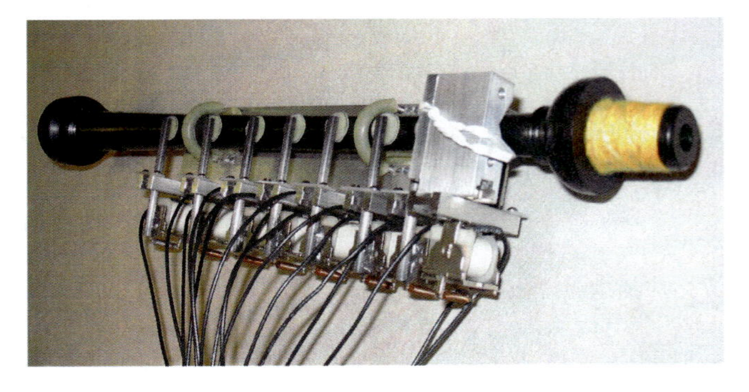

Fig. 10.4 Chanter is mounted on aluminum block along with electromagnetic coils that open and close sound holes using rubber pads at the end of lightweight plastic tubes.

Grace notes, which are fast notes played between the notes of a melody (see "Ornamentation," below), are traditionally played by simply lifting one finger when possible, taking advantage of alternate fingerings, but since McBlare has very fast and precisely coordinated "fingers," we use standard fingerings for all notes including grace notes. In principle, we could send special MIDI commands to control individual fingers to achieve the same fingerings used by human pipers.

10.4 McBlare in Practice

McBlare is supported by a lightweight tripod that folds (see Figure 10.5) and the entire robot fits into a special airline-approved case for the pump and a suitcase for the remainder, making travel at least manageable. The chanter control works extremely well. The speed allows for authentic-sounding grace notes and some very exciting computer-generated sequences. The maximum measured rate is 16ms per up/down finger cycle, which allows 125 notes per second in the worst case. The chanter control is also compact, with the mechanism attached directly to and supported by the chanter.

We developed a small laptop-based program to play useful sequences for tuning and pressure adjustment. The user can then select and play a tune from a MIDI file. The program can also record sequences from a keyboard and add ornamentation as described below.

As might be expected, there is considerable mechanical noise generated by the air compressor. In addition, the electro-mechanical chanter "fingers" make clicking sounds. However, the chanter is quite loud, and few people notice the noise once the chanter begins to play! We attempted to quantify this with a sound level meter. At 1m, McBlare generates an SPL of about 102dB outdoors, whereas the pump alone generates about 76dB. Thus, the pump noise is about 26dB down from the continuous bagpipe sounds. The bagpipe SPL varies a few dB with direction, pitch, and perhaps phasing among the drones, so this should be taken only as a rough estimate.

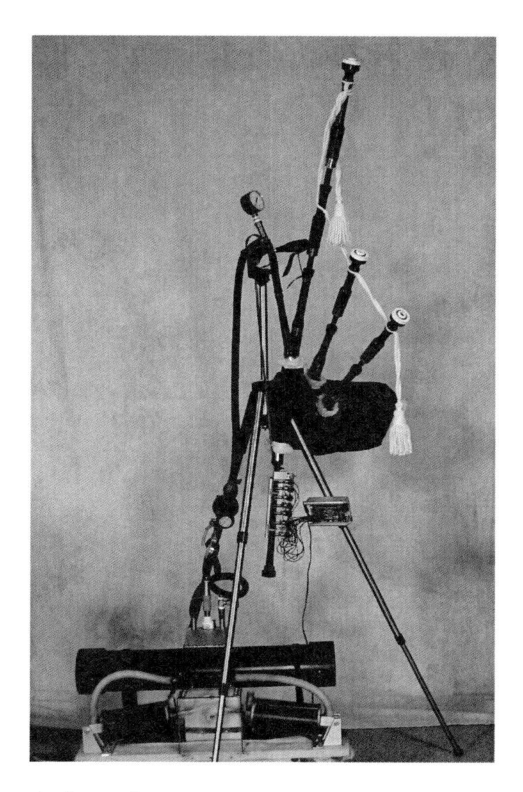

Fig. 10.5 McBlare ready for performance.

In our original work, we reported difficulty covering the full range of pitches from G_4 to A_5 [2]. More recently, we have found that the combination of good pressure regulation, eliminating even the slightest leak from closed tone holes, and a good reed (all three being critical) enable good performance across the full pitch range. A method to humidify the air has been strongly suggested by a number of bagpipe players. Although we have tried various approaches, we have been unable to achieve any solid improvements by raising the humidity, supporting a conclusion that humidity is at most of secondary importance after pressure regulation, the reeds, and sealing the tone holes. It should be noted, however, that humidity is hard to control and study, so we cannot rule out the importance of humidity. In particular, we suspect that humidity might affect the timbral quality of the chanter. Note also that another category of bagpipe is played by bellows and hence is not subject to the naturally humid breath of player, indicating that "dry" playing is at least in the realm of "normal" bagpipe playing. The adjustment of reeds to play well and reliably in the resulting dry environment is the subject of much discussion. Perhaps McBlare can someday serve as a testbed for comparing reeds in dry vs. humid playing conditions.

10.4.1 Ornamentation

The use of quick flourishes of notes ("grace notes") between longer notes of a melody (ornamentation) is a characteristic of bagpipe music. Without ornaments, all bagpipe tunes would be completely "legato," lacking any strong rhythm. In particular, if a melody contains two or more repeated notes, ornamentation is essential: since the chanter never stops sounding, there is no other way to signal a separation between the two notes. Ornaments are also used for rhythmic emphasis.

There are some basic principles used for ornamenting traditional highland bagpipe tunes, so a hand-coded, rule-based approach might allow ornaments to be added automatically to a given melody. Instead, we have implemented a simple case-based approach using a small database of existing bagpipe melodies in standard MIDI file format, complete with ornaments. Typical ornament sequences are automatically extracted from the database and then inserted into new melodies using the following procedure.

The first step is to build a collection of typical ornaments. Ornaments are defined as a sequence of one or more notes with durations less than 0.1s bounded by two "melody" notes with durations greater than 0.1s. A table is constructed indexed by the pitches of the two longer, or "melody" notes. For example, there is one entry in the table for the pair (E_4, D_4). In this entry are all of the ornaments found between melody pitches E_4 and D_4. The database itself comes from standard MIDI files of bagpipe performances collected from the Web. These appear to be mostly produced by music editing software, although actual recordings from MIDI chanters could be used instead.

The second step uses the table to obtain ornaments for a new, unornamented melody. For each note in the melody (except for the last), the pitch of the note and the following note are used to find an entry in the table. If no entry is found, no ornaments are generated. If an entry exists, then it will be a list of ornaments. An element of the list, which is a sequence of short notes, is chosen at random. The melody note is shortened by the length of the ornament sequence (something that human players do automatically to maintain the rhythm) and the ornament notes are inserted between the melody note and the next note.

There are many obvious variations on this approach. For example, the ornament could be chosen based also on the length of the melody note so that perhaps shorter ornaments would be chosen for shorter notes. One option in our system is to choose ornaments of maximal length to exaggerate the ornamentation. (In traditional practice, longer ornaments, or "doublings," are often used to create a stronger rhythmic emphasis.)

10.4.2 Gestural Control

The use of MIDI control makes it possible to adapt all sorts of controllers to McBlare, including keyboards (which are very useful for experimentation), novel sensors, or even MIDI bagpipe controllers [6]. In our exploration of robot performance practice with McBlare, we wrote real-time software to enable the pipes to be played using a Nintendo Wii game controller (see Figure 10.6). The Wii controller contains a 3-axis accelerometer and a variety of buttons. The accelerometers can sense rapid acceleration in any direction. Because gravitation provides an absolute reference, the Wii controller can also sense orientation in 2 dimensions: left to right rotation (roll) and up to down rotation (elevation). The Wii controller provides multiple degrees of freedom, discrete buttons as well as continuous controls, wireless operation, and low cost, but certainly other controllers and interfaces could be developed.

One mode of control uses orientation to provide two parameters to a music generation algorithm. The roll axis controls note density, and the elevation axis controls interval size. The generation algorithm creates notes that fall on equally spaced rhythmic boundaries. At every boundary, a new note is generated with a probability determined by the roll parameter. As the controller is rotated clockwise from left to right, the probability of a new note increases, so the density of notes increases. At the extreme ranges of roll, the tempo is slowly decreased or increased. Each new note has a pitch determined as a random offset from the current pitch. The random offset is scaled by the elevation axis so that larger intervals tend to be generated with higher elevation. (Of course, the next pitch is also limited to the fixed range of the bagpipes.) This gives the user (performer) the ability to create and control a variety of melodic textures at virtuosic speeds.

After each new note is generated in this mode, we automatically insert ornamentation as described in the previous section. The ornamentation adds to the virtuosity and gives the performance a more idiomatic character.

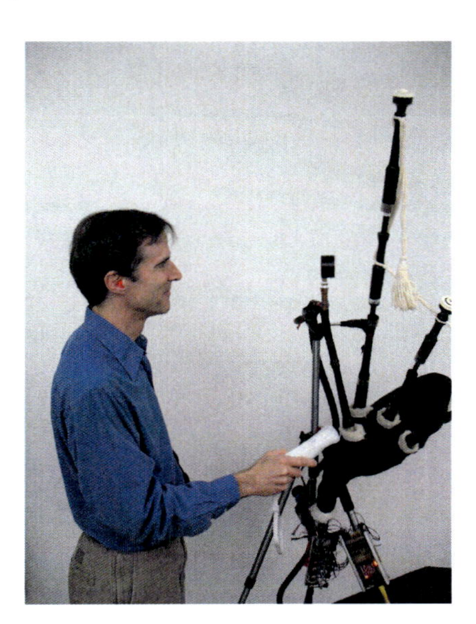

Fig. 10.6 McBlare operated by the first author using a Nintendo Wii controller.

A second mode of operation simply maps the elevation to pitch, allowing the user to run up and down scales and even play melodies in a "theremin-like" manner.

Finally, a third mode integrates the side-to-side acceleration sensor and maps the integral to pitch. The integral is clamped to a minimum and maximum to keep it in range. This allows pitch change to be directed by the user and correlated closely to the user's gestures. In addition, rotating the controller slightly right or left (the roll axis) will bias the accelerometer positively or negatively with gravity, causing the integral (and pitch) to drift upward or downward, respectively. There is no absolute position reference, so this mode does not allow the user to play a specific melody with any accuracy.

Buttons on the controller allow the performer to switch modes at any time. The combination of modes gives the performer access to a variety of musical textures and mappings of physical gesture to control. Although this control is not suitable for traditional music (and it is hard to imagine a better interface for traditional bagpipe music than human fingers and tone holes), the approach does offer new modes of music generation and interaction that would be extremely difficult or impossible using traditional means.

10.5 Conclusion and Future Work

McBlare is interesting for both scientific and artistic reasons. From the scientific perspective, McBlare allows for careful study of the behavior of bagpipes. For example, we have found that there is a very narrow range of pressure that allows the chanter to play its full range properly. This would explain the tendency for pipers to boost the pressure slightly for higher notes, but it also confirms the possibility of playing with constant pressure as advocated by expert players. McBlare offers a controlled environment for examining the effects of reed adjustments, humidity, and adjustments to tone holes. We have also recorded McBlare's chanter playing all 256 possible fingerings. Further analysis of these recordings may uncover some interesting new timbral and microtonal opportunities for bagpipe players.

Artistically, McBlare (and robotic instruments in general) offers a way for computers to generate or control music without loudspeakers. The three-dimensional radiation patterns of acoustic instruments, the sheer loudness of highland bagpipes, and visibility of the means of sound production are important differences between McBlare and sound synthesis combined with loudspeakers. Aside from these physical differences, there is something about robotic instruments that captures the imagination in a way that must be experienced to be appreciated. The human fascination with automatons and the ancient tradition of bagpipes combine powerfully in McBlare, which has been featured not only in concerts but as a museum installation. Interactive control of McBlare leads to a unique and fascinating instrument.

One obvious difference between McBlare and human pipers is that humans can use their arms to rapidly apply pressure to the bag to start the pipes singing and release the pressure quickly to stop. McBlare, on the other hand, takes time to build up pressure. The chanter typically will not start until the optimum pressure is reached, but a chanter that is not in oscillation offers less air resistance, which in turn causes a pressure drop. The pressure drop inhibits the chanter from starting. This feedback process makes the bagpipes somewhat unstable and reluctant to start: until the chanter starts sounding, the lowered pressure will inhibit the chanter from starting. Usually, the (human) McBlare operator intervenes and speeds up the process by temporarily raising the system pressure until the chanter starts. At this point, one or more drones might be overblowing and need to be manually restarted. This all takes less than a minute, but is something a human can accomplish in seconds.

A more advanced system might sense when the chanter is sounding and automatically raise the pressure to restart the chanter when it stops. One could then go even further by automatically adjusting the pressure to eliminate "gurgling" on low notes (pressure is too high) or stopping vibration on high notes (pressure is too low). Since all of this would add weight and complexity, we will probably keep McBlare in its current configuration.

Bagpipes and drums are a traditional combination, and we plan to work on a robotic drum to play along with McBlare. With computer control, hyper-virtuosic pieces, complex rhythms, and super-human coordination will be possible. Examples include playing 11 notes in the time of 13 drum beats or speeding up the drums while simultaneously slowing down the bagpipes, ending together in phase. In order to explore the musical possibilities, we hope to create a website where composers can upload standard MIDI files for McBlare. We will then record performances and post them for everyone to enjoy.

Acknowledgments. The School of Computer Science, which includes the Robotics Institute, supported this work by providing funds for equipment and purchasing bagpipes. Additional support was provided by Ivan Sutherland, who also suggested the original concept. McBlare would not be possible without enormous efforts by Garth Zeglin, who designed, built, and programmed a series of microcontrollers for McBlare. We would like to thank Alasdair Gillies, Chris Armstrong, Patrick Regan, Soshi Iba, and Marek Michalowski for teaching us about the bagpipe and being subjects for our tests. Roddy MacLeod and the National Piping Centre in Glasgow supported a visit to Piping Live! – Glasgow International Piping Festival where many fruitful exchanges occurred (none actually involved flying fruits or vegetables). Thanks to Carl Disalvo for helping with McBlare's construction and debut and to Richard Dannenberg for his photograph in Figure 6. This chapter extends work that was originally presented at the NIME conference in Vancouver, B.C, Canada, in 2005 [2].

References

1. BBC News UK Edition. Robot trumpets Toyota's know-how, BBC (2004), `http://news.bbc.co.uk/1/hi/technology/3501336.stm` (accessed April 2005)
2. Dannenberg, R., Brown, B., Zeglin, G., Lupish, R.M.: A Robotic Bagpipe Player. In: Proceedings of the International Conference on New Interfaces for Musical Expression, pp. 80–84. University of British Columbia, Vancouver (2005)
3. Guillemain, P.: A Digital Synthesis Model of Double-Reed Wind Instruments. EURASIP Journal on Applied Signal Processing 2004, 990–1000 (2004)

4. Jorda, S.: Afasia: the Ultimate Homeric One-man-multimedia-band. In: Proceedings of New Interfaces for Musical Expression, Dublin, Ireland (2002)
5. Lenz, A.: Andrew's Tips: Making a Water Manometer (2004), `http://www.bagpipejourney.com/articles/manometer.shtml` (accessed April 2005)
6. Music Thing. Burns Night Special: MIDI Bagpipes are everywhere! (2005), `http://musicthing.blogspot.com/2005/01/burns-night-special-midi-bagpipes-are.html` (accessed April 2005)
7. Ohta, H., Akita, H., Ohtani, M.: The Development of an Automatic Bagpipe Playing Device. In: Proceedings of the 1993 International Computer Music Conference, Tokyo, Japan, pp. 430–431. International Computer Music Association, San Francisco (1993)
8. Roads, C.: Sequencers: Background. In: The Computer Music Tutorial, pp. 662–669. MIT Press, Cambridge (1996)
9. Sekiguchi, K., Amemiya, R., Kubota, H.: The Development of an Automatic Drum Playing Device. In: Proceedings of the 1993 International Computer Music Conference, Tokyo, Japan, pp. 428–429. International Computer Music Association, San Francisco (1993)
10. Singer, E., Larke, K., Bianciardi, D.: LEMUR GuitarBot: MIDI Robotic String Instrument. In: Proceedings of the 2003 International Conference on New Interfaces for Musical Expression (NIME 2003), pp. 188–191. McGill University, Montreal (2003)
11. Solis, J., Chida, K., Taniguchi, K., Hashimoto, S.M., Suefuji, K., Takanishi, A.: The Waseda Flutist Robot WF-4RII in Comparison with a Professional Flutist. Computer Music Journal 30(4), 12–27 (2006)
12. Takanishi, A., Maeda, M.: Development of Anthropomorpic Flutist Robot WF-3RIV. In: Proceedings of the 1998 International Computer Music Conference, Ann Arbor, MI, pp. 328–331. International Computer Music Association, San Francisco (1998)
13. Tosa, N.: The Nonsense Machines. Maywa Denki, Japan (2004)
14. Vergez, C., Rodet, X.: Comparison of Real Trumpet Playing, Latex Model of Lips and Computer Model. In: 1997 International Computer Music Conference, Thessaloniki, Greece, pp. 180–187. International Computer Music Association, San Francisco (1997)

Chapter 11
Violin Playing Robot and *Kansei*

Koji Shibuya

Abstract. *Kansei* is a Japanese word similar in meaning to "sensibility," "feeling," "mood," etc. Although *kansei* seems to affect musical instrument playing greatly, many musical instrument playing robots do not utilize *kansei* very well. Violin playing has been chosen as an example because it is very difficult, and it seems that *kansei* affects it more clearly than other musical instruments. First, in this chapter, a violin playing-robot is introduced and the sounds produced by the robot are analyzed and discussed. The robot consists of an anthropomorphic right arm offering 7 degrees of freedom (DOF) with a simple hand that grasps a bow. By using the arm, the robot can produce various sounds. The left hand fingers of the robot which are under development are presented next. After that, the information flow of a violin-playing robot from musical notes to sounds considering *kansei* is proposed. As a first approach, in the flow, timbre is regarded as *kansei* information, and *kansei* mainly affects processes that determine sound data from musical notes. Then, human violinists' *kansei* is analyzed based on the flow. It has been found that bow force and sounding point play important roles in determining timbre that results from human *kansei*.

11.1 Introduction

Violin playing is a very difficult task not only for humans but also for robots. Human violinists have to practice the violin for a very long period, usually beginning in childhood. To become a good violinist, two skills are required. One is the physical skill usually called "technique." The other is *kansei* (or *kansei* skill). *Kansei*, explained later, is a Japanese word and similar in meaning to "sensibility," "feeling," "mood," etc. Both skills are equally important for musical instrument playing. As one approach to understand these skills, the implementation of

Koji Shibuya
Ryukoku University, Dept. of Mechanical and Systems Engineering, Faculty of Science and Technology, 1-5 Yokotani Seta-Oe, Otsu, Shiga, 520-2194, Japan
e-mail: koji@rins.ryukoku.ac.jp

J. Solis and K. Ng (Eds.): Musical Robots and Interactive Multimodal Systems, STAR 74, pp. 179–193.
springerlink.com

computer music approaches have been demonstrated to be capable of generating high-quality human-like performances based on examples of human performers [1]. However, all of these systems have tested only by computer systems or MIDI-enabled instruments which limited the unique experience of a live performance.

To investigate physical skills, the author proposed to develop a violin-playing robot. If a robot is to play the violin very well, which requires dexterous arm and hand motion, robot technology must progress greatly. There are some studies on violin-playing robots. Kajitani built a violin-playing robot, a recorder-playing robot, and a cello-playing robot and created an ensemble with them [2]. Also Sobh and Wange built a unique violin-playing mechanism with two bows for one violin [3]. The right arms of the two robots mentioned above are not anthropomorphic. Furthermore, their left hands use many solenoids for fingering. Shimojo built a violin-playing robot for bowing using an industrial robot with 7-DOF (Mitsubishi Heavy Industries Ltd., PA-10) [4]. The robot is slightly larger than a human. Although Toyota Motor Corporation built a violin-playing robot, its arm has only 6-DOF [5]. In robotics, it is said that humans have seven joints. Thus, Toyota's robot is not anthropomorphic in the strict sense of the word. However, from a technical point of view, the author believes that Toyota's robot is the best violin-playing robot at present. However, Toyota's robot does not focus on human *kansei*. Their aim in building such a robot is only to demonstrate their technology. They wanted to build service and nursing robots with the technology developed for the violin-playing robot. Instead the proposed robot in this research has a right arm with 7-DOF for bowing, and left-hand fingers for fingering. Because a violin fits a human body, the robot must be anthropomorphic.

The purpose of building the violin-playing robot is not only to study the physical skill but also to reveal the role of *kansei* in the process of determining body motion. Because many readers may not be familiar with the word "*kansei*," it shall be briefly explained here. Recently, many Japanese researchers in many fields such as philosophy, psychology, medical science, and robotics have focused on *kansei*. The Japan Society of Kansei Engineering (JSKE) was established in 1998 [6]. The society publishes an English journal ("*Kansei Engineering International*"), in which some articles discuss the relationship between robots and *kansei*. In particular, in 2009, a special issue on "KANSEI Robotics" was published [7], in which *kansei* technology and human-robot interaction were discussed [8, 9].

Because *kansei* is a vague word, in this chapter the author defined *kansei* as follows: *Kansei* is the human ability to process and express complex information such as human facial expressions or voice. Suppose that two violinists play the same musical notes. The produced sounds may be different from each other because the target sound that is determined through their *kansei* is different. This means that bowing parameters such as bow force and bow speed to produce the sound are also different. Thus, the role of *kansei* in the process of determining human body motion must be considered. Therefore, a set of experiments were carried out to reveal the role of *kansei* in determining human motion planning.

This chapter is organized as follows. Firstly, features of the violin and violin playing are introduced. Secondly, the violin-playing robot developed in my laboratory is introduced. Then, *kansei* in violin playing is discussed. To investigate *kansei* in a motion planning, an information flow in violin playing is proposed. Based on the flow, some experiments have been conducted, and the relationship between *kansei* and violin playing is discussed.

11.2 Features of the Violin and Violin playing

Figure 11.1 shows the names of each part of a violin and a bow. A violin has four strings: "G," "D," "A," and "E." The ends of each string are connected to a peg and a tail piece, and the bridge supports the strings. Violinists usually put their chin on the chin rest and pin the violin between their chin and shoulder.

A bow consists of three parts: stick, hair and frog. The hair is horse hair. We can adjust the tension of the hair by rotating the screw, which changes the position of the frog. To regulate friction between hair and strings, violinists must apply an appropriate quantity of rosin to the hair.

We can divide violin playing into three tasks: bowing, fingering and holding a violin. Bowing and fingering are much more important than holding because those tasks affect sounds directly. In bowing; bow force, bow speed and sounding point are significant parameters that determine the parameters of produced sounds, such as volume and timbre. The sounding point is the position at which the bow touches a string.

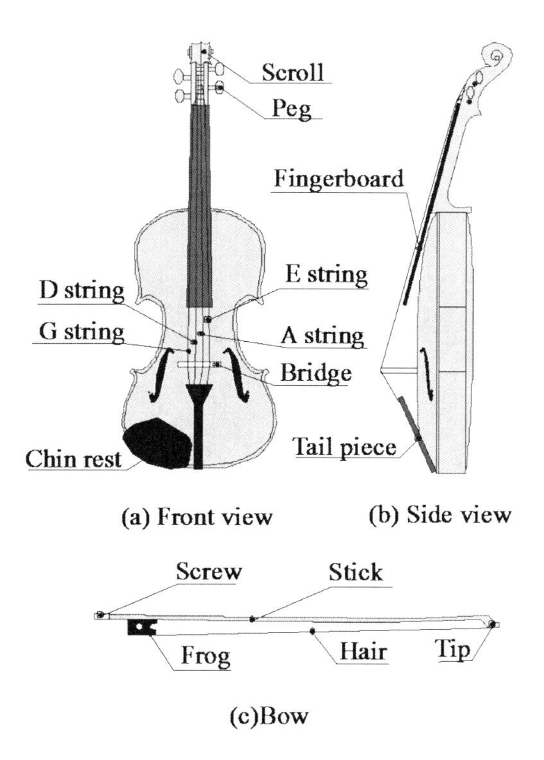

Fig. 11.1 Names of each part of a violin and a bow.

11.3 Violin-Playing Robot

In order the proposed research, an anthropomorphic robot arm for bowing was developed. Shimojo et al. use a commercial 7-DOF manipulator [7], but the robot is larger than typical human size. The author suspects that there are some difficulties in performing bowing motion because of the size differences. Therefore, the author aims to build a human sized robot. Also, a left hand with three fingers has been developed recently. In this section, the robot arm and the left hand are introduced.

11.3.1 Hardware of the Right Arm for Bowing

Figure 11.2 shows the right robot arm for bowing. The number of degrees of freedom in the shoulder, elbow, and wrist joint is three, one and three, respectively, as shown in Fig. 11.3. In the figure, circles, pairs of triangles and rectangles represent the joints of the robot. A circle represents a top view of a cylinder. A pair of triangles represents a side view of two circular cones. Also a rectangle represents a side view of a cylinder. All the icons represent directions of the rotation as shown in Fig. 11.3. Therefore, this robot has 7-DOF, and the movement ranges of all joints were determined based on those of humans, as shown in Table 11.1. Also, Table 11.1 shows directions of each joint. The lengths between shoulder and elbow joints and between elbow and wrist joints are approximately 262 and 304 mm, respectively. Therefore, the size of the robot is almost the same as an adult human. A DC motor with an encoder drives each joint, and a personal computer and servo amplifiers control all joint movements. The robot arm is controlled in position and velocity (force control is not implemented as no force sensors are used).

Fig. 11.2 Photograph of right robot arm.

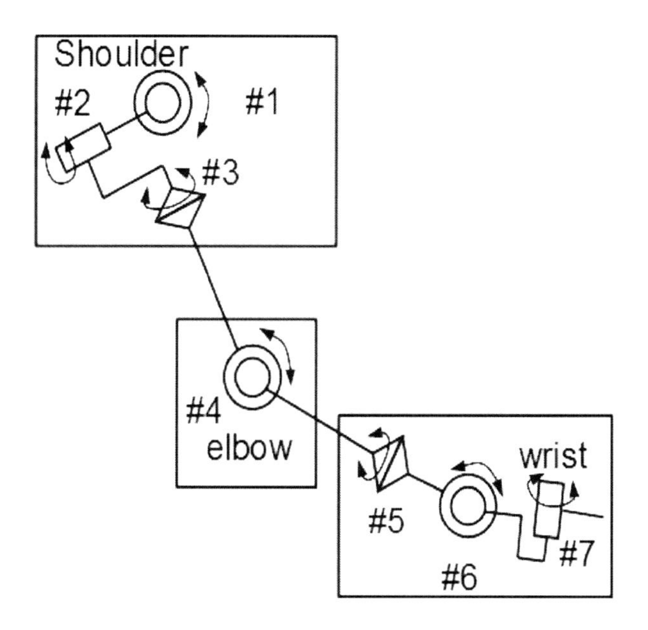

Fig. 11.3 Link model of right arm.

Table 11.1 Movement ranges of all joints.

Joint	Joit Number	Direction of Rotation	Angle [deg]
Shoulder	#1	Flexion	145
		Extension	50
	#2	Abduction	90
		Adduction	20
	#3	Inner rotation	90
		Outer rotation	45
Elbow	#4	Flexion	145
		Extension	5
Wrist	#5	Pronation	90
		Supination	90
	#6	Flexion	90
		Extension	70
	#7	Abduction	30
		Adduction	55

The proposed robot uses photoelectric switches to obtain absolute joint angles. Figure 11.4 shows an example of the sensor arrangement in joint #3. The top of the upper arm axis was cut in a semicircular shape. This part will interrupt or pass light based on the specific absolute joint angle. The robot includes photoelectric sensors with similar mechanisms on all joints. Every time after the power is switched on; the control system uses these sensors to obtain the correct absolute joint angles.

Also, the robot is includes a hand and attached to the end of the arm. The hand is a simple gripper that only grips a bow, as shown in Fig. 11.5. Although the author believes that hands for controlling bow movement should be anthropomorphic, it is too difficult to make a five-fingered hand with many joints. Therefore, the robot includes a simple gripper.

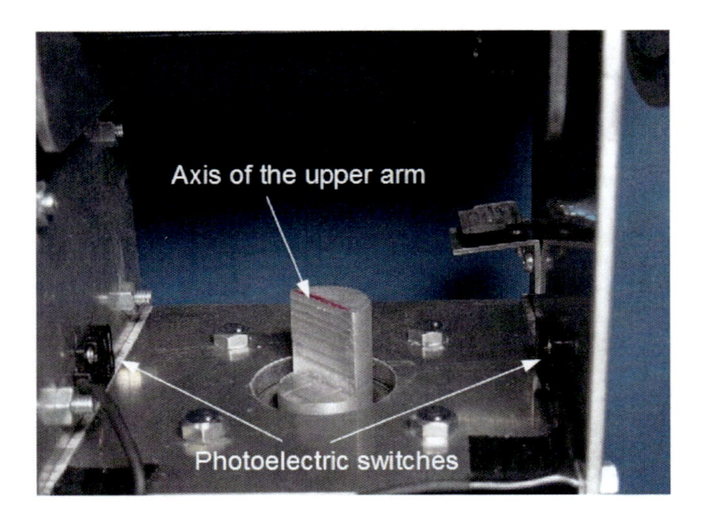

Fig. 11.4 Photoelectric switch installed in the #3 joint.

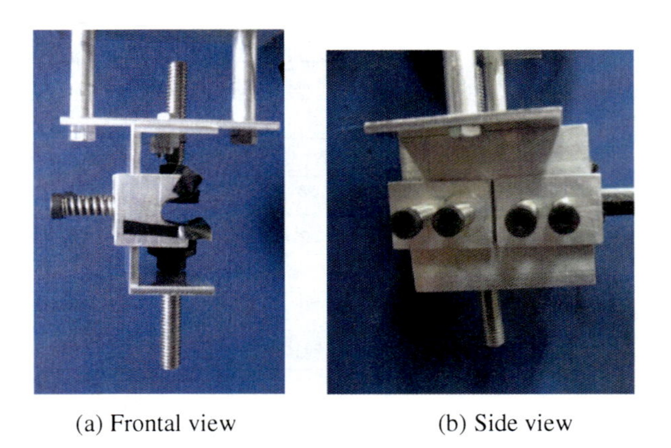

Fig. 11.5 Photograph of hand for grasping bow.

11.3.2 Bowing Motion by Right Arm

The robot can produce a single sound without fingering. The path of its hand is a straight line orthogonal to the strings. We can obtain reference angles by solving the robot's inverse kinematics. To solve it, a constraint is needed due to a redundant degree of freedom. The results of our previous analyses on human violinist's movements clarified that the movements of joint #3 of the human violinists were smaller than the other joints. Therefore, the author considered fixing the joint angular velocity of #3, when solving inverse kinematics. From the solved data, the patterns of al angular velocities are computed and then sent to the servo amplifiers.

Also, after playing one string, the robot can play another string except the E string. For example, this robot can play the D string in up bow after playing the G string in down bow. This motion is calculated as follows. First, two straight lines for playing two strings are calculated. Then, the end point of one line and the starting point of the other line are connected in a straight line. Obtained lines are the trajectory of the hand. Finally, all the angular velocities of the arm joints are calculated by the previously mentioned method.

11.3.3 Sound Analysis

(1) Single sound

Figure 11.6 shows the sound data of the D string. Because undulations cannot be observed, it is possible to conclude that the sound is relatively good. Also, Figure 11.7 shows the sound data of the G string and D string. It is found that the robot plays the two strings correctly.

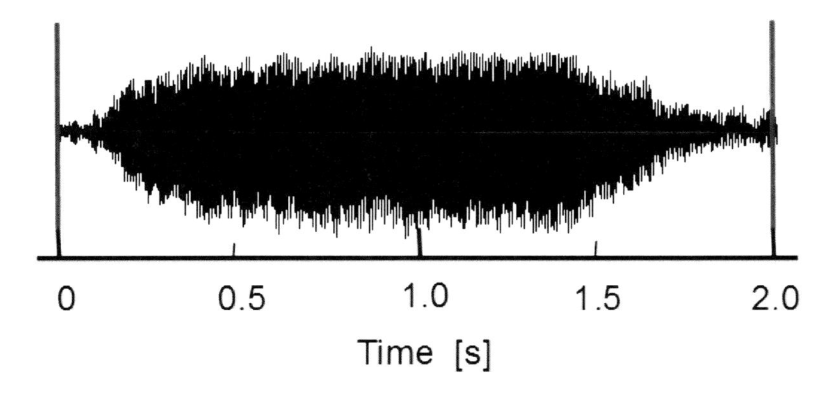

Fig. 11.6 Sound data of D string.

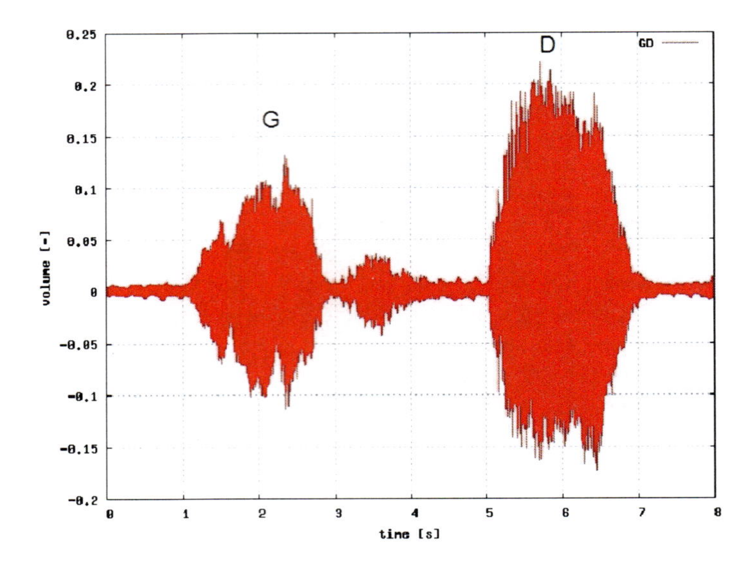

Fig. 11.7 Sound data of G and D strings.

11.3.4 Hardware of the Left Hand with Three Fingers for Fingering

Recently, the author is developing a left hand with three fingers for fingering. Figure 11.8 shows a photograph of the hand. This left hand has three fingers. Usually, human violinists use four fingers, except the thumb, for pushing the strings. However, a robotic four-fingered left hand is too big. Therefore, the left hand has three fingers, which are installed oblique the strings, as human violinists usually do.

Each finger has two joints and they are actuated by D.C. motors and rack-and-pinion mechanisms, as shown in Fig. 11.9. Motors 1 and 2 drive the arms 1 and 2 respectively. By driving motor 1, rack 1 is moved, which makes pinions 1 and 3 move. The gear ratio of pinions 1 and 3 is 2:1. As a result, the finger tip moves downward or upward to push or release the string. Also, motor 2 drives rack 2 and pinion 2, which is separated from pinion 1. This makes the fingertip move forward or backward and can change the strings to be pushed. The fingertips are covered with rubber.

Therefore, by using this hand, the robot can play simple scales or simple music. At present, fingerings of a few bars of "Mary Had a Little Lamb," and "Twinkle, Twinkle, Little Star" are achieved with bowing by human arm. The latter have to use two different stings (E and A). Unfortunately, some pitches are not strictly correct. To address this problem, we have to develop a left arm that can move the hand. Also, at the present due to its complexity, the control of the right arm and left hand are not coordinated.

Fig. 11.8 Photograph of the left-hand fingers.

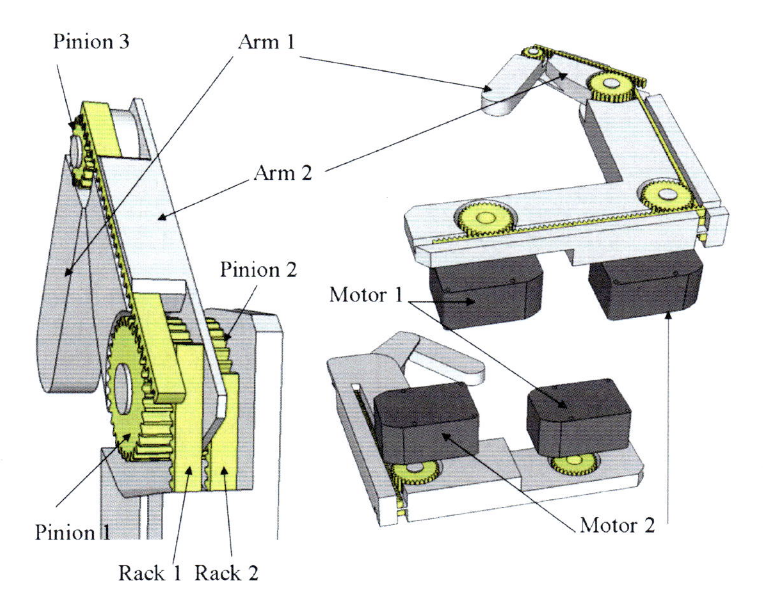

Fig. 11.9 Mechanism of a single finger.

11.4 Violin Playing and *Kansei*

This section the relationship between *kansei* and human body motion is explained. It is a key concept of this study.

The author believes that professional musicians, including violinists, change their performance based on the timbre or sound color that they want to produce because producing good timbre is very important. For instance, when they want to produce "bright" sounds, they choose such parameters as bow force, bow speed, and sounding point, which can represent such timbre. Humans evaluate timbre through their sensibility, in other words, *kansei*. From the above discussion, after concluding that *kansei* affects violin playing, *kansei* has been taken into account when developing the violin-playing robot.

11.5 Information Flow in Violin Playing Considering *Kansei*

This section, discusses how *kansei* affects human motion and the information flow from musical notes to sounds in violin playing shown in Fig. 11.10. The flow is divided into three parts: task planning, motion planning, and playing. Each of the parts is discussed in the following sections.

11.5.1 Task Planning

The task planning determines bowing and fingering parameters such as bow force, bow speed, sounding point, positions of the left-hand fingers and so on.

In almost all cases, musical notes are the fundamental information for musical instrument performance. Musical notes contain information about notes that should be produced. From the musical notes, the task planning derives images of sounds. "Images" means mental impression of the sounds for the musical notes, which can be expressed only by words or sentences.

Then, the task planning determines note information and timbre from the images. Note information consists of pitch, length and strength of a sound, which can be expressed numerically by physical parameters, such as frequency and time. We can consider timbre as *kansei* information because its expression is difficult by physical parameters. Because musical notes lack information on timbre, violinists need to create that through their *kansei*. Also, human evaluate timbre through their *kansei*. Finally, violinists determine bowing and fingering parameters from them.

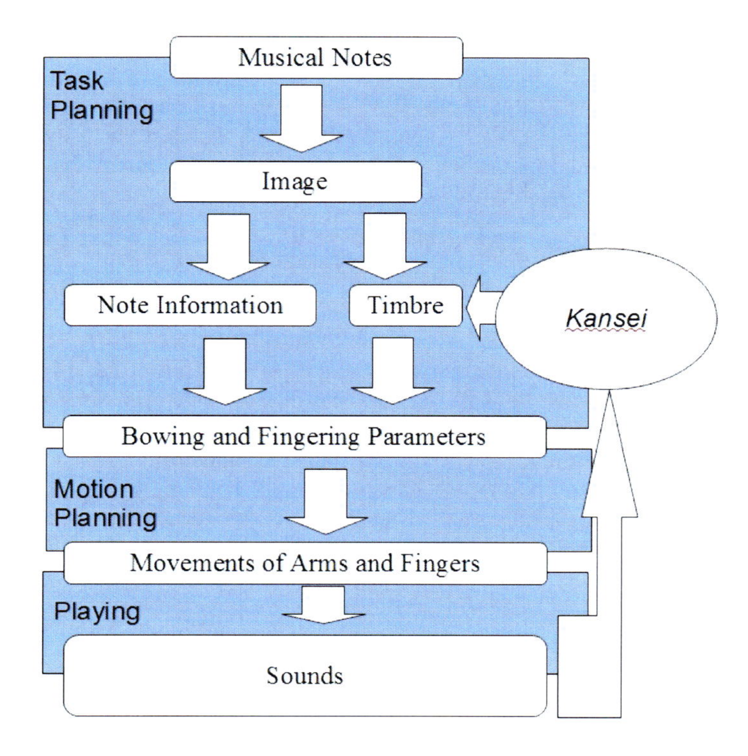

Fig. 11.10 Information flow in violin playing.

11.5.2 Motion Planning

Motion planning determines body movements, which must satisfy bowing and fingering parameters determined in the task planning. The author believes that the kinematics and dynamics of the human body play a key role in this process.

11.5.3 Playing

In the playing process, violinists or violin-playing robot move their body based on the information derived from the motion planning to produce sounds. Violinists evaluate the sounds and change timbre and bowing and fingering parameters through *kansei*.

From the above discussion, *kansei* will mainly affect task planning because timbre is only used in task planning, which was partially confirmed through analysis of human playing [10].

11.6 Analysis of Human *Kansei*

Based on the proposed information flow (Fig. 11.10), the author analyzed human *kansei* [11-13]. In those analyses, the relationship between timbre and bowing parameters represents the main focus. This section discusses one of the examples of the analyses.

11.6.1 Experiment

The objective of the analysis is to clarify the relationships between timbre and bowing and fingering parameters. In the experiment, three professional violinists played a melody shown in Fig. 11.11. They imagined eleven timbres shown in Table 11.2 before playing, and produced sounds according to these images. We performed the experiment three times to ensure reproducibility.

To measure the bow force, the author mounted strain gauges on the bow. Bow speed, and sounding point were measured by a 3D video tracker system that could calculate 3D positions of specified points. Fingering motion was also observed from videos.

Table 11.2 Timbres used in experiment

Bright	Dark
Lucid	
Powerful	Feeble
Abundant	Barren
Heavy	Light

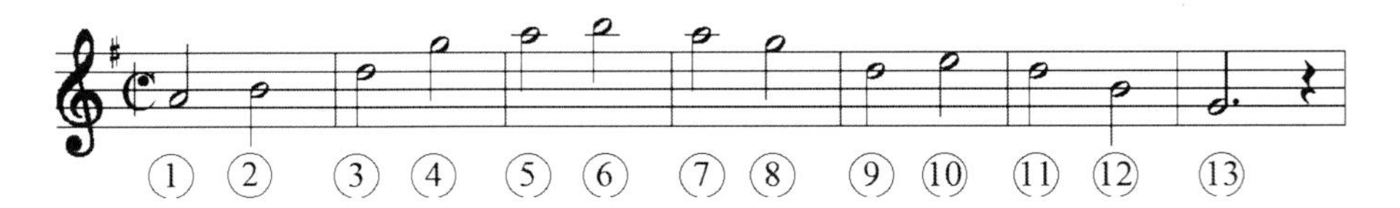

Fig. 11.11 Melody used in experiment.

11.6.2 Results

The author analyzed the relationships between timbre and three bowing parameters using averaged bow force, bow speed and sounding point data. A clear relationship among timbre and bow force and sounding point was found. However, no relationship between bow speed and timbre were found.

Then, a more detailed analysis of the relationship between bow force and sounding point in each timbre was done. Figure 11.12 shows the results. The vertical axis represents averaged sounding point and the horizontal axis represents averaged bow force. From this figure, we can see that the directions of the lines that connect plots of opposite timbres, such as "bright" and "dark," are markedly downward except for a few timbres. This means that when professional violinists use strong bow force, the sounding point is near the bridge. It can be concluded that bow force has a strong relationship with sounding point, and violinists change those parameters according to timbre.

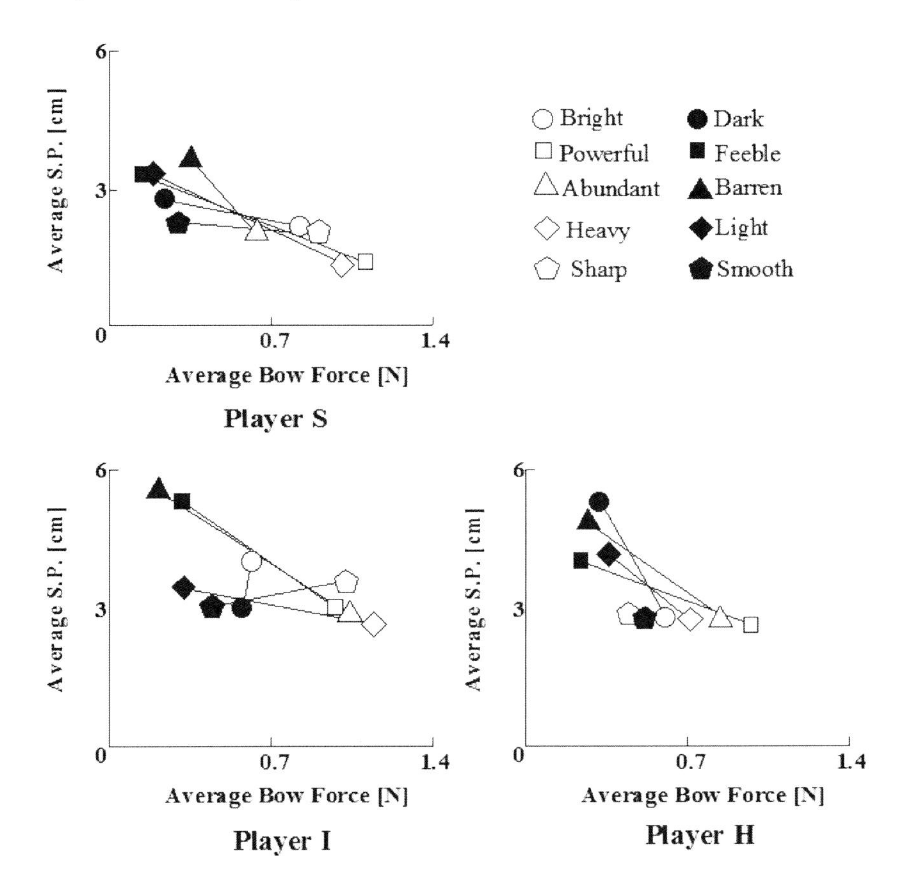

Fig. 11.12 Relationships between sounding point and bow force.

Lastly, the author analyzed the fingering pattern and found three fingering patterns. However, the change of fingering pattern for each violinist was not the same. Particularly, one violinist used only one fingering pattern among the three, while the other two subjects changed fingering pattern according to timbre. This suggests that violinists change fingering pattern according to timbre, but the strategy of changing fingering patterns is different for each violinist.

From the above results, we may conclude that violinists change bow force, sounding point and fingering pattern mainly according to the timbre they imagine.

11.7 Conclusion and Future Works

In this chapter, first, the author introduced the right arm and left hand of the violin-playing robot developed in my laboratory and its sound data were shown. Then, an information flow of violin playing considering *kansei* has been proposed. Based on the flow, the author analyzed the relationship between timbre and bowing parameters. As a result, it was found that that timbre affects bow force and sounding point. Although these two discussions seem to be disconnected, these will be connected tightly in future work.

Also, the following points should be addressed as future works. First, the left arm of the robot with seven joints should be built and coordinated motions of both arms and left fingers should be planned to achieve producing more expressive sounds. Second, an algorithm to convert musical notes to robot motion, which includes *kansei* information, should be constructed. From the results, the author want to consider the roles of *kansei* in motion planning of both human and robots.

References

1. Canazza, S., De Poli, S., Drioli, C., Vidolin, A.: Modeling and Control of Expressiveness in Music Performance. Proceedings of the IEEE 92(4), 686–701 (2004)
2. Kajitani, M.: Development of Musician Robots. Journal of Robotics and Mechatronics 1(3), 254–255 (1989)
3. Sobh, T.M., Wange, B.: Sobh and Bei Wange Experimental Robot Musicians. Journal of Intelligent and Robotic Systems 38, 197–212 (2003)
4. Kuwabara, H., Seki, H., Sasada, Y., Aiguo, M., Shimojo, M.: The development of a violin musician robot. In: IEEE/RSJ International Conference on Intelligent Robots and Systems (IRO S2006) Workshop: Musical Performance Robots and Its Applications, Beijing, pp. 18–23 (2006)
5. Kusuda, Y.: Toyota's violin-playing robot. Industrial Robot: An International Journal 35(6), 504–506 (2008)
6. `http://www.jske.org/` (in Japanese)
7. Special Issue on KANSEI Robotics. Kansei Engineering International 8(1), 1–83 (2009)
8. Hashimoto, S.: Kansei Technology and Robotics Machine with a Heart. Kansei Engineering International 8(1), 11–14 (2009)
9. Cho, J., Kato, S., Kanoh, M., Ito, H.: Bayesian Method for Detecting Emotion From Voice for Kansei Robotics. Kansei Engineering International 8(1), 15–22 (2009)
10. Shibuya, K., Fukatsu, K., Komatsu, S.: Influences of Timbre on Right Arm Motion in Bowing of Violin Playing. Journal of Biomechanisms Japan 28(3), 146–154 (2004) (in Japanese)

11. Shibuya, K., Ssugano, S.: The Effect of KANSEI Information on Human Motion - Basic Model of KANSEI and Analysis of Human Motion In Violin Playing. In: Proceedings of the 4th IEEE International Workshop on Robot and Human Communication (RO-MAN 1995), Tokyo (1995)
12. Shibuya, K., Tanabe, H., Asada, T., Sugano, S.: Construction of KANSEI Model for the Planning of Bowin. In: Violin Playing. Proceedings of the 5th IEEE International Workshop on Robot and Human Communication (RO-MAN 1996), Tsukuba, pp.422-427 (1996)
13. Shibuya, K.: Analysis of Human KANSEI and Development of a Violin Playing Robot. In: IEEE/RSJ International Conference on Intelligent Robots and Systems (IROS 2006) Workshop: Musical Performance Robots and Its Applications, pp. 13–17 (2006)

Chapter 12
Wind Instrument Playing Humanoid Robots

Jorge Solis and Atsuo Takanishi

Abstract. Since the golden era of automata (17th and 18th centuries), the development of mechanical dolls served as a mean to understand how the human brain is able of coordinating multiple degrees of freedom in order to play musical instruments. A particular interest was given to wind instruments as a research approach since this requires the understanding of human breathing mechanisms. Nowadays, different kinds of automated machines and humanoid robots have been developed to play wind instruments. In this chapter, we will detail some issues related to the development of humanoid robots and the challenges in their design, control and system integration for playing wind instruments.

12.1 Introduction

The development of wind instrument playing automated machine and humanoid robots has interested the researchers since the golden era of automata up to today. As an example, we may find some classic examples of automata displaying human-like motor dexterities to play instruments such as the "Flute Player" [18]. In addition, we find the first attempt to develop an anthropomorphic musical robot, the WABOT-2. The WABOT-2 was capable of playing a concert organ, built by the late Prof. Ichiro Kato [6]. In particular, Prof. Kato argued that the artistic activity such as playing a keyboard instrument would require human-like intelligence and dexterity. Compared to other kinds of instruments (i.e. piano, violin, etc.), the research on wind instruments have interested researchers from the point of view of human science (i.e. study of breathe mechanism), motor learning control (i.e. coordination and synchronization of several degrees-of-freedom), musical

---

Jorge Solis
Faculty of Science and Engineering, Waseda University/Humanoid Robotics Institute, Waseda University, 3-4-1 Ookubo, Shinjuku-ku, Tokyo, Japan
e-mail: solis@ieee.org

Atsuo Takanishi
Faculty of Science and Engineering, Waseda University/Humanoid Robotics Institute, Waseda University, 3-4-1 Ookubo, Shinjuku-ku, Tokyo, Japan
e-mail: contact@takanishi.mech.waseda.ac.jp

J. Solis and K. Ng (Eds.): Musical Robots and Interactive Multimodal Systems, STAR 74, pp. 195–213.
springerlink.com

engineering (i.e. modeling of sound production.), etc. In this chapter, an overview of different kinds of automated machines and humanoid robots designed for playing wind instruments (flute and saxophone) will be presented. In particular, their mechanism design principles and the control strategies implemented for playing musical instruments will be detailed.

12.2 Flute and Saxophone Sound Production

The sound of wind instruments is a self-excited oscillation [1]. The sound production system comprises three major elements: energy source, generator, and sound resonator. The sound generator includes the dynamics of reed vibration and air flowing through a reed aperture for reed woodwind instruments. The sound resonator is related to the air-column resonance of the instruments. In a single reed woodwind instrument, a flow modulated by the reed in the generator enters the resonator and excites an oscillation of the air column. As a response, the resonator generates sound pressure at the entrance and it acts as an external force on the reed and influences the oscillation. In this manner, the sound production system forms a feedback loop. If the loop gain becomes positive and overcomes losses such as the acoustic radiation, the system yields a self-excited oscillation or sounding. In order to develop musical performance humanoid robots, we are required to understand in detail the principle of sound production of the instrument as well as the mechanism of humans to control different kinds of properties of the sound. In particular, in this section, we will provide a general overview of the differences in the principles of sound production between the flute and the saxophone.

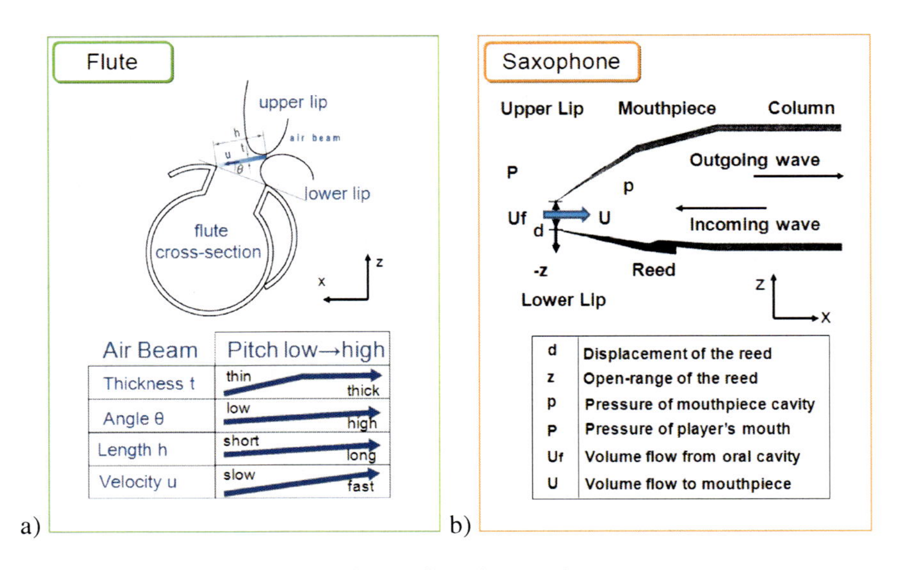

Fig. 12.1 Principle of sound production: a) flute; b) saxophone.

The flute is an air reed woodwind which, due to the absence of a reed, is driven by an air beam characterized by the length, thickness, angle and velocity [2]. Slight changes of any of these parameters are reflected in the pitch, volume, and tone of the flute sound (Figure 12.1a). On the other hand, acoustically speaking, the saxophone is very similar to the clarinet, except that the conical bore of the saxophone resonates at multiples of the lowest resonance frequency whereas the clarinet resonates at odd multiples [11]. When a player blows on a saxophone; the reed acts as a pressure-operated valve in such a manner that the flow of air into the mouthpiece is increased and decreased as the acoustical pressure difference between the mouthpiece cavity and the player's mouth rises and falls. This pressure is then adjusted to suit the note being played. In Fig. 12.1b, the principle of sound production of single reed instruments is shown; where d and z represent the displacement and open-range of the reed respectively. P and p are the pressure of player's mouth and pressure mouthpiece cavity. Finally, U_f and U are the volume flow at the mouth piece and after it respectively.

12.3 Wind Instrument-Playing Anthropomorphic Robots

12.3.1 Flute-Playing Robots

During the golden era of automata, the "Flute Player" developed by Jacques de Vaucanson was designed and constructed as a means to understand the human breathing mechanism [4]. Vaucanson presented "The Flute Player" to the Academy of Science in 1738 (Figure 12.2a). For this occasion he wrote a lengthy report carefully describing how his flutist can play exactly like a human. The design principle was that every single mechanism corresponded to every muscle [18]. Thus, Vaucanson had arrived at those sounds by mimicking the very means by which a man would make them. Nine bellows were attached to three separate pipes that led into the chest of the figure. Each set of three bellows was attached to a different weight to give out varying degrees of air, and then all pipes joined into a single one, equivalent to a trachea, continuing up through the throat, and widening to form the cavity of the mouth. The lips, which bore upon the hole of the flute, could open and close; and move backwards or forwards. Inside the mouth was a moveable metal tongue, which governed the air-flow and created pauses.

More recently, the "Flute Playing Machine" developed by Martin Riches was designed to play a specially-made flute somewhat in the manner of a pianola, except that all the working parts are clearly visible [9]. The Flute Playing Machine (Figure 12.2b) is composed of an alto flute, blower (lungs), electro-magnets (fingers) and electronics. The design principle is basically transparent in a double sense. The visual scores can be easily followed so that the visual and acoustic information is synchronized. The pieces it plays are drawn with a felt tip pen on long transparent music roll which are then optically scanned by the photo cells of a reading device. The machine has a row of 15 photo cells which read felt-tip pen markings on a transparent roll. Their amplified signals operate the 12 keys of the flute and the valve which controls the flow of air into the embouchure. The two remaining tracks may be used for regulating the dynamics or sending timing signals to a live performer when performing a duet.

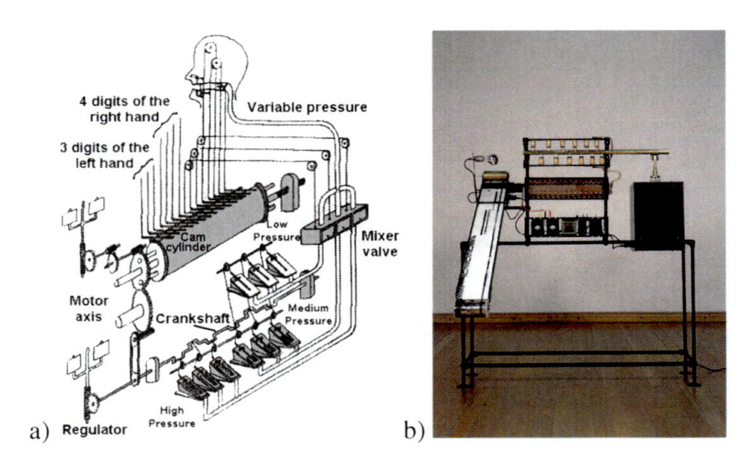

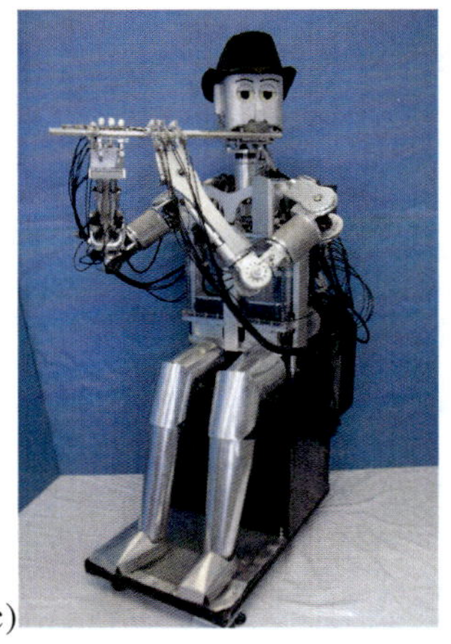

Fig. 12.2 Examples of automated machines and humanoid robots for emulating the flute playing: a) Flute Player (adapted version from the illustration done by Doyon and Liaigre); b) Flute-Playing Machine (courtesy by Martin Riches[1]); c) Anthropomorphic Waseda Flutist Robot.

The authors have developed an Anthropomorphic Waseda Flutist Robot to understand the human motor control from an engineering point of view (Figure 12.2c). In the next sub-sections, the technical details of the mechanism design and musical performance control strategies will be detailed.

[1] Photography by Hermann Kiessling, collection of the Berlinische Galerie, State Museum of Modern Art, Photography, and Architecture.

12.3.1.1 Anthropomorphic Waseda Flutist Robot

Since 1990, the research on the development of the anthropomorphic Waseda Flutist Robot has been focused on mechanically emulating the anatomy and physiology of the organs involved during the flute playing. As a result of this research, in 2007, the Waseda Flutist Robot No. 4 Refined IV (WF-4RIV) was developed. The WF-4RIV has a total of 41-DOFs and it is composed of the following simulated organs [13]: lungs, lips, tonguing, vocal cord, fingers, and other simulated organs to hold the flute (i.e. neck and arms). The WF-4RIV has a height of 1.7 m and a weight of 150 kg. In particular, the WF-4RIV improved the mechanical design of the artificial lips (to produce more naturally the shape of human lips so that more natural sounds can be produced), the tonguing mechanism (to reproduce the double tonguing so that smoother transitions between notes can be done), the vibrato (to add more natural vibrations to the air beam), and the lung system (to enhance the air flow efficiency). In the following sub-sections, we will focus in providing the technical design details of such mechanisms and the implementation of the musical performance control system.

12.3.1.2 Mechanical Design of the WF-4RIV

One of the most complicate organs to mechanically reproduce is the shape of the lips. In fact; the previous version of the lips mechanism was composed of 5-DOFs which basically were used to control the parameters of the air stream. Moreover, the artificial lips were made of EPDM rubber (ethylene-propylenediene-monomer rubber). However, the EPDM rubber could not simulate the elasticity properties of human lips. From our discussions with a professional flutist, a more accurate control of the shape of lips and higher elasticity of the artificial lips are important issues that may contribute in improving the robot's playing. For that reason, the lips mechanism of WF-4RIV was simplified and designed by 3-DOFs (Figure 12.3a) to realize an accurate control of the motion of the superior lip (control of airstream's thickness), inferior lip (control of airstream's angle) and sideway lips (control of airstream's length). The artificial lip is made of a thermoplastic rubber named "Septon" (Kuraray Co. Ltd., Japan). The Septon was selected due to its high stiffness (19.85 kPa measured at a measuring temperature of 23°C at a tensile rate of 500 mm/min according to JIS K-6251) and low hardness (can be easily designed between 30 A and 80 A hardness). In order to change the shape of the artificial lips, an array of pins were placed on four places (top, bottom and sideways).

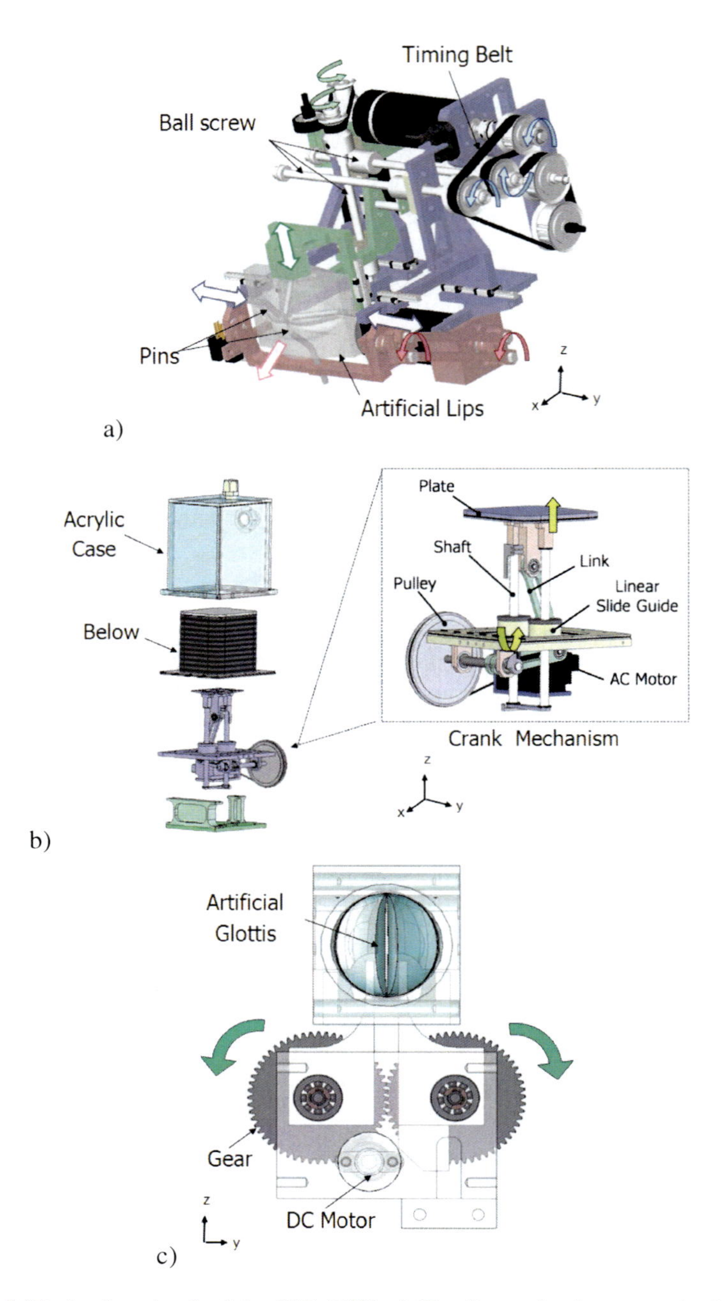

Fig. 12.3 Mechanism detail of the WF-4RIV: a) The lip mechanism controls the length, thickness and angle of the airstream; b) The lung mechanism controls the speed of the airstream; c) The artificial vocal cord adds vibrations to the airstream.

On the other hand, the reproduction of the lung system was previously implemented by using two vane mechanisms which were controlled by an AC motor. The breathing process was controlled by a couple of valve mechanisms which were located behind the robot. The use of vane mechanisms effectively reduced the mechanical noise during the breathing process; however, we detected low air conversion efficiency (51%) due to the loss of air coming from the lungs to the lip. This problem also inhibits the production of a natural vibrato as little air arrives to the vibrato mechanism (vocal cord). Therefore, we designed a new lung mechanism for the WF-4RIV which is more air-tight during the breathing process. The lung system of the WF-4RIV is composed by two acrylic cases, which are sealed (Figure 12.3b). Each of the cases contains a bellow which is connected to an independent crank mechanism. The crank mechanism is controlled by using an AC motor so that the robot can breathe air in into the acrylic cases and breathe air out from them by controlling the speed of motion of the bellow. Moreover, the oral cavity of WF-4RIV is composed by a clamped plate (located at the front) and a coupler (located at the rear). The clamped plate is where the artificial lips of the robot are attached, and the coupler connects to a tube with the air flow coming from the throat mechanism. Inside the oral cavity, a simulated tongue was installed.

The reproduction of the vibrato mechanism was previously implemented by a voice coil motor which presses directly on a tube to add vibrations to the air beam passing through this mechanism. However, human uses a more complicated mechanism to produce a vibrato. In fact, by observing the laryngeal movement while playing a wind instrument using a laryngo-fiberscope, it has been demonstrated that the shape of the vocal cord of the flutist differs according to the level of expertise [10]. Therefore, we believe that the aperture control of the glottis plays a key role in producing a human-like vibrato which will help in producing a performance with expressiveness. As a result, the vocal cord of the WF-4RIV is composed by 1-DOF and the artificial glottis was also made of Septon. In order to add vibration to the incoming air stream, a DC motor linked to a couple of gears was used (Figure 12.3c). The gears are connected to each both sides of the vocal folds by links. This design enables the control of the amplitude and frequency of the aperture of the glottis (after the addition of vibrations to the air stream, it is then directed to the robot's oral cavity through a tube).

12.3.1.3 Musical Performance Control of the WF-4RIV

As it has been previously detailed, different organs have several functions during flute playing. The motor control of such organs are basically the result of an accurate coordination of contracting muscles and the correct positioning of the embouchure hole of the flute with respect to the lips. Human beings have a large and complex set of muscles that can produces dynamic changes of those organs. Due to such a complexity, human players perform different scores several times to improve the accuracy control and synchronization of the organs. For each single note, humans listen and evaluate the sound quality. If the quality is not acceptable, they adjust some parameters until the produced sound is acceptable.

Inspired by this principle, the control strategy for the WF-4RIV has been implemented by defining an Auditory Feedback Control System (AFCS) [14]. The AFCS it is composed by three main modules (Figure 12.4): Expressive Music Generator (*ExMG*), Feed Forward Air Pressure Control System (*FFAiPC*) and Pitch Evaluation System (*PiES*). The *ExMG* aims to output musical information required to produce an expressive performance from a nominal score. For this purpose, a set of musical performance rules (which defines the deviations introduced by the performer) are defined (offline). The process of modeling the expressiveness features of the flute performance from the performance of a professional flutist is done by means of Artificial Neural Networks (ANN) [3]. In particular, three different musical parameters were considered: duration rate, vibrato frequency and vibrato amplitude. For each of the networks, a number of inputs were defined based on the music computer research as follows: *Duration rate* (14 inputs); *Vibrato duration* (19 inputs) and *Vibrato frequency* (18 inputs). We have experimentally determined different numbers of hidden layers for each of them. In particular, seven units were defined for the duration rate, five units for the vibrato frequency and 17 units for the vibrato amplitude. In order to train the ANN, the back-propagation algorithm was considered. This kind of supervised learning incorporates an external teaching signal (the performance of a professional flutist has been used).

During learning, the weight vectors (W_i) are updated using Eq. (12.1); where $E(t)$is the error between the actual output y_k and the teaching signal d_k computed as Eq. (12.2), and η is the learning rate (which affects network learning speed). There are a few techniques to select this parameter so that we have experimentally determined it (η =0.75). The ANN was trained to learn the extracted performance rules obtained from the analysis of the professional flutist performance. Based on the previous setting of the ANN, the duration rate converges at 148 steps, the vibrato frequency at 72 steps and the vibrato amplitude at 49 steps. The output data from the *ExMG* (i.e. note duration, vibrato frequency, vibrato duration, attacking time and tonguing) is then sent directly to the robot's control system by sending MIDI messages from a sequencer device (personal computer). As a result, the flutist robot is capable of performing a musical performance with expressiveness.

$$w_i(t+1) = w_i(t) - \eta\left(\frac{\delta E(t)}{\delta w_i}\right) \tag{12.1}$$

$$E(t) = \frac{1}{2}\sum_{k=1}^{N}\left(y_k(t) - d_k(t)\right)^2 \tag{12.2}$$

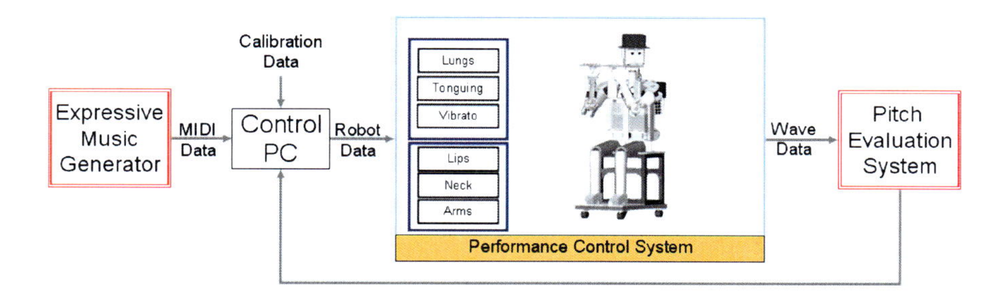

Fig. 12.4 Block diagram of the auditory feedback control system implemented for the WF-4RIV.

However, it is important to consider that music is not defined by a set of independent events sent trough the MIDI data. In fact, musical sounds are continuously produced by taking into consideration adjacent notes. Human players may need to adjust some dynamic parameters during the performance. Therefore, the *FFAiPC* uses a feed-forward control system to control the air pressure coming out from the lungs. For this purpose, the inverse model of the lung system is computed to control of the air pressure during the attack time. In order to implement the feed-forward control system of the air pressure from the lungs, the Feedback Error Learning approach has been implemented which it is also based on the use of an ANN. The feedback error learning is a computational theory of supervised motor learning [7], which is inspired by the central nervous system. In the case of the WF-4RIV, there are 9 input layers units, 5 hidden layer units and a single output layer unit. The output signal is the air pressure and the teaching signal is the data collected from a pressure sensor placed inside the artificial lungs. In order to produce the inverse model, a total of 179 learning steps were required.

Finally, in order to evaluate the flute sound produced by the robot, the PES has been implemented in order to detect both the pitch of the flute sound as well evaluation its quality. The *PES* is designed to estimate the pitch or fundamental frequency of a musical note. In the case of the WF-4RIV, the Cepstrum method has been implemented because it is the most popular pitch tracking method in speech and as it can be computed in real-time [5]. The Cepstrum is calculated by taking the Short-Fourier Transform (STFT) of the log of the magnitude spectrum of sound frame (frame size of 2048, 50% of overlapping at 44.1Hz). However, this method presents the problem of deciding how to divide the frequency. Therefore, the MIDI data of the score is used along to synchronize the output obtained from the implemented Cepstrum method.

After the detection of the pitch, it is possible to evaluate the quality of the sound by using the Eq. (12.3); which it is based on the experimental results done by Ando [2]. The weighting coefficients w_1 and w_2 have been experimentally determined (1.0 and 0.5 respectively). Basically, by using the proposed PES, the WF-4RIV is able of autonomously detect when a note was incorrectly played. Therefore, during a performance, when a note is incorrectly played, the AFS automatically puts a mark to indicate to the performance control system to adjust the required parameters.

In particular, the lip's shape and lung velocity are modified based on the "General Position" proposed and detailed in [12], which is an algorithm that automatically finds the optimal parameter values by adjusting (based on an orthogonal table) the lip's shape and lung's velocity while continuously blowing a simple etude until all the notes are produced with a uniform sound quality.

$$EvalF = \frac{w_1 \cdot (M - H) + w_2 \cdot (L_e - L_o)}{Volume} \tag{12.3}$$

M: Harmonic level [dB] L_e: Even-harmonics level [dB] H: Semi-Harmonics level [dB] L_o: Odd-harmonics level [dB] $Volume$: Volume [dB] w_1, w_2: Weighting Coefficients

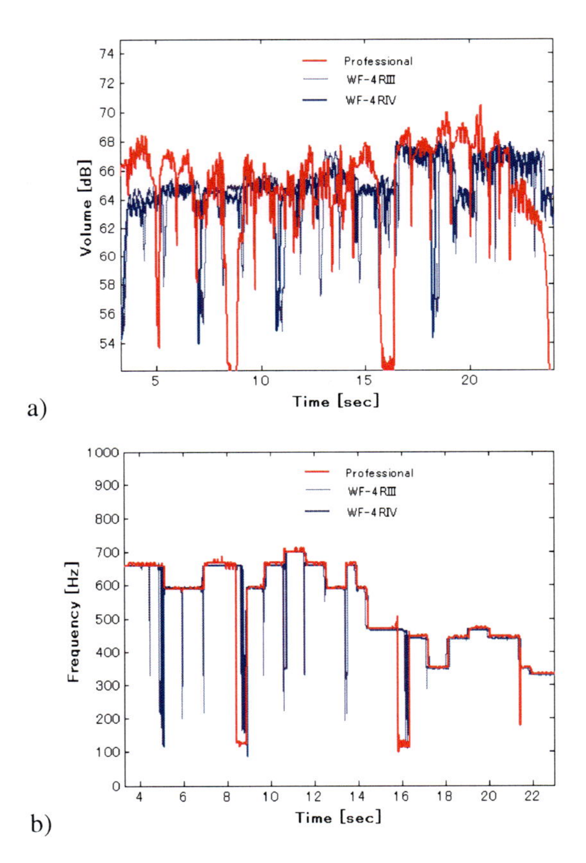

Fig. 12.5 Experimental results to compare the musical performance among the robot versions and a professional flautist: a) The sound volume; b) the pitch.

12.3.1.4 Musical Performance Evaluation of the Flutist Robot

In this experiment, we have verified the improvements of the musical performance of the Waseda Flutist Robot. For this purpose, we have programmed the previous version (WF-4RIII) and the improved one (WF-4RIV) to perform the traditional folklore Japanese song "Sakura", and compare with the performance of a professional flutist player in terms of sound volume and pitch. The experimental results are shown in Figure 12.5. As we may observe, the dynamic changes in the sound volume are considerably improved on the performance of the WF-4RIV thanks to the mechanical improvements done on the lip and tonguing mechanisms (Figure 12.5a). Furthermore, the improved control system implemented for the WF-4RIV enabled the robot to produce more stable tones compared with the WF-4RIII (Figure 12.5b).

However; even though the WF-4RIV is able to reproduce more naturally the vibrato, there are still considerable differences both in terms of sound volume and pitch (especially during note transitions) with respect to the professional human player. This issue may be related to the way the proposed control system has been implemented (see Figure 12.4). Actually, the *PES* has been designed to evaluate exclusively steady tones so that during note transitions, the proposed control could not adjust accurately the mechanical parameters (i.e. lip's shape, lung velocity, etc.).

12.3.2 Saxophone-Playing Robots

One of the first attempts to develop a saxophone-playing robot was done by Takashima at Hosei University [17]. His robot; named APR-SX2, is composed of three main components (Figure 12.6a): mouth mechanism (as a pressure controlled oscillating valve), the air supply mechanism (as a source of energy), and fingers (to make the column of air in the instrument shorter or longer). The artificial mouth consisted of flexible artificial lips and a reed pressing mechanism. The artificial lips were made of a rubber balloon filled with silicon oil with the proper viscosity. The air supplying system (lungs) consists of an air pump and a diffuser tank with a pressure control system (the supplied air pressure is regulated from 0.0 MPa to 0.02 MPa). The APR-SX2 was designed under the principle that the instrument played by the robot should not be changed. A total of twenty-three fingers were configured to play the saxophone's keys (actuated by solenoids), and a modified mouth mechanism was designed to attach it to the mouthpiece, no tonguing mechanism was implemented (normally reproduced by the tongue motion). The control system implemented for the APR-SX2 is composed by one computer dedicated to the control of the key-fingering, air pressure and flow, pitch of the tones, tonguing, and pitch bending. In order to synchronize all the performance, the musical data was sent to the control computer through MIDI in real-time. In particular, the SMF format was selected to determine the status of the tongue mechanism (on or off), the vibrato mechanism (pitch or volume), and pitch bend (applied force on the reed).

More recently, an Anthropomorphic Saxophonist Robot developed by the authors, increased the understanding of the human motor control, from an engineering point of view, by mechanically reproducing the human organs involved during saxophone playing (Figure 12.6b). In the next sub-sections, the technical details of the mechanism design and control strategies will be detailed.

12.3.2.1 Anthropomorphic Waseda Saxophonist Robot

In 2009, the Waseda Saxophonist Robot No.2 (WAS-2) was developed at Waseda University [15]. The WAS-2 is composed by 22-DOFs that reproduce the physiology and anatomy of the organs involved during the saxophone playing as follows (Figure 12.6b): 3-DOFs to control the shape of the artificial lips, 16-DOFs for the human-like hand, 1-DOF for the tonguing mechanism, and 2-DOFs for the lung system. In the following sub-sections, we will focus in providing the technical design details of such mechanisms and the implementation of the musical performance control system.

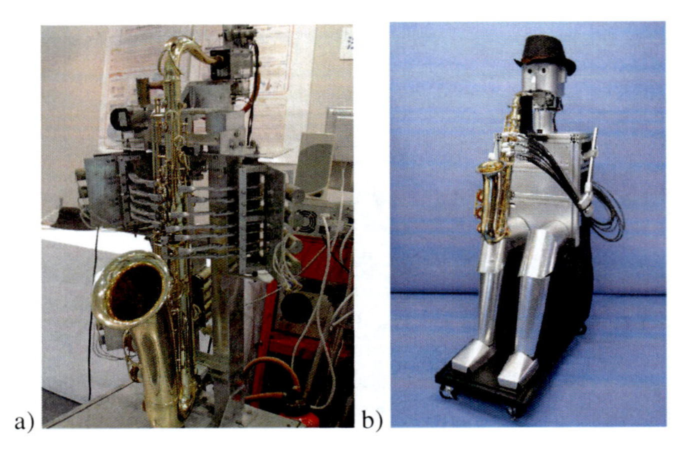

Fig. 12.6 Examples of saxophone playing automated machines and humanoid robots: a) APR-SX2 (courtesy by Suguru Takashima); b) WAS-2.

12.3.2.2 Mechanical Design of the WAS-2

The previous mouth mechanism was designed with 1-DOF in order to control the vertical motion of the lower lip. Based on the up/down motion of the lower lip, it became possible to control the pitch of the saxophone sound. However, it is difficult to control the sound pressure by means of 1-DOF. Therefore, the mouth mechanism of the WAS-2 consists of 2-DOFs designed to control the up/down motion of both lower and upper lips (Figure 12.7a). In addition, a passive 1-DOF has been implemented to modify the shape of the side-way lips. The artificial lips were also made of Septon. In particular, the arrangement configuration of the lip mechanism is as follows: upper lip (rotation of the motor axis is converted into vertical motion by means of a timing belt and ball screw to avoid the leak of air flow), lower lip (a timing belt and ball screw so that the rotational movement of the motor axis is converted into vertical motion to change the amount of pressure on the reed) and sideway lip. On the other hand, the tonguing mechanism is shown in Fig. 12.7b. The motion of the tongue tip is controlled by a DC motor which is connected to a link attached to the motor axis. In such a way, the air flow can be blocked by controlling the motion of the tongue tip. Thanks to this tonguing mechanism of the WAS-2, the attack and release of the note can be reproduced.

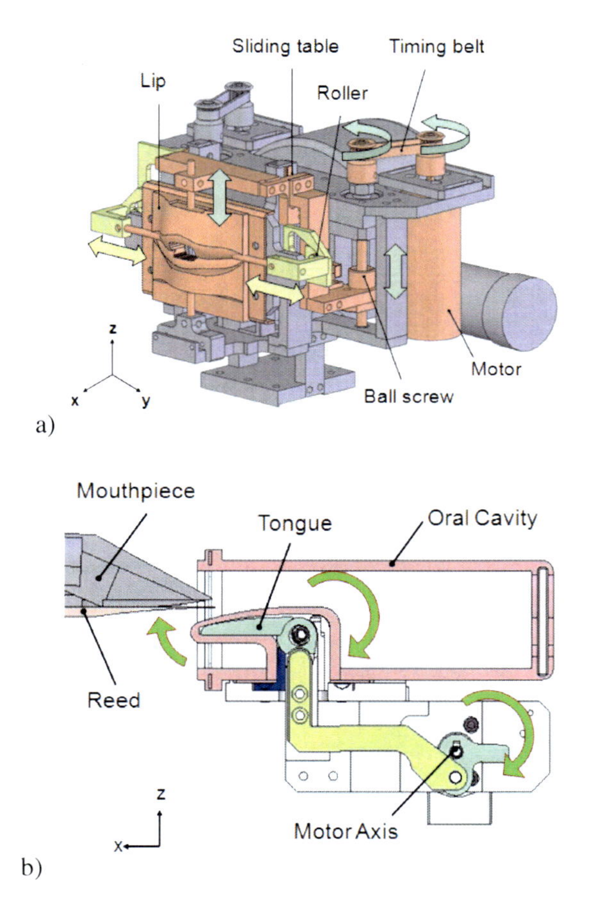

Fig. 12.7 Mechanism details of the WAS-2: a) Mouth mechanism; b) Tonguing mechanism.

Regarding the WAS-2 air source, a DC servo motor has been used to control the motion of the air pump diaphragm; which it is connected to an eccentric crank mechanism (Figure 12.8a). This mechanism has been designed to provide a minimum 20 L/min air flow and a minimum pressure of 30kPa. In addition, a DC servo motor has been designed to control the motion of an air valve so that the delivered air by the air pump is effectively rectified. Finally, the finger mechanism of the WAS-2 is composed of 16-DOFs to push the correspondent keys from A#2 to F#5 (Figure 12.8b). In order to reduce the weight on the hand part, the actuation mechanism uses a wire and pulley attached to the RC motor axis. RS-485 communication protocol has been used to control the motion of each single finger.

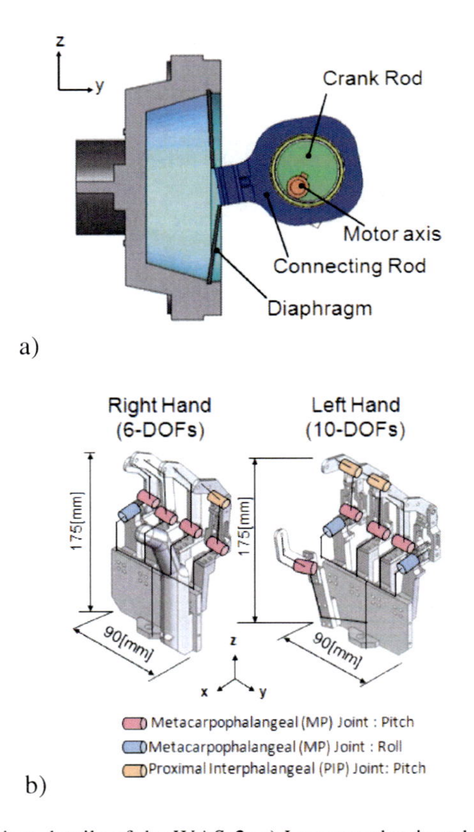

Fig. 12.8 Mechanism details of the WAS-2: a) Lung mechanism; b) Human-like hand.

12.3.2.3 Musical Performance Control of the WAS-2

The control system implemented on the WAS-2 is also integrated by a PC Control and a PC Sequencer (Figure 12.9). The PC Control is used to acquire and process the information from each of the degrees of freedom of the saxophonist robot as well as controlling the air flow/pressure to produce the desired sound. The PC Control has as inputs the MIDI data and Music Pattern Generator (calibration data). The Music Pattern Generator is designed to output the calibration parameters required in order to produce the desired saxophone sound. Inspired on the principle of sound production of single-reed instruments, the WAS-2 requires the control of the following parameters [16]: lower lip's position, valve closing rate, air flow, and pressure. In particular, the accurate control of the lower lip's position ˉd air pressure is required during saxophone performance.

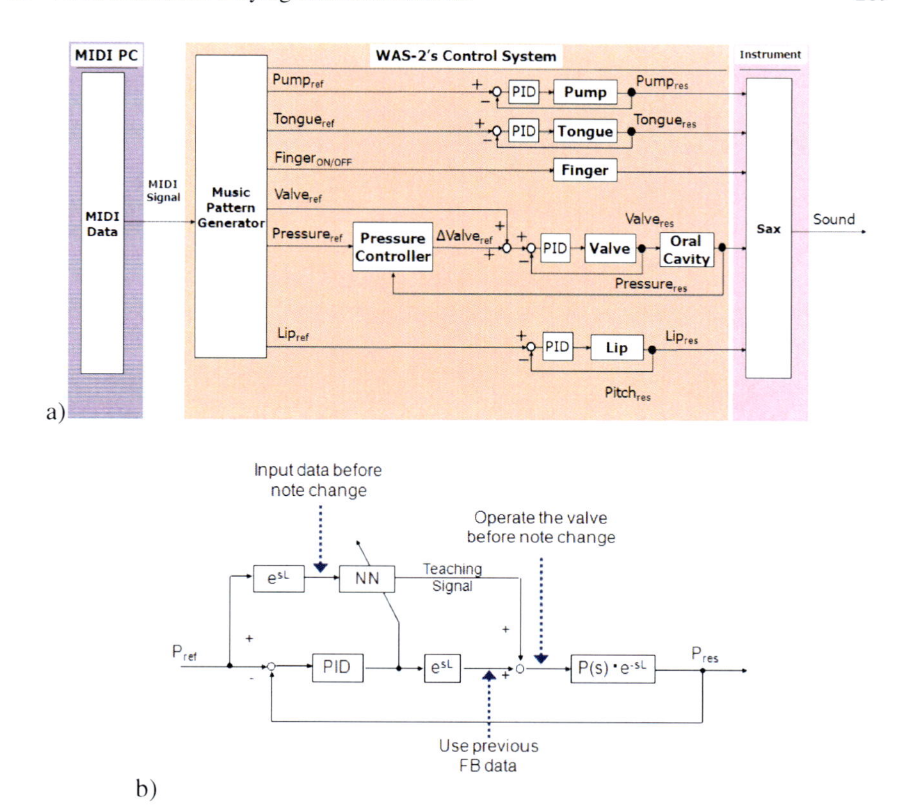

Fig. 12.9 The control system implemented for the WAS-2: a) The block diagram implemented for the WAS-2; b) Detail of the air pressure controller.

The previously implemented control system for the robot was based on a cascade feedback control system to assure the accuracy of the air pressure during a musical performance. Basically, air pressure is controlled based on the measurements of the pressure sensed at the output of the air pump and the position of the lower lips, the air pressure was been controlled. However; during the attack time, the target air pressure is reached around 100ms later during a musical performance. Basically, the signal of the note to be played is sent to the control system through a MIDI message. As soon as message of a note change is received, the air pressure, as well as the position of the lower lips is adjusted. Thus, a delay on the control of the air pressure (during the attack time) was observed.

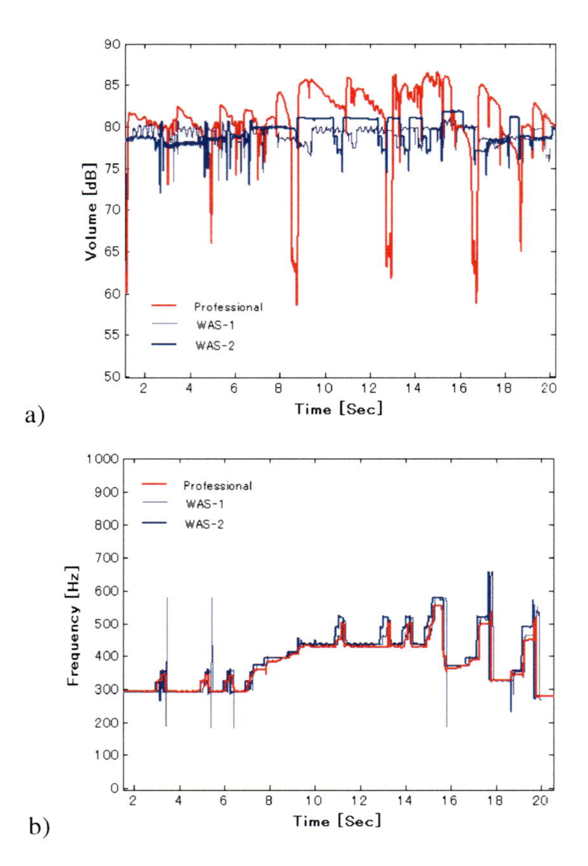

Fig. 12.10 Experimental results to compare the musical performance among the robot versions and a professional saxophonist: a) The sound volume; b) The pitch.

Actually, if we analyze the performance of a human playing the saxophone, the distance between the lungs and the oral cavity are a few dozens of centimeters. This distance provokes the existence of dead-time. However, musicians when playing a musical performance, in order to avoid any delay on the adjustment of the air pressure located inside the oral cavity, control the required parameters of the lungs and the mouth before the notes change. In order to assure the accurate control of the air pressure, an improved control system has been implemented for the WAS-2 (Figure 12.9a). In particular, a feed-forward error learning control system with dead-time compensation was implemented (Figure 12.9b). The inputs of the ANN are defined as follows: pressure reference, note, and lower/upper lips position. For this case, a total of six hidden units were used (experimentally determined). The output is the position of the air valve. The system is trained to produce the required air pressure to blow a sound. In addition, a dead-time factor (referred as e^{sL}; which is an element to predict how changes made now by the controller will affect the controlled variable in the future) is introduced to compensate the delay during the attack time [8].

12.3.2.4 Musical Performance Evaluation

In this experiment, we have focused on verifying the musical performance improvements of the Waseda Saxophonist Robot. For this purpose, we have programmed the previous version (WAS-1) and the improved one (WAS-2) to perform the Moonlight Serenade composed by Glenn Miller and compared them with the performance of a professional saxophone player in terms of sound volume and pitch. The experimental results are shown in Figure 12.9. As we may observe, the dynamic changes on the sound volume have been relatively improved on the performance of WAS-2 thanks to the mechanical improvements done in the mouth and finger mechanisms (Figure 12.10a). Furthermore, the improved control system implemented on the WAS-2 enabled the robot to produce more stable tones compared with the WAS-1 (Figure 12.10b).

However, if we compare the musical performance of the WAS-2 to that of the professional player, we can still observe significant differences (particularly in the dynamic changes on the sound volume). This issue may be related to the way the mechanical parameters are controlled (see Figure 12.9). In fact, we may notice that the air pressure, lip positioning and fingering activation are controlled separately and synchronized by means of the MIDI clock signal.

12.4 Conclusions and Future Work

The mechanism design and performance control of wind instrument-playing of automated machines and humanoid robots pose different kind of challenging issues from the point of view of motor control learning and music technology. In this chapter, different approaches were introduced for the development of robots able to play the flute and saxophone. More recently, the authors have focused on analyzing the internal motion of the human organs while playing wind instruments by means of imaging data (i.e. MRI, CT-Scan, etc.). From this, the mechanical design of the anthropomorphic wind playing robots could be improved to enable the robot to produce a more natural sound. Furthermore, intelligent control system strategies will be tested and improved to enhance the musical expressiveness of the robot.

Regarding the development of automated machines, we expect that novel ways of expression will be introduced thanks to the simplicity of their mechanical design and the implementation of music engineering and artificial intelligence. The development of humanoid robots will certainly contribute not only to the better understanding of the mechanisms of human musical performance, but also in developing higher-skilled robots that can reproduce different kinds of human-skills (i.e. playing the flute as well as the saxophone, Musician-Humanoid Interaction (MHI), etc.).

Acknowledgments. Part of this research was done at the Humanoid Robotics Institute (HRI), Waseda University and at the Center for Advanced Biomedical Sciences (TWINs). This research is supported (in part) by a Gifu-in-Aid for the WABOT-HOUSE Project by Gifu Prefecture. This work is also supported (in part) by Global COE Program "Global Robot Academia" from the Ministry of Education, Culture, Sports, Science and Technology of Japan.

References

1. Adachi, S.: Principles of sound production in wind instruments. Acoustical Science and Technology 25(6), 400–405 (2004)
2. Ando, Y.: Drive conditions of the flute and their influence upon harmonic structure of generated tone. Journal of the Acoustical Society of Japan, 297–305 (1970) (in Japanese)
3. Bishop, C.M.: Neural Networks for Pattern Recognition, pp. 116–121. Oxford University Press, Great Britain (2004)
4. Doyon, A., Liaigre, L.: Jacques Vaucanson: Mecanicien de Genie. In: PUF (1966)
5. Gerhard, D.: Pitch extraction and fundamental frequency: History and current techniques. Technical Report TR-CS, Dept. of Computer Science, University of Regina, pp. 1–22 (2003)
6. Kato, I., Ohteru, S., Kobayashi, H., Shirai, K., Uchiyama, A.: Information-power machine with senses and limbs. In: Proc. of the CIS-IFToMM Symposium Theory and Practice of Robots and Manipulators, pp. 12–24 (1973)
7. Kawato, M., Gomi, H.: A computational model of four regions of the cerebellum based on feedback-error-learning. Biological Cybernetics 68, 95–103 (1992)
8. Kim, H., Kim, K., Young, M.: On-Line Dead-Time Compensation Method Based on Time Delay Control. IEEE Transactions on Control Systems Technology 11(2), 279–286 (2003)
9. Klaedefabrik, K.B.: Martin Riches - Maskinerne / The Machines, pp. 10–13. Kehrer Verlag (2005)
10. Mukai, S.: Laryngeal movement while playing wind instruments. In: Proceedings of International Symposium on Musical Acoustics, pp. 239–241 (1992)
11. Schumacher, R.T.: Ab Initio Calculations of the Oscillations of a Clarinet. Acustica 48(2), 71–85 (1981)
12. Solis, J., Chida, K., Suefuji, K., Takanishi, A.: The Development of the anthropomorphic flutist robot at Waseda University. International Journal of Humanoid Robots 3(2), 127–151 (2006)
13. Solis, J., Taniguchi, K., Ninomiya, T., Takanishi, A.: Understanding the Mechanisms of the Human Motor Control by Imitating Flute Playing with the Waseda Flutist Robot WF-4RIV. Mechanism and Machine Theory 44(3), 527–540 (2008)
14. Solis, J., Taniguchi, K., Ninomiya, T., Petersen, K., Yamamoto, T., Takanishi, A.: Implementation of an Auditory Feedback Control System on an Anthropomorphic Flutist Robot Inspired by the Performance of a Professional Flutist. Advanced Robotics Journal 23, 1849–1871 (2009)
15. Solis, J., Takeshi, N., Petersen, K., Takeuchi, M., Takanishi, A.: Development of the Anthropomorphic Saxophonist Robot WAS-1: Mechanical Design of the Simulated Organs and Implementation of Air Pressure. Advanced Robotics Journal 24, 629–650 (2009)

16. Solis, J., Petersen, K., Yamamoto, T., Takeuchi, M., Ishikawa, S., Takanishi, A., Hashimoto, K.: Design of New Mouth and Hand Mechanisms of the Anthropomorphic Saxophonist Robot and Implementation of an Air Pressure Feed-Forward Control with Dead-Time Compensation. In: Proceedings of the International Conference on Robotics and Automation, pp. 42–47 (2010)
17. Takashima, S., Miyawaki, T.: Control of an automatic performance robot of saxophone: performance control using standard MIDI files. In: Proc. of the IEEE/RSJ Int. Conference on Intelligent Robots and Systems - Workshop: Musical Performance Robots and Its Applications, pp. 30–35 (2006)
18. de Vaucanson, J.: Le Mécanisme du Fluteur Automate; An Account of the Mechanism of an Automation: or, Image Playing on the German-Flute. The Flute Library, First Series No. 5, F. Vester. (ed.) Intro. David Lasocki. Buren (GLD. Uit-geverij Frits Knuf, The Netherlands (1979)
19. Wood, G.: A Magical History of the Quest for Mechanical Life, Faber & Faber ed., p. 278 (2002)

Chapter 13
Multimodal Techniques for Human/Robot Interaction

Ajay Kapur

Abstract. Is it possible for a robot to improvise with a human performer in real-time? This chapter describes a framework for interdisciplinary research geared towards finding solutions to this question. Custom built controllers, influenced by the Human Computer Interaction (HCI) community, serve as new interfaces to gather musical gestures from a performing artist. Designs on how to modify a sitar, the 19-stringed North Indian string instrument, with sensors and electronics are described. Experiments using wearable sensors to capture ancillary gestures of a human performer are also included. A twelve-armed solenoid-based robotic drummer was built to perform on a variety of traditional percussion instruments from around India. The chapter describes experimentation on interfacing a human sitar performer with the robotic drummer. Experiments include automatic tempo tracking and accompaniment methods. This chapter shows contributions in the areas of musical gesture extraction, musical robotics and machine musicianship. However, one of the main novelties was completing the loop and fusing all three of these areas together into a real-time framework.

13.1 Introduction

There are many challenges in interfacing a human with a mechanical device controlled by a computer. Many methods have been proposed to address this problem, usually including sensor systems for human perception and simple robotic mechanisms for actuating the machine's physical response. Conducting this type of experiments in the realm of music is obviously challenging, but fascinating at the same time. This is facilitated by the fact that music is a language with traditional

Ajay Kapur
California Institute of the Arts, Valencia, California, USA;
New Zealand School of Music, Victoria University of Wellington, New Zealand,
24700 McBean Parkway, Valencia CA 91355, USA
e-mail: ajay@karmetik.com

J. Solis and K. Ng (Eds.): Musical Robots and Interactive Multimodal Systems, STAR 74, pp. 215–232.
springerlink.com

rules, which must be obeyed to constrain a machine's response. Therefore the evaluation of successful algorithms by scientists and engineers is feasible. More importantly, it is possible to extend the number crunching into a cultural exhibition, building a system that contains a novel form of artistic expression that can be used on stage.

The goal of this research is to make progress towards a system for combining human and robotic musical performance. We believe that in order for such a system to be successful it should combine ideas from the frequently separate research areas of music robotics, hyperinstruments, and machine musicianship.

The art of building musical robots has been explored and developed by musicians and scientists [1-6] . A recent review of the history of musical robots is described in Kapur [7]. The development of hyperinstruments plays a crucial role in obtaining data from the human's performance. Work and ideas by Machover [8] and Trueman [9] have greatly influenced the development of the interface and sensors described in this chapter. The area of machine musicianship is another part of the puzzle. Robert Rowe (who also coined the term machine musicianship) describes a computer system which can analyze, perform and compose music based on traditional music theory [10]. Other systems which have influenced the community in this domain are Dannenberg's score following system [11], George Lewis's Voyager [12], and Pachet's Continuator [13].

There are few systems that have closed the loop to create a real live human/robotic performance system. Audiences who experienced Mari Kimura's recital with the LEMUR GuitarBot [14] can testify to its effectiveness. Gil Weinberg's robotic drummer Haile [15] continues to grow in capabilities to interact with a live human percussionist [16]. Trimpins performance with Kronos Quartet as portrayed in the documentary "Trimpin: The Sound of Invention" was one of the greatest success stories in the field.

This chapter describes a human-robot performance system based on North Indian classical music, drawing theory from ancient tradition to guide aesthetic and design decisions. Section 13.2 describes the revamped hyperinstrument, known as the Electronic Sitar (*ESitar*). Section 13.3 describes the design of wearable sensors for multimodal gesture extraction. Section 13.4 describes the building of the robotic Indian drummer, known as *MahaDeviBot*. Section 13.5 describes experimentation and algorithms toward "intelligent" multimodal machine musicianship. Section 13.5 discusses the system used live in various musical performance scenarios. Section 13.6 contains conclusions and future work.

13.2 The Electronic Sitar

Sitar is *Saraswati's* (the Hindu Goddess of Music) 19-stringed, gourd shelled, traditional North Indian instrument. Its bulbous gourd (Figure 13.1), cut flat on the top, is joined to a long-necked, hollowed, concave stem that stretches three feet long and three inches wide. The typical sitar contains seven strings on the upper bridge, and twelve sympathetic strings below, all tuned by tuning pegs. The upper strings include rhythm and drone strings, known as *chikari*. Melodies, which are primarily performed on one of the upper-most strings, induce sympathetic resonant vibrations in the corresponding sympathetic strings below. The sitar can have up to 22 moveable frets, tuned to the notes of a *Raga* (the melodic mode, scale, order, and rules of a particular piece of Indian classical music) [18, 19]. The sitar is a very sophisticated and subtle instrument, which can create vocal effects with incredible depths of feeling, making it a challenging digital controller to create. The initial work on transforming the sitar into a hyperinstrument is described in [17], which serves as a source to gain a more detailed background on traditional sitar performance technique, the sitar's evolution with technology, and initial experimentation and design in building a controller out of a sitar.

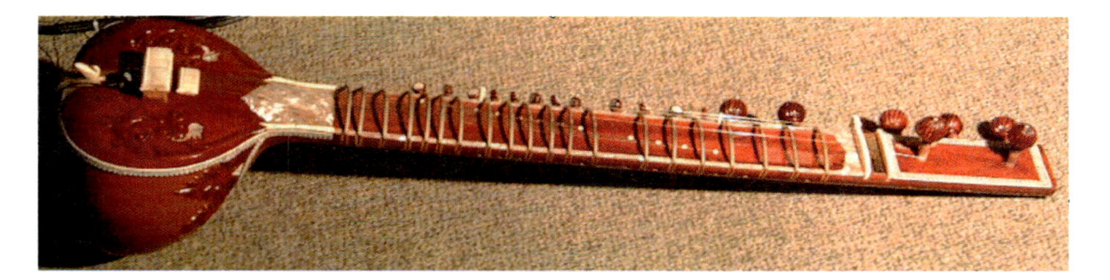

Fig. 13.1 A traditional Sitar.

13.2.1 Traditional Sitar Technique

It is important to understand the traditional playing style of the sitar to comprehend how our controller captures its hand gestures. It should be noted that there are two main styles of sitar technique: Ustad Vilayat Khan's system and Pandit Ravi Shankar's system. The main differences between the styles are that Ustad Vilayat Khan performs melodies on the higher octaves, eliminating the lowest string from the instrument, whereas Pandit Ravi Shankar's style has more range, and consequently melodies are performed in the lower octaves [20]. The *ESitar* is modeled on the Vilayat Khan's system or *gharana*.

A performer generally sits on the floor in a cross-legged fashion. Melodies are performed primarily on the outer main string, and occasionally on the copper string. The sitar player uses his left index finger and middle finger, as shown in Figure 13.2a, to press the string to the fret for the desired note. In general, a pair of frets are spaced a half-step apart, with the exception of a few that are spaced by a whole. The frets are elliptically curved so the string can be pulled downward, to bend to a higher note. This is how a performer incorporates the use of *shruti* (microtones).

On the right index finger, a sitar player wears a ring like plectrum, known as a *mizrab,* shown in Figure 13.2b. The right hand thumb remains securely on the edge of the *dand* (shaft of instrument) as shown on Figure 13.2c, as the entire right hand gets pulled up and down over the main seven strings, letting the *mizrab* strum the desired melody. An upward stroke is known as *Dha* and a downward stroke is known as *Ra* [19, 20].

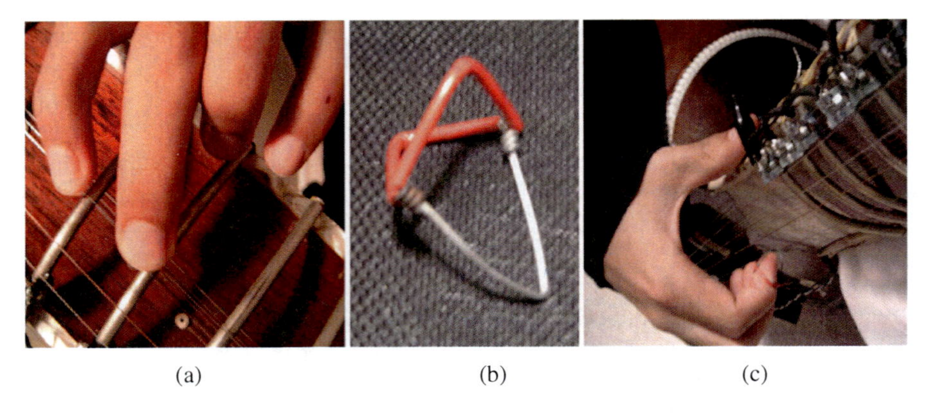

(a) (b) (c)

Fig. 13.2 Traditional Sitar Playing Technique.

13.2.2 The MIDI Sitar Controllers

With the goal of capturing a wide variety of gestural input data, the *ESitar* controller combines several different families of sensing technology and signal processing methods. Three *ESitar*'s were constructed: *ESitar* 1.0 in 2003, *ESitar* 2.0 in 2006, and ESitar 3.0 in 2009. The methods used in all three will be described including microcontroller platforms, different sensors systems and algorithms.

Each version of the ESitar used a different Microcontroller from the Atmel[1] AVR [22], PIC[2] microchip, and the Arudino MEGA[3]. The first *ESitar* was encased in a controller box as seen in Figure 13.3, with three switches, shaft encoders, and potentiometers used to trigger events, toggle between modes, and fine tune settings. The box also has an LCD to display controller data and settings to the performer, enabling him/her to be completely detached from the laptops running sound and graphic simulations. In the second version, a major improvement was encasing the microchip, power regulation, sensor conditioning circuits, and MIDI out device in a box that fits behind the tuning pegs on the sitar itself. This reduces the number of wires, equipment, and complication needed for each performance. This box also has two potentiometers, six momentary buttons, and four push buttons for triggering and setting musical parameters. In the third version, the entire microchip was embedded inside the sitar itself, with USB access and a ¼ inch jack for audio output.

[1] http://www.atmel.com/ (January 2010)
[2] http://www.microchip.com/ (January 2010)
[3] http://arduino.cc/en/Main/ArduinoBoardMega (Feb 2010)

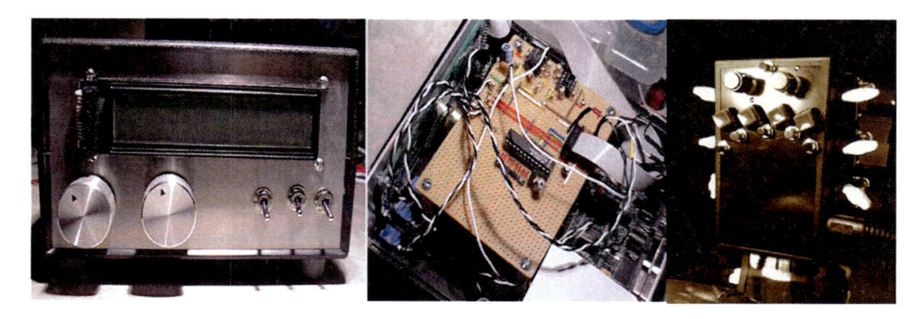

Fig. 13.3 Atmel Controller Box Encasement of *ESitar* 1.0 (left, middle). PIC Controller Box Encasement on *ESitar* 2.0 (right).

Gesture Capturing

The controller captures gesture data including the depressed fret number, thumb pressure, 3 axes of the performer's head tilt, and 3 axes of the sitar's tilt.

Fret Detection

The currently played fret is deduced using an exponentially distributed set of resistors which form a network interconnecting in series each of the frets on the *ESitar* (Figure 13.4). When the fingers of the left hand depress the string to touch a fret (as shown in Figure 13.2a), current flows through the string and the segment of the resistor network between the bottom and the played fret. The voltage drop across the in-circuit segment of the resistor network is digitized by the microcontroller. Using a lookup table it maps that value to a corresponding fret number and sends it out as a MIDI message. This design is inspired by Keith McMillan's Zeta Mirror 6 MIDI Guitar [23].

The *ESitar* used a modified resistor network for fret detection based on more experimentation. Military grade resistors at 1% tolerance were used in this new version for more accurate results. Soldering the resistors to the pre-drilled holes in the frets provided for a more reliable connection that does not have to be re-soldered at every sound check!

As mentioned above, the performer may pull the string downward, bending a pitch to a higher note (for example play a *Pa [5*th*]* from the *Ga [major 3*rd*]* fret). To capture this additional information that is independent of the played fret, we fitted the instrument with a piezo pick-up whose output was fed into a pitch detector. For initial experiments, the pitch detector was implemented in a *pure data* [24] external object using an auto-correlation based method [25]. The pitch detection is bounded below by the pitch of the currently played fret and allows a range of eight semi-tones above. This system was later ported to a real time system in *ChucK* programming language.

Fig. 13.4 The network of resistors on the frets of the *ESitar* 1.0 (left, middle). The *ESitar* 2.0 full body view (right).

Mizrab Pluck Direction

We are able to deduce the direction of a *mizrab* stroke using a force sensing resistor (FSR), which is placed directly under the right hand thumb, as shown in Figure 13.5. As mentioned before, the thumb never moves from this position while playing. However, the applied force varies based on *mizrab* stroke direction. A *Dha* stroke (upward stroke) produces more pressure on the thumb than a *Ra* stroke (downward stroke). We send a continuous stream of data from the FSR via MIDI, because this data is rhythmic in time and can be used compositionally for more then just deducing pluck direction.

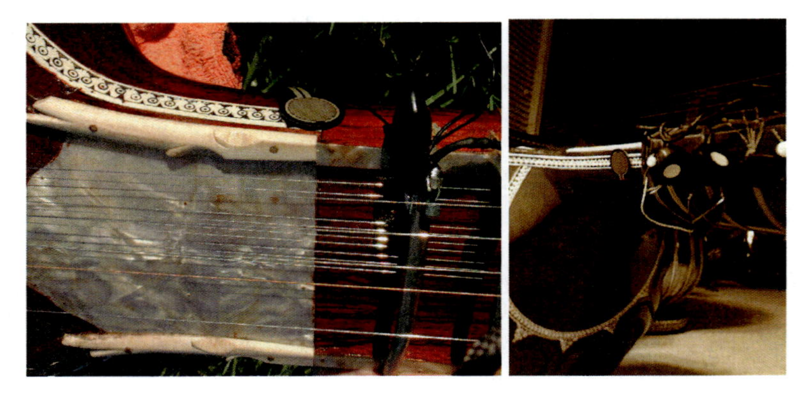

Fig. 13.5 FSR sensor used to measure thumb pressure on *ESitar* 1.0 (left) and *ESitar* 2.0 (right).

3-axes Sitar Tilt

In the *ESitar* 2.0, there is a 3-axis accelerometer embedded in the controller box at the top of the neck, to capture ancillary sitar movement, as well as serve as yet another means to control synthesis and audio effect parameters. This sensor can be used to derive data for performer's posture with their instrument, as well as intricacies about playing technique such as jerk detection to help evaluate the beginning and end of melodic phrasing. In *ESitar 3.0*, the accelerometer is embedded inside the body of the instrument itself, achieving similar results.

13.3 Wearable Sensors

The motion of the human body is a rich source of information, containing intricacies of musical performance which can aid in obtaining knowledge about intention and emotion through human interaction with an instrument. Proper posture is also important in music performance for musician sustainability and virtuosity. Building systems that could aid as pedagogical tools for training with correct posture is useful for beginners and even masters. This section explores a variety of techniques for obtaining data from a performing artist by placing sensors on the human body. There are two ways in which we have tried to gather data from the human musician: (1) a wearable sensor package to obtain acceleration data, (2) a wireless sensor package system that obtains orientation data.

The design of the *KiOm* [28] (Figure 13.6) was built using a Kionix KXM52-1050[4] three-axis accelerometer. The three streams of analog gesture data from the sensor is read by the internal ADC of the Microchip PIC[5]. These streams are converted to MIDI messages for capturing gesture signals.

Fig. 13.6 The *KiOm* Circuit Boards and Encasement (left, middle). Wireless Inertial Sensor Package (*WISP*) (right).

Our paradigm is to keep traditional instrument performance technique, while capturing both the amplified acoustic signal and the gesture sensor data. A similar paradigm is that of the hyperinstrument [8, 17] where an acoustic instrument is augmented with sensors. In our approach, any performer can wear a low-cost sensor while keeping the acoustic instrument unmodified, allowing a more accessible and flexible system. This technique is particularly useful in sitar performance. A *KiOm* is attached to the head of the sitar player as an easy way to control and trigger different events in the performance [26]. This would be a useful addition to almost any controller as a replacement for foot pedals, buttons, or knobs. It is particularly useful in this system as a sitar player's hands are always busy, and cannot use his/her feet due to the seated posture.

The Wireless Inertial Sensor Package (*WISP*) [29], designed by Bernie Till and the Assistive Technology Team at University of Victoria, is a miniature Inertial Measurement Unit (IMU) specifically designed for the task of capturing human body movements. It can equally well be used to measure the spatial orientation of

[4] http://www.kionix.com/ (February 2005)
[5] http://www.microchip.com/ (February 2005)

any kind of object to which it may be attached. Thus the data from the *WISP* provides an intuitive method to gather data from a musical performer. The *KiOm*'s has the disadvantage of being heavy and having wires that connect to the computer, certainly putting constraints on a musician. With the wireless *WISP*, the performer is free to move within a radius of about 50m with no other restrictions imposed by the technology such as weight or wiring. The *KiOm* and *WISP* are used in multimodal experiments discussed later in the chapter.

13.4 MahaDeviBot – Robotic Percussionist

Mechanical systems for musical expression have developed since the 19th Century. Before the phonogram, player pianos and other automated devices were the only means of listening to compositions, without the presence of live musicians. The invention of audio recording tools eliminated the necessity and progression of these types of instruments. In modern times, with the invention of the microcontroller and inexpensive electronic actuators, mechanical music is being revisited by many scholars and artists.

Musical robots have come at a time when tape pieces and laptop performances have left some in the computer music audiences wanting more interaction and physical movement from the performers [31]. The research in developing new interfaces for musical expression continues to bloom as the community is now beginning to focus on how actuators can be used to raise the bar even higher, creating new mediums for creative expression. Robotic systems can perform tasks not achievable by a human musician. Speakers, no matter how many directions they point, can never replace the sound of a bell being struck on stage with its acoustic resonances with the concert hall. The use of robotic systems as pedagogical implements is also proving to be significant. Indian classical students practice to a Tabla box with pre-recorded drum loops. The use of robotic strikers, performing real acoustic drums gives the students a more realistic paradigm for concentrated rehearsal.

The development of the *MahaDeviBot* as a paradigm for various types of solenoid-based robotic drumming is described. The *MahaDeviBot* serves as a mechanical musical instrument that extends North Indian musical performance scenarios, while serving as a pedagogical tool to keep time and help portray complex rhythmic cycles to novice performers in a way that no audio speakers can ever emulate. This section describes the design strategies for the *MahaDeviBot*, including four different methods for using solenoids for rhythmic events. There are four different designs proposed, and appropriately named by the inventor: Kapur Fingers, Singer Hammer, Trimpin Hammer and Trimpin BellHop are described.

13.4.1 Design

Kapur Fingers

The Kapur Fingers involve modifications of a push solenoid. One issue with the off-the-shelf versions of the solenoids is that during use they click against themselves making lots of mechanical sound. A key goal for a successful music robotic system is to reduce the noise of its parts so it does not interfere with the desired musical sound. Thus the push solenoids were reconfigured to reduce noise. The shaft and inner tubing were buffed with a wire spinning mesh using a *Dremel* hand power tool. Then protective foam was placed toward the top of the shaft to stop downward bounce clicking. Rubber grommets were attached in order to prevent upward bounce-back clicking (Figure 13.7). The grommets were also used to simulate the softness of the human skin when striking the drum as well as to protect the drum skin.

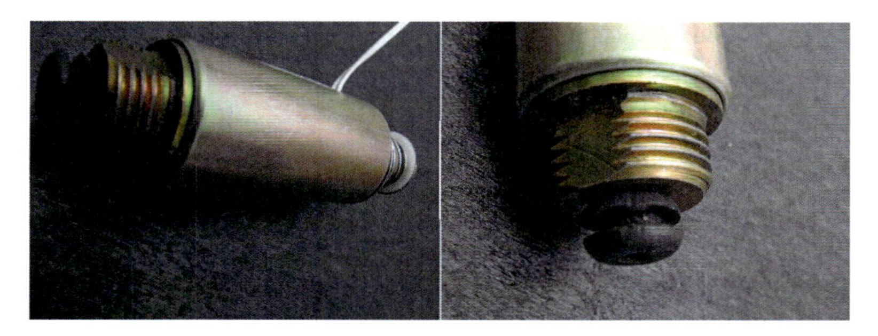

Fig. 13.7 Kapur Finger using a grommet and padding.

Singer Hammer

The Singer Hammer is a modified version of the Eric Singer's *ModBot* [32]. The mechanism strikes a drum using a steel rod and ball (Figure 13.8). A pull solenoid is used to lift a block to which the rod is attached. A ball joint system was added to connect the solenoid to the bar for security and reliability of strokes. The trade-off was that it added some mechanical noise to the system. The *MahaDeviBot* has four Singer Hammers striking a variety of frame drums.

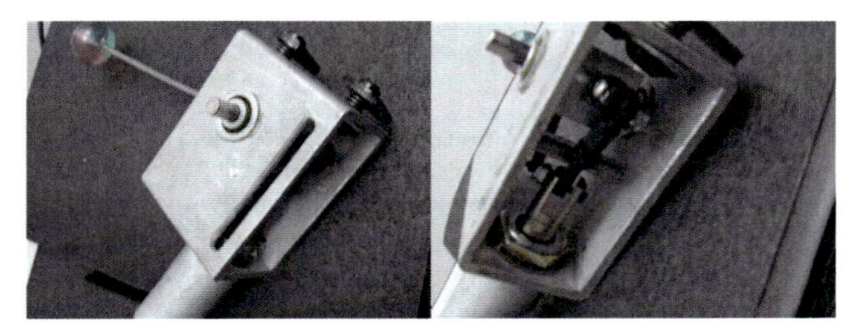

Fig. 13.8 Singer Hammer with added ball-joint striking mechanism.

Trimpin Hammer

The Trimpin Hammer is a modified version of Trimpin's variety of percussion instruments invented over the last 20 years [33]. Its key parts include female and male rod ends, and shaft collars. This is a very robust system that involves using a lathe to tap the shaft of the solenoid so a male rod end can be secured. This is a mechanically quiet device, especially with the added plastic stopper to catch the hammer on the recoil. These devices are used to strike frame drums, gongs, and even bells as shown in Figure 13.9.

Fig. 13.9 Trimpin Hammer modified to fit the *MahaDeviBot* schematic.

Trimpin BellHop

The Trimpin BellHop is a modified version of technology designed for Trimpin's ColoninPurple, where thirty such devices were used to perform modified xylophones suspended from the ceiling of a gallery. These are made by modifying a pull solenoid by extending the inner tubing so that the shaft can be flipped upside down and triggered to hop out of the front edge and strike a xylophone or Indian bell (as shown in Figure 13.10). These, too, are mechanically quiet and robust.

Fig. 13.10 Trimpin BellHop outside shell tubing (left) and inside extended tubing (middle). Trimpin BellHops used on MahaDeviBot (right).

Head

The headpiece of the *MahaDeviBot* is a robotic head that can bounce up and down at a given tempo (Figure 13.11). This was made using a pull solenoid attached to a pipe. Two masks are attached to either side and recycled computer parts from ten-year old machines that have no use in our laboratories anymore visualize the brain. In performance with a human musician, the head serves as a visual feedback cue to inform the human of the machine-perceived tempo at a given instance in time.

Fig. 13.11 The bouncing head of *MahaDeviBot (left)*. *MahaDeviBot* hanging in REDCAT Walt Disney Concert Hall, Los Angeles 2010 (right).

This section described in detail the various design strategies used to build the final version of the robotic Indian drummer. As an example of how these tradeoffs can influence robotic design for musical performance, the four designs are integrated into *MahaDeviBot* in the following ways: The Kapur Fingers are added to a drum with the Singer Hammer to allow large dynamic range and quick rolls from one frame drum. The Trimpin Hammer is used to perform drum rolls and is used for robotic Tabla performance. The Trimpin BellHop is used to strike bells and other instruments where volume is important and which will not be struck at high rates. A solo melodic artist can now tour the world with a robotic drummer to accompany if software is "intelligent" enough to keep the interest of the audience. The next section discusses our pursuits in this direction.

13.5 Machine Musicianship

The "intelligence" of interactive multimedia systems of the future will rely on capturing data from humans using multimodal systems incorporating a variety of environmental sensors. Research on obtaining accurate perception about human action is crucial in building "intelligent" machine response. This section describes experiments testing the accuracy of machine perception in the context of music performance. The goal of this work is to develop an effective system for human-robot music interaction. We look at two methods in this section: (1) Multimodal Tempo Tracking and (2) Multimodal Rhythmic Accompaniment.

13.5.1 Multimodal Tempo Tracking

This section describes a multimodal sensor capturing system for traditional sitar performance. As described before, sensors for extracting performance information are placed on the instrument. In addition wearable sensors are placed on the human performer. The MahaDeviBot is used to accompany the sitar player. In this section, we ask the question: How does one make a robot perform in tempo with the human sitar player?

Analysis of accuracy of various methods of achieving this goal is presented. For each signal (sensors and audio) we extract onsets that are subsequently processed by Kalman filtering [34] for tempo tracking [35]. Late fusion of the tempo estimates is shown to be superior to using each signal individually. The final result is a real-time system with a robotic drummer changing tempo with the sitar performer in real-time.

The goal of this section is to improve tempo tracking in human-machine interaction. Tempo is one of the most important elements of music performance and there has been extensive work in automatic tempo tracking on audio signals [36]. We extend this work by incorporating information from sensors in addition to the audio signal. Without effective real-time tempo tracking, human-machine performance has to rely on a fixed beat, making it sound dry and artificial. The area of machine musicianship is the computer music communities' term for machine perception. Our system evolves the state-of-the-art different as it involves a multimodal sensor design to obtain improved accuracy for machine perception.

There are four major processing stages in our system. A block diagram of the system is shown in Figure 13.12. In the following subsection we describe each processing stage from left to right. In the acquisition stage performance information is collected using audio capture, two sensors on the instrument and a wearable sensor on the performer's body. Onsets for each separate signal are detected after some initial signal conditioning. The onsets are used as input to four Kalman filters used for tempo tracking. The estimated beat periods for each signal are finally fused to provide a single estimate of the tempo.

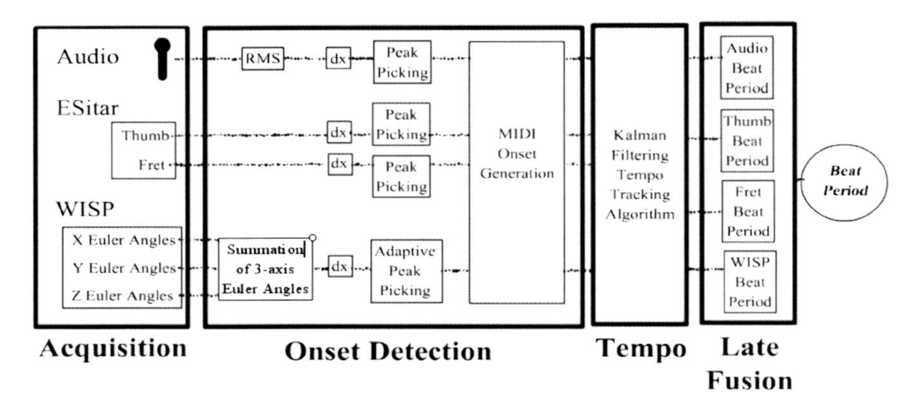

Fig. 13.12 Multimodal Sensors for Sitar Performance Perception.

For our experiments we recorded a data set of a performer playing the *ESitar* with a *WISP* on the right hand. Audio files were captured at a sampling rate of 44100 Hz. Thumb pressure and fret sensor data synchronized with audio analysis windows were recorded with *Marsyas* at a sampling rate of 44100/512 Hz using MIDI streams from the *ESitar*. Orientation data for the Open Sound Control (OSC) [30] streams of the *WISP* were also recorded.

While playing, the performer listened to a constant tempo metronome through headphones. 104 trials were recorded, with each trial lasting 30 seconds. Trials were evenly split into 80, 100, 120, and 140 BPM, using the metronome connected to the headphones. The performer would begin each trial by playing a scale at a quarter note tempo, and then a second time at double the tempo. The rest of the trial was an improvised session in tempo with the metronome.

Figure 13.13 shows the percentages of frames for which the tempo was correctly estimated. Tempo estimates are generated at 86Hz resulting in approximately 2600 estimates/30 second clip in the dataset. From the percentages of Fig. 13.13, we can conclude that when using a single acquisition method, the *WISP* obtained the best results at slower tempos, and the audio signal was best for faster tempos. Overall, the audio signal performed the best as a single input, whereas the fret data provided the least accurate information.

When looking carefully through the detected onsets from the different types of acquisition methods, we observed that they exhibit outliers and discontinuities at times. To address this problem we utilize a late fusion approach where we consider each acquisition method in turn for discontinuities. If a discontinuity is found, we consider the next acquisition method, and repeat the process until either a smooth estimate is obtained or all acquisition methods have been exhausted. When performing late fusion the acquisition methods are considered in the order listed on bottom half of Figure 13.13.

Signal	Tempo (BPM)			
	80	100	120	140
Audio	46%	85%	86%	80%
Fret	27%	27%	57%	56%
Thumb	35%	62%	75%	65%
WISP	50%	91%	69%	53%
LATE FUSION:				
Audio/*WISP*/Thumb/Fret	45%	83%	89%	84%
Audio/*WISP*/Thumb	55%	88%	90%	82%
Audio/ *WISP*	58%	88%	89%	72%
Audio/Thumb	57%	88%	90%	80%
WISP/Thumb	47%	95%	78%	69%

Fig. 13.13 Comparison of Acquisition Methods.

By fusing the acquisition methods together, we are able to get more accurate results. At 80 BPM, by fusing the information from *WISP* and the audio streams, the algorithm generates more accurate results then either signal on its own. When all the sensors are used together, the most accurate results are achieved at 140 BPM, proving that even the fret data can improve accuracy of tempo estimation. Overall, the information from the audio fused with the thumb sensor was the strongest.

13.5.2 Rhythmic Accompaniment

Traditional Indian Classical music has been taught with the aid of an electronic Tabla box, where students can practice along with pre-recorded *theka loops* (rhythmic cycles). This allows the performer to select any time cycle and rehearse at a variable tempo. The main problem with this system is that one beat repeats over and over again, which is boring and not realistic for a true performance situation. This motivated the work explored in this section of generating an interactive rhythm accompaniment system that would evolve based on human input. We will present a software framework to generate machine driven performance using a database structure for storing "memory" of "what to perform". This application introduces a new level of control and functionality to the modern North Indian musician with a variety of flexible capabilities, and a new performance scenarios using custom written software designed to interface the *ESitar* with *MahaDeviBot*.

The Music Information Retrieval (MIR) community[6] inspired our initial framework and experimentation for this approach. The goal of this system is to generate a variety of rhythmic accompaniment that evolves over time based on human performance by using sensors to query databases of pre-composed beats. To achieve this, symbolic event databases (shown in Figure 13.14) for each robotic instrument were filled with rhythmic phrases and variations. During performance, at any given time, queries are generated by sensor data captured from the human performer. As this software is written in *ChucK* [27], it was easy for the databases to be time and tempo locked to each other to allow for multiple permutations and combinations of rhythm. Figure 13.14 shows an example of how the system can be mapped. In this case, thumb pressure from the *ESitar* queries what rhythm robotic instrument 1 (*Dha* strokes) will mechanically play on the low frame drum. It is possible to generate a large number of combinations and permutations of rhythms by accessing patterns in each database. This proved to be a successful technique for performances on stage[7].

---

[6] http://www.ismir.net/ (January 2007)
[7] Videos Available at: http://www.karmetik.com (Technology → Robotics Department)

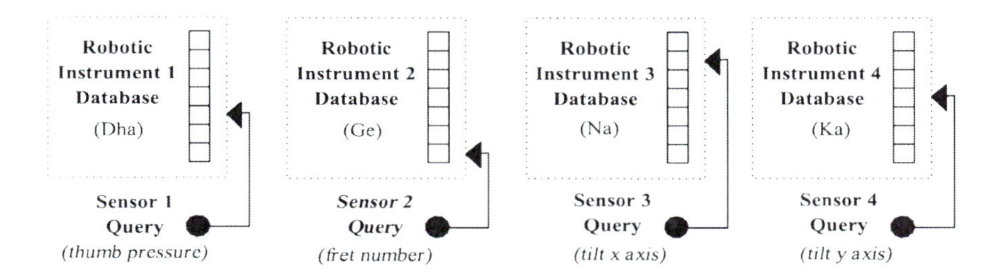

Fig. 13.14 Symbolic MIR-based approach showing how *ESitar* sensors are used as queries to multiple robotic drum rhythm databases.

One issue to address is how the queries are generated. In order to provide additional information for queries the derivatives and second derivatives of each sensor data stream are also utilized. Also there are more advanced feature extraction methods, for example obtaining interonset interval values between peaks of the thumb pressure data. There are many algorithms that can be explored; however, the main philosophical question is whether the human should have full control of the machine's performance (Figure 13.15).

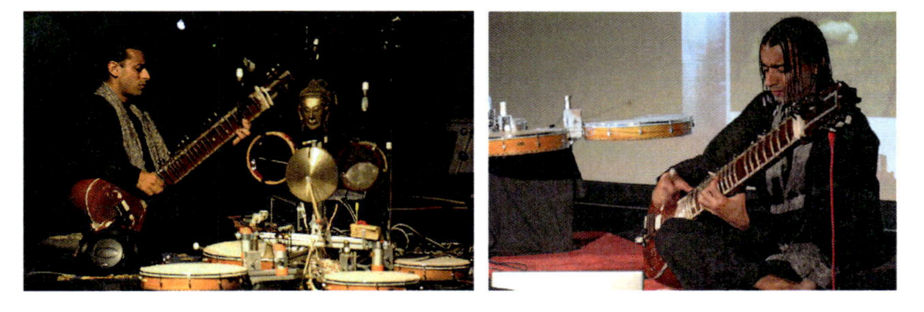

Fig. 13.15 *MahaDeviBot* being controlled with the *ESitar* at NUS Arts Festival in Singapore and NIME 2007 in New York City.

13.6 Conclusion and Future Work

The body of work described in this chapter was truly an artistic venture calling on knowledge from a variety of engineering disciplines, musical traditions, and philosophical practices. It also called on the collaborations and expertise of many professionals, professors and students. The goal of the work was to preserve and extend North Indian musical performance using state of the art technology including multimodal sensor systems, machine learning and robotics. The process of achieving our goal involved strong laboratory practice with regimented experiments with large data sets, as well as a series of concert performances showing how the technology can be used on stage to make new music, extending the tradition of Hindustani music.

Overall, our main motivation in using a robotic system was our discontent for hearing rhythmic accompaniment through a set of speakers. In is our opinion, triggering pre-recorded samples does not have enough expressiveness, and physical models to-date do not sound real enough. However, having a machine perform acoustic instruments using mechanical parts has its disadvantages. We must tune the machines' instruments for every show, which would not be necessary if we were just triggering "perfect" samples. Also, because of the nature of any mechanical system, there are imperfections in event timings based on varying spring tension, speed and strength of previous strikes. However, this produces more realistic rhythms, as humans also have imperfections when actually "grooving".

Our experimentation with robotic systems for musical performance brought many familiar yet new challenges to working with sensors. A set of allen wrenches, screw drivers, plyers, a Calliper and a Dremel are carted to each performance along with a box set of extra springs, screws, washers, and spare parts. Our first designs had frameworks made of wood. This obviously is too heavy a material, and using aluminum is ideal because of its sturdiness and lightweight. However, we learned from our initial prototypes that welding anything would be a mistake. All parts should be completely modular to allow for changes in the future. Thus designing our robots out of 20/20 T-slotted aluminum was a perfect material to accomplish all our goals of sustainability, modularity, mobility and professional appearance.

This research lays the ground work for future work in building a Machine Orchestra, with multiple robotic instruments on stage performing with multiple electronic musicians using multimodal sensor enhanced instruments. Algorithms will need to generalized to work with a variety of interfaces and musicians and robots will all need to be networked together in order to communicate with one another.

It is our belief that eventually, as ubiquitous computing prevails, every musical instrument will be enhanced with sensors and every musician will use a computer to amplify or manipulate their sound, and have automatic analysis of their real-time performance. Though musical robotics as presented in this chapter have a very particular aesthetic motivation, we do believe that more common entertainment robotic systems will become omnipresent for moving speakers, microphones, lights, projections and even digital displays for reading music.

References

1. Trimpin.: SoundSculptures: Five Examples. Munich MGM MediaGruppe, Munchen (2000)
2. Solis, J., Bergamasco, M., Isoda, S., Chida, K., Takanishi, A.: Learning to Play the Flute with an Anthropomorphic Robot. In: International Computer Music Conference, Miami, Florida (2004)
3. Jorda, S.: Afasia: the Ultimate Homeric One-Man-Multimedia-Band. In: International Conference on New Interfaces for Musical Expression, Dublin, Ireland (2002)
4. MacMurtie, C.: Amorphic Robot Works
5. Baginsky, N.A.: The Three Sirens: A Self Learning Robotic Rock Band
6. Raes, G.W.: Automations by Godfried-Willem Raes

7. Kapur, A.: A History of Robotic Musical Instruments. In: International Computer Music Conference, Barcelona, Spain (2005)
8. Machover, T., Chung, J.: Hyperinstruments: Musically Intelligent and Interactive Performance and Creativity Systems. In: International Computer Music Conference, pp. 186–190 (1989)
9. Trueman, D., Cook, P.R.: BoSSA: The Deconstructed Violin Reconstructed. In: Proceedings of the International Computer Music Conference, Beijing, China (1999)
10. Rowe, R.: Machine Musicianship. MIT Press, Cambridge (2004)
11. Dannenberg, R.B.: An On-line Algorithm for Real-Time Accompaniment. In: International Computer Music Conference, Paris, France, pp. 193–198 (1984)
12. Lewis, G.: Too Many Notes: Computers, Complexity and Culture in Voyager. Leonardo Music Journal 10, 33–39 (2000)
13. Pachet, F.: The Continuator: Musical interaction with Style. In: International Computer Music Conference, Goteborg, Sweden (2002)
14. Singer, E., Larke, K., Bianciardi, D.: LEMUR GuitarBot: MIDI Robotic String Instrument. In: International Conference on New Interfaces for Musical Expression, Montreal, Canada (2003)
15. Weinberg, G., Driscoll, S., Parry, M.: Haile - A Preceptual Robotic Percussionist. In: International Computer Music Conference, Barcelona, Spain (2005)
16. Weinberg, G., Driscoll, S., Thatcher, T.: Jam'aa - A Middle Eastern Percussion Ensemble for Human and Robotic Players. In: International Computer Music Conference, New Orleans, pp. 464–467 (2006)
17. Kapur, A., Lazier, A., Davidson, P., Wilson, R.S., Cook, P.R.: The Electronic Sitar Controller. In: International Conference on New Interfaces for Musical Expression, Hamamatsu, Japan (2004)
18. Menon, R.R.: Discovering Indian Music. Somaiya Publications Pvt. Ltd, Mumbai (1974)
19. Vir, R.A.: Learn to Play on Sitar. Punjab Publications, New Delhi (1998)
20. Bagchee, S.: NAD: Understanding Raga Music. Ceshwar Business Publications, Inc, Mumbai (1998)
21. Sharma, S.: Comparative Study of Evolution of Music in India & the West. Pratibha Prakashan, New Delhi (1997)
22. Wilson, R.S., Gurevich, M., Verplank, B., Stang, P.: Microcontrollers in Music Education - Reflections on our Switch to the Atmel AVR Platform. In: International Conference on New Interfaces for Musical Expression, Montreal, Canada (2003)
23. Wait, B.: Mirror-6 MIDI Guitar Controller Owner's Manual. Zeta Music Systems, Inc., Oakland (1989)
24. Puckette, M.: Pure Data: Another Integrated Computer Music Environment. In: Second Intercollege Computer Music Concerts, Tachikawa, Japan, pp. 37–41 (1996)
25. Zolzer, U.: DAFX: Digital Audio Effects. John Wiley and Sons, Ltd. England (2002)
26. Merrill, D.: Head-Tracking for Gestural and Continuous Control of Parameterized Audio Effects. In: International Conference on New Interfaces for Musical Expression, Montreal, Canada (2003)
27. Wang, G., Cook, P.R.: ChucK: A Concurrent, On-the-fly Audio Programming Language. In: International Computer Music Conference, Singapore (2003)
28. Kapur, A., Yang, E.L., Tindale, A.R., Driessen, P.F.: Wearable Sensors for Real-Time Musical Signal Processing. In: IEEE Pacific Rim Conference, Victoria, Canada (2005)

29. Till, B.C., Benning, M.S., Livingston, N.: Wireless Inertial Sensor Package (WISP). In: International Conference on New Interfaces for Musical Expression, New Yark City (2007)
30. Wright, M., Freed, A., Momeni, A.: OpenSound Control: State of the Art 2003. In: International Conference on New Interfaces for Musical Expression, Montreal, Canada (2003)
31. Schloss, A.W.: Using Contemporary Technology in Live Performance: The Dilemma of the Performer. Journal of New Music Research 32, 239–242 (2003)
32. Singer, E., Feddersen, J., Redmon, C., Bowen, B.: LEMUR's Musical Robots. In: International Conference on New Interfaces for Musical Expression, Hamamatsu, Japan (2004)
33. Trimpin, Portfolio, Seattle, Washington
34. Brown, R.G., Hwang, P.Y.C.: Introduction of Random Signals and Applied Kalman Filtering. John Wiley & Sons Inc., Chichester (1992)
35. Kasteren T.V.: Realtime Tempo Tracking using Kalman Filtering. In: Computer Science. University of Amsterdam, Masters Amsterdam (2006)
36. Gouyon, F., Klapuri, A., Dixon, S., Alonso, M., Tzanetakis, G., Uhle, C., Cano, P.: An Experimental Comparison of Audio Tempo Induction Algorithms. IEEE Transactions on Speech and Audio Processing 14 (2006)

Chapter 14
Interactive Improvisation with a Robotic Marimba Player

Guy Hoffman and Gil Weinberg

Abstract. Shimon is an improvisational robotic marimba player that listens to human co-players and responds musically and choreographically based on analysis of musical input. The paper discusses the robot's mechanical and motion control and presents a novel interactive improvisation system based on the notion of physical gestures. Our system uses anticipatory action to enable real-time improvised synchronization with the human player. It was implemented on a full-length human-robot Jazz duet, displaying coordinated melodic and rhythmic human-robot joint improvisation. We also describe a study evaluating the effect of visual cues and embodiment on one of our call-and-response improvisation module. Our findings indicate that synchronization is aided by visual contact when uncertainty is high. We find that visual coordination is more effective for synchronization in slow sequences compared to faster sequences, and that occluded physical presence may be less effective than audio-only note generation.

14.1 Introduction

This paper describes Shimon, an interactive robotic marimba player. Shimon improvises in real-time while listening to, and building upon, a human pianist's performance. We have built Shimon as a new research platform for Robotic Musicianship (RM). As part of this research, we use the robot to evaluate some of the core claims of RM. In particular, we test the effects of embodiment, visual contact, and acoustic sound on musical synchronization and audience appreciation.

Guy Hoffman
Interdisciplinary Center Herzliya, P.O.Box 167, Herzliya 46150, Israel
e-mail: hoffman@idc.ac.il

Gil Weinberg
Georgia Institute of Technology, 840 McMillan Street, Atlanta, GA, USA
e-mail: gil.weinberg@coa.gatech.edu

J. Solis and K. Ng (Eds.): Musical Robots and Interactive Multimodal Systems, STAR 74, pp. 233–251.
springerlink.com

We also introduce a novel robotic improvisation system. Our system uses a gesture-centric framework, based on the belief that musicianship is not merely a sequence of notes, but choreography of movements. These movements result in musical sounds, but also perform visually and communicatively with other band-members and the audience. Our system was implemented on a full-length human-robot Jazz duet, displaying highly coordinated melodic and rhythmic human-robot joint improvisation. We have performed with the system in front of a live public audience.

14.1.1 Robotic Musicianship

We define RM to extend both the tradition of computer-supported interactive music systems, and that of music-playing robotics [19]. Most computer-supported interactive music systems are hampered by not providing players and audiences with physical cues that are essential for creating expressive musical interactions. For example, in humans, motion size often corresponds to loudness, and gesture location to pitch. These cues provide visual feedback and help players anticipate and coordinate their playing. They also create a more engaging experience for the audience by providing a visual connection to the sound. Most computer-supported interactive music systems are also limited by the electronic reproduction and amplification of sound through speakers, which cannot fully capture the richness of acoustic sound [15]. On the other hand, research in musical robotics focuses mostly on the physics of sound production, and rarely addresses perceptual and interactive aspects of musicianship, such as listening, analysis, improvisation, or interaction. Most such devices can be classified into two groups: robotic musical instruments, which are mechanical constructions that can be played by live musicians or triggered by pre-recorded sequences [3, 16]; or anthropomorphic musical robots that attempt to imitate human musicians [17, 18]. Some systems use the human's performance as a user-interface to the robot's performance [14]; and only a few attempts have been made to develop perceptual, interactive robots that are controlled by autonomous methods [1]. In contrast, in previous work, we have developed a perceptual and improvisatory robotic musician in the form of Haile, a robotic drummer [19]. However, Haile's instrumental range was percussive and not melodic, and its motion range was limited to a small space relative to the robot's body, hindering the effectiveness of visual cues. We have addressed these limitations with Shimon, presented here, a robot that plays a melodic instrument—a marimba—and does so by covering a larger and visible range of movement [20]. We build on these traits, developing an expressive motion-control system as well as a gesture-based improvisation framework, as described in this paper.

14.2 Physical Structure

Several considerations informed the physical design of Shimon: we wanted large movements for visibility, as well as fast movements for virtuosity. Another goal was to allow for a wide range of sequential and simultaneous note combinations. The resulting design was a combination of fast, long-range, linear actuators, and two sets of rapid parallel solenoids, split over both registers of the instrument.

Fig. 14.1 Overall view of the robotic marimba player Shimon.

The physical robot is comprised of four arms, each actuated by a voice-coil linear actuator at its base, and running along a shared rail, in parallel to the marimba's long side. The robot's trajectory covers the marimba's full 4 octaves (Fig 14.1). The linear actuators are based on a commercial product by IAI and are controlled by a SCON trajectory controller. They can reach an acceleration of 3g, and—at top speed—move at approximately one octave per 0.25 seconds. The arms are custom-made aluminum shells housing two rotational solenoids each. The solenoids control mallets, chosen with an appropriate softness to fit the area of the marimba that they are most likely to hit. Each arm contains one mallet for the bottom-row ("white") keys, and one for the top-row ("black") keys. Shimon was designed in collaboration with Roberto Aimi of Alium Labs.

14.3 Motor Control

A standard approach for musical robots is to handle a stream of MIDI notes and translate them into actuator movements that produce those notes. In Shimon's case, this would mean a note being converted into a slider movement and a subsequent mallet strike. Two drawbacks of this method are (a) an inevitable delay between activation and note production, hampering truly synchronous joint musicianship, and (b) not allowing for expressive control of gesture-choreography, including tonal and silent gestures. We have therefore separated the control for the mallets and the sliders to enable more artistic freedom in the generation of musical and choreographic gestures, without compromising immediacy and safety. This section describes the two control systems designed for safe artistic expression.

14.3.1 Mallets

The mallets are struck using rotational solenoids responding to a MIDI interface. Eight MIDI notes are mapped to the eight mallets, and the MIDI NOTE_ON and NOTE_OFF messages are used to activate and deactivate the solenoid. Given this electro-mechanical setup, we want to be able to achieve a large dynamic range of striking intensities. We also want to be able to strike repeatedly at a high note rate. Since we can only control the solenoids in an on/off fashion, the striking intensity is a function of two parameters: (a) the velocity gained from the distance traveled; and (b) the length of time the mallet is held on the marimba key. We therefore need to maintain a model of the mallet position for each striker. In order to do so, we have empirically sampled sound intensity profiles for different solenoid activation lengths, and used those to build a model for each striker. This model includes four parameters: (i) the mean travel time from the rest position to contact with the key; (ii) the mean travel time from the down position back to the rest position; (iii) the hold duration that results in the highest intensity note for that particular mallet; and (iv) the duty cycle that results in the highest intensity note for that mallet, when it starts from the resting position.

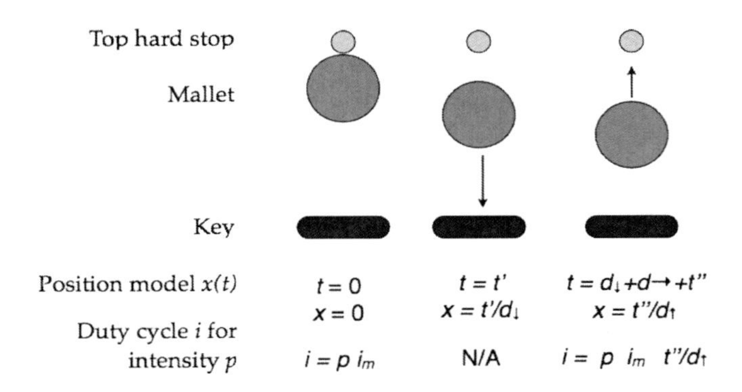

Fig. 14.2 Duty-Cycle computation based on mallet position model.

Using this model, each of the eight mallet control modules translates a combination of desired auditory intensity and time of impact into a solenoid duty cycle. Intuitively—the lower a mallet is at request time, the shorter the duty cycle needs to be to achieve impact, and to prevent muting of the key through a prolonged holding time. An estimated position is thus dynamically maintained based on the triggered solenoid commands, and the empirical mallet model (Fig. 14.2). The described system results a high level of musical expressivity, since it (a) maintains a finely adjustable dynamic striking range, and (b) allows for high-frequency repetitions for the same mallet, during which the mallet does not travel all the way up to the resting position.

14.3.2 Sliders

The horizontally moving sliders are four linear carriages sharing a rail and actuated through voice coil actuators under acceleration- and velocity-limited trapezoid control. There are two issues with this control approach. (a) a mechanical and non-expressive ("robotic", so to speak) movement quality associated with the standard fire-and-forget motion control approach, and (b) collision-avoidance, since all four arms share one rail. To tackle these issues, we chose to take an animation approach to the gesture control. Based on our experience with other expressive robots [5, 6] we use a high-frequency controller that updates the absolute position of each slider at a given frame rate. This controller is fed position data for all four arms at a lower frequency, based on higher-level movement considerations. This approach has two main advantages: (a) for each of the robotic arms, we are able to generate a more expressive spatio-temporal trajectory than just a trapezoid, as well as add animation principles such as ease-in, ease-out, anticipation, and follow-through [11]; and (b) since the position of the sliders is continuously controlled, collisions can be avoided at the position request level. An intermediate layer handles the slider position requests and generates the positions for each of the four sliders, while maintaining collision safety. Responding to position requests, it uses a combination of Proportional Integral Derivative (PID) control for each slider, with a simulated spring system between sliders, to update the position of all four sliders during an update cycle (Fig 14.3). For each position request, we calculate the required PID force, and add a force exerted by "virtual springs", which helps prevent collisions and moves unlocked sliders out of the way. The result of this control approach is a system that is both safe—carriages will never collide and push each other out of the way—and expressive.

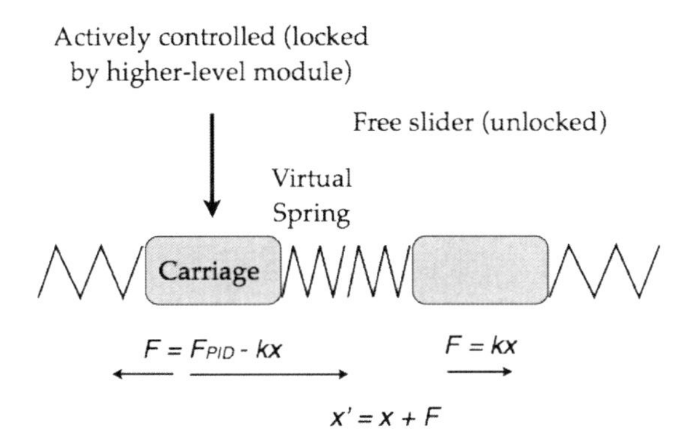

Fig 14.3. Interaction between PID control and simulated spring model.

14.4 Gestures and Anticipation

A main innovation of this paper is modeling interactive musical improvisation as gestures. Using gestures as the building blocks of musical expression is appropriate for robotic musicianship, as it puts the emphasis on physical movement instead of on the sequencing of notes. This is in line with an embodied view of human-robot interaction [4]. Gestures separately control the timing of the mallet strikers and the movement of the sliders. They can take many forms, from a simple "go-to and play" gesture for a single note, to a more complex improvisation gesture, as described below.

14.4.1 Anticipatory Action

In order to allow for real-time synchronous non-scripted playing with a human counterpart, we also take an anticipatory approach, dividing gestures into preparation and follow-through. This principle is based on a long tradition of performance, such as ensemble acting [13], and has been explored in our recent work, both in the context of human-robot teamwork [7], and for human-robot joint theater performance [6]. By separating the potentially lengthy preparatory movement from an almost instant follow-through, we can achieve a high level of synchronization and beat keeping without relying on a full-musical-bar delay of the system. This separation will also enable us, in the future, to explore different anticipatory strategies for synchronized playing.

14.5 Improvisation

Implementing this gesture-based approach, we have developed a Jazz improvisation system, which we employed in a human-robot joint performance. In our system, a performance is made out of interaction modules, each of which is an independently controlled phase in the performance. It is continuously updated until the part's end condition is met. This is usually a perceptual condition, but can also be a pre-set amount of bars to play.

Fig 14.4 shows the general structure of an interaction module. It contains a number of gestures, which are either triggered directly, or registered, to play based on the current beat, as managed by the beat keeper. Gestures are selected and affected by information coming in from percepts, which analyze input from the robot's sensory system. These percepts can include, for example, a certain note density, or the triggering of a particular phrase or rhythm. While there are a number of sensory modules possible, we are using a MIDI sensory input, responding to the notes from a MIDI-enabled electric piano. On top of this sensor, we developed several perceptual modules described later in this section. Common to all parts, and continuously running is the Beat Keeper module, which serves as an adjustable metronome that can be dynamically set and reset during play. The Beat Keeper interacts with the system by calling registered callback functions in the modules and gestures making up the performance.

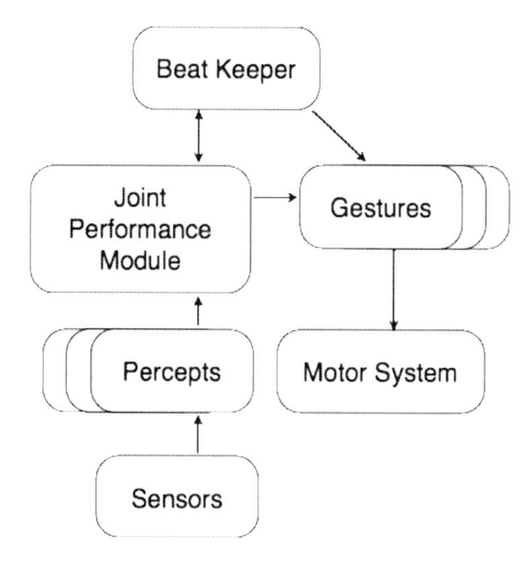

Fig 14.4 Schematic interaction module for each phase of the performance.

14.5.1 Interaction Module I: Call-and Response

The first interaction module is the call-and-response module. In this module, the system responds to a musical phrase with a set-beat chord sequence. The challenge is to be able to respond in time and play on a synchronized beat to that of the human player. This module makes use of the anticipatory structure of gestures. During the sequence detection phase, the robot prepares the chord gesture. When the phrase is detected, the robot can strike the response almost instantly, resulting in a highly meshed musical interaction. This module includes two kinds of gestures: (i) Simple chord gestures—select an arm configuration based on a given chord during the preparation stage, and strike the prepared chord in the follow-through stage; and (ii) Rhythmic chord gestures—which are similar to the simple chord gestures in preparation, but during follow-through will strike the mallets in a preset pattern. This can be an arpeggiated sequence, or any other rhythmic structure. The robot adapts to the call phrase using a simultaneous sequence spotter and beat estimator percept. Using an on-beat representation of the sequences that are to be detected, we use a Levenshtein distance metric [12] with an allowed distance of d=1 to consider a phrase detected. Naturally, we do not allow the last note in the phrase to be deleted for the purposes of comparison, as this would invalidate the synchronization. At that stage, the beat estimator will estimate both the played beat based on the duration of the sequence, and the beat synchronization based on the time of the last note played. These are transmitted to the beat keeper, which will execute a sequence of simple and rhythmic chords, as beat callbacks. The result is an on-sync, beat-matched call-and-response pattern, a common interaction in a musical ensemble.

14.5.2 Interaction Module II: Opportunistic Overlay Improvisation

A second type of interaction module is the opportunistic overlay improvisation. This interaction is centered on the choreographic aspect of movement with the notes appearing as a side effect of the performance. The intention of this module is to play a relatively sparse improvisation that is beat-matched, synchronized, and chord-adaptive to the human's playing. The central gesture in this module is a rhythmic movement gesture that takes its synchronization from the currently active beat in the beat keeper module. This beat is updated through a beat detection percept tracking the beat of the bass line in the human playing. In parallel, a chord classification percept is running, classifying the currently played chord by the human player. Without interrupting the choreographic gesture, this interaction module attempts to opportunistically play notes that belong to the currently detected chord, based on a preset rhythmic pattern. If the rhythmic pattern is in a "beat" state, and one or more of the mallets happen to be in a position to play a note from the detected chord, those mallets strike. Since both the choreographic gesture and the rhythmic strike pattern are activated through a shared beat keeper, the result is a confluence of two rhythms and one chord structure, resulting in a novel improvisational gesture which is highly choreographic, can only be conceived by a machine, and is tightly synchronized to the human's playing.

14.5.3 Interaction Module III: Rhythmic Phrase-Matching Improvisation

The third interaction module that we implemented is a rhythmic phrase-matching improvisation module. As in the previous section, this module supports improvisation that is beat- and chord-synchronized to the human player. In addition, it attempts to match the style and density of the human player, and generate improvisational phrases inspired by the human playing. Beat tracking and chord classification is done in a similar fashion as the in the opportunistic overlay improvisation: The timing and pitch of the bass notes are used for detecting the beat, for synchronizing the downbeats of the human's playing, and for chord classification. In addition, this module uses a decaying-history probability distribution to generate improvisational phrases that are rhythm-similar to phrases played by the human. The main gesture of this part selects—in each bar—one of the arm positions that correspond to the currently classified chord. This is the gesture's anticipatory phase. When in position, the gesture then plays a rhythmic phrase tempo- and sync-matched to the human's performance. Each arm plays a different phrase. Specifically, arm i plays a phrase based on a probabilistic striking pattern, which can be described as a vector of probabilities

$$\mathrm{p}_i = \{p^i_0 \; p^i_0 \; \cdots \; p^i_k\} \tag{14.1}$$

where k is the number of quantizations made. E.g.—on a 4/4 beat with 1/32 note quantization, k=32. Thus, within each bar, arm i will play at time j with a probability of p^i_j. This probability is calculated based on the decayed history of the human player's quantized playing patterns of the human player. The system listens to the human player's last beat's improvisation, quantizes the playing into k bins, and then attempts to cluster the notes in the phrase into the number of arms which the robot will use. This clustering is done on a one-dimensional linear model, using only the note pitch as the clustering variable. Once the clusters have been assigned, we create a human play vector

$$h_i = \{h^i_k\} = \{\text{1 if a note in cluster I was played at time k, 0 otherwise}\} \quad (14.2)$$

The probability p^i_j is then updated inductively as follows, where lambda is the decay parameter

$$p^i_0 = h^i_0 \lambda + p^i_j (1 - \lambda) \quad (14.3)$$

Fig 14.5 A live performance using Shimon's gesture-based improvisation system described was held on 4/17/09 in Atlanta, GA, USA.

The result is an improvisation system, which plays phrases influenced by the human player's rhythm, phrases, and density. For example, if the human plays a chord rhythm, then the vectors h_i would be identical or near identical for all clusters, resulting in a robot improvisation that will be close to a chord rhythm. However, there is variance in the robot's playing since it is using the human phrases as a probability basis, therefore changing the pattern that the human plays. Also, since the arm positions change according to the current harmonic lead of the human, and the robot's exploration of the chord space, the phrases will never be a precise copy of the human improvisation but only rhythmically inspired.

Moreover, as the probability vectors mix with data from earlier history, the current playing of the robot is always a combination of all the previous human plays. The precise structure of the robot's memory depends on the value of lambda. Another example would be the human playing a 1–3–5 arpeggio twice in one bar. This would be clustered into three clusters, each of which would be assigned to one of the arms of the robot, resulting in a similar arpeggio in the robot's improvisation. An interesting variation on this system is to re-assign clusters not according to their original note-pitch order. This results in the maintenance of the rhythmic structure of the phrase but not the melodic structure. In the performance described below, we have actually used only two clusters and assigned them to crossover arms, i.e. cluster 0 to arms 0 and 2 and cluster 1 to arms 1 and 3. Note that this approach maintains our focus on gestures as opposed to note sequences, as the clustering records the human's rhythmic gestures, matching different spatial activity regions to probabilities, which are in turn used by the robot to generate its own improvisation. Importantly—in both improvisation modules—the robot never maintains a note-based representation of the keys it is about to play. This is in line with our embodied music approach.

14.6 Evaluation: Embodiment in Robotic Musicianship

In our laboratory, we use Shimon as a research platform to evaluate core hypotheses of Robotic Musicianship (RM). As mentioned in the Introduction, one of the potential benefits of RM over other computer-supported interactive music systems is the generation of music-related physical and visual cues to aid joint musicianship. This could, for example, enable better synchrony through the use of anticipation of the robot's moves on the human's part. In addition, embodiment in human-robot interaction has been explored and usually been shown to have a significant effect on non-temporal interaction and subjects' reported perception of the robot [2. 10]. Similarly, a robot musician's physical presence could inspire human musicians to be more engaged in the joint activity. The robot's physical movement could also have choreographic and aesthetic effects on both players and audience. And the acoustic sound produced by the robot could similarly contribute to the enjoyment of the musical performance. We tested some of these hypotheses in a number of experiments using Shimon as an experimental platform. In this paper, we discuss the effects of physical embodiment and visual contact on joint synchronization and audience appreciation.

Fig 14.6 Experimental setup showing the human pianist on the right, and the robotic marimba player Shimon on the left.

14.6.1 Embodiment and Synchronization

In the human-robot musical synchronization case, we predict that human musicians take advantage of the physical presence of the robot in order to anticipate the robot's timing, and thus coordinate their playing with that of the robot. However, due to the auditory and rhythmic nature of music, human musicians have also been known to be able to play with no visual cues, and without any physical co-presence. We thus tested to what extent robot embodiment aids in synchronization, and to what extent this effect can be related to the visual connection between the human and the robot.

14.6.1.1 Experimental Design

We conducted a preliminary 3x2 within-subject study manipulating for level of embodiment and robot accuracy. Six experienced pianists from the Georgia Tech Music Department were asked to play the call-and-response segment from "Jordu" described above, jointly with a robotic musician. The interaction starts by the pianist playing the 7-note introductory phrase on a grand piano. The robot detects the tempo and bar sync of the phrase and responds in a rhythmic three-chord pattern on the marimba. The pianists were asked to synchronize a single bass note with each of the robot's chord, as best as they could. Each pianist repeated the sequence 90 times. They were asked to play at a variety of tempos, without specifying the precise tempo to play in. The timing of the human's playing was recorded through a MIDI interface attached to the grand piano, and the robot's playing time was also recorded, both to millisecond precision. MIDI delays were accounted for. *Manipulation I: Precision*: In the first half of the sequences (the PRECISE condition), the robot was programmed to play its response in the precise tempo of the human's call phrase. In this condition, the pianists were informed that the robot will try to match their playing precisely. In second half of the sequences (the IMPRECISE condition), the robot was programmed to play its response either on tempo and on-beat to the human's call phrase, slightly slower than the human's introduction phrase, or slightly faster. The pianists were informed that the robot might play slightly off their proposed beat, but that its response will be consistent throughout each individual response sequence. Also, the pianists were asked to try to synchronize their playing with the actual notes of the robot, and not with their own "proposed" tempo and beat. *Manipulation II: Embodiment:* Within each half of the trials—for a third of the interaction sequences (the VISUAL condition), the pianists were playing alongside the robot to their right, enabling visual contact with the robot (as shown in Figure 14.6). In another third of the interaction sequences (the AUDITORY condition), the robot is physically present, but separated from the human musician by a screen. In this condition, the human player can hear the robot move and play, but not see it. The remaining third of the interaction sequences (the SYNTH condition), the robot does not move or play. In this condition, the human player hears a synthesized marimba play over a set of headphones. In both the AUDITORY and the SYNTH condition there is no visual contact with the robot; in both the VISUAL and the AUDITORY condition there is an acoustic note effect indicating the presence of a physical instrument and a physical player, and in addition, the robot's motor noise can indicate to the pianist that the robot is in motion. The order of the conditions was randomized for each subject.

14.6.1.2 Results

We analyze the (signed) delay and (absolute) offset error between pianist and robot for each of the response chords. In particular, we analyze the offset separately for the first, second, and third chord. This is due to the different musical role each chords plays: the first chord occurs an eighth beat after the introductory phrase, so that the pianists can easily synchronize with the robot by simply playing according to their original tempo. The second chord reveals the robot's perceived tempo, and its temporal placement may vary, in particular in the IMPRECISE condition. Since all three chords play at a fixed tempo, the temporal placement of the third chord can be implied by the interval between the first and the second chord, in which case the synchronization can, again, be implied by rhythm alone. We thus expect that the first chord will be the most synchronized, with the second chord being the most difficult to synchronize. Testing this hypothesis we find that the absolute delays for each of the three chords are indeed significantly different, at a confidence of $p < 0.001$, across all six conditions. The offsets for the first chord are the lowest (49.35ms), those of the second chord are significantly more offset (116.16ms), and the offsets for the third chord are lower than the second, but not as low as the first (67.79ms). *PRECISE Condition:* Comparing the mean offset between pianist and robot in the PRECISE condition, we find no significant difference between the three embodiment conditions, and in particular, we find no advantage to visual contact with the robot. That said, the AUDITORY condition has a slightly higher error rate, as well as scatter, than either of the other two conditions.

IMPRECISE Condition: In contrast, when the robot intentionally changes the detected tempo of the introductory phrase, we expect to detect a larger difference in synchronization between embodiment conditions. Fig 14.7 shows the mean and standard error for all trials in the IMPRECISE condition. For the first and third chord, we see no difference between the embodiment conditions, indicating that, indeed, the human musicians can use the auditory and rhythmic cues to synchronized these two chords. In particular, it is notable that the first two chords are enough for the subjects to synchronize the third chord based on the same interval. However, for the second chord—the timing of which has some uncertainty—the offset is smaller for the VISUAL condition compared to both non-visual conditions. This difference is nearly significant: VISUAL: 129.32 +/- 10.01 ms; other conditions: 162.80 +/- 13.62 ms, $T(182)=-1.66$, $p=0.09$, suggesting that visual cues are indeed used to synchronize the unpredictably-timed event. The "offset" discussed above is the absolute error between the human's key-hit and the robot's marimba-strike. The effect of visual contact is more apparent when looking at the sign of the error: evaluating the signed delay, we find a significant difference between the VISUAL and the other two conditions (Fig 14.8): VISUAL: 16.78 +/- 19.11 ms; other conditions: -75.21 +/- 18.95 ms, $T(182)=3.10$, $p < 0.01$ **. In particular, we find that trials in the VISUAL condition to be delayed with respect to the robot, whereas the trials in the non-visual conditions pre-empt the robot's playing, indicating that pianists react to the robot's movement when they can see it, but try to anticipate the robot's timing when they cannot see it.

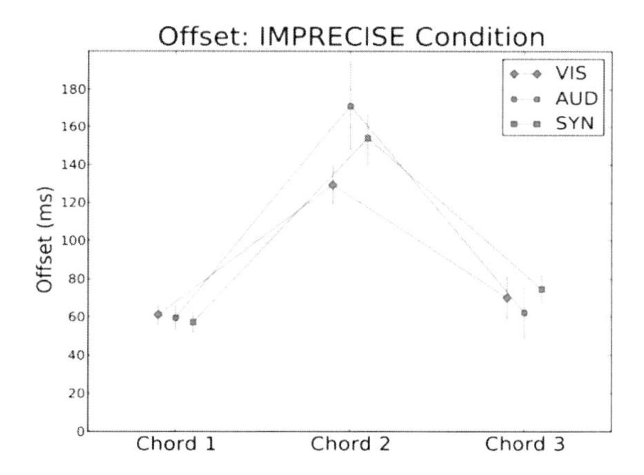

Fig. 14.7 Mean offset in milliseconds between pianist and robot in IMPERCISE condition.

Effects of tempo: We also find that the benefits of the visual connection increase at slower playing tempos. While the AUDITORY condition is significantly more error-prone in trials under 100 BPM, than in trials over 100 BPM (234.54 +/- 56.25 ms [slow] vs. 131.25 +/- 10.75 ms [fast]; $T(57)=2.14$, $p < 0.05$ *), the errors in the VISUAL condition is not affected by the decrease in tempo (138.59 +/- 17.62 ms [slow] vs. 119.43 +/- 11.54 ms [fast]; $T(60)=0.94$). As above, the effect on the SYNTH condition is similar to the AUDITORY condition, but less pronounced. In addition, we find that the significant advantage for the VISUAL condition over the SYNTH condition is mainly due to the trials in which the robot played slower than expected. In those cases (Figure 14.9), the visual condition caused a significantly lower error on the human's part: VISUAL: 121.42 +/- 5.39 ms; SYNTH: 181.56 +/- 16.57 ms; T(57)=2.66, p < 0.01 **.

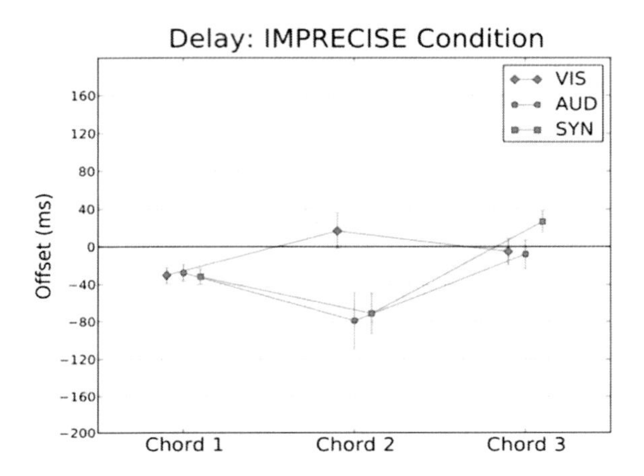

Fig. 14.8 Mean delay in milliseconds between pianist and robot in IMPERCISE condition.

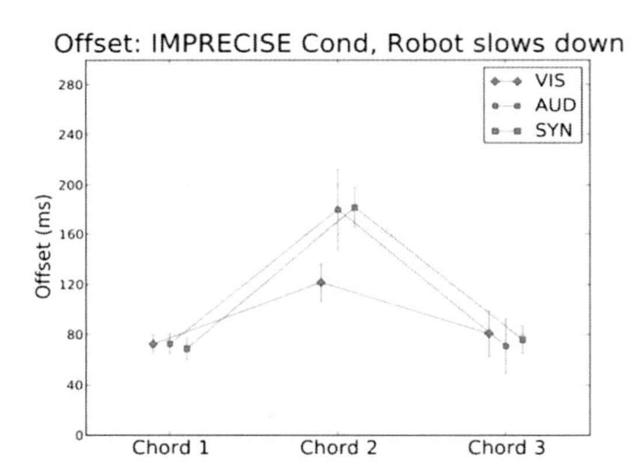

Fig. 14.9 Mean offset in milliseconds between pianist and robot in IMPECISE condition for trials in which the robot slowed down compared to the human's "call" phrase.

14.6.1.3 Discussion

In our preliminary tests, we find that visual contact with the robot contributes only partially to the degree of synchronization in a call-and-response interaction. The effect of embodiment without visual contact, compared to a disembodied musician, seems to be even less pronounced, and sometimes detrimental. In the case where the robot does not intentionally change the tempo, we see no difference between any of the three conditions. We believe that this is due to the fact that the robot precisely follows the pianist's rhythm, allowing for perfect synchronization simply by playing along in the same rhythm, without any input from the robot. In the case where the robot slightly alters the tempo in response to the human's playing, we find that the pianists' ability to synchronize with the robot is significantly reduced. For the second chord (in which there is an uncertainty in timing), visual contact reduces the error compared to the auditory and synthesized condition. In particular, visual contact allows the pianists to react to the robot instead of preempting the timing of their playing. This indicates that the pianists use the visual cues to time their playing. By the third chord, the human players seem to be able to achieve a high level of synchronization regardless of the embodiment of the robot. This may indicate that they resort again to an internal rhythmic cue based on the first two chords. We also find that visual contact is more crucial during slow trials, and during trials in which the robot slows down, possibly suggesting that visual cues are slow to be processed and do not aid much in fast sequences. For example, it may be that during fast sequences, the pianists did not have time to look at the robot. Another explanation is that their internal beat keeping is more accurate over short time spans. In general, it seems that pianists use visual information when they can, but can resort to rhythmic and auditory cues when necessary and possible. Interestingly, it seems that the synthesized condition is less error-prone than the present-but-screened (AUDITORY) condition. This may be due to the fact that the pianists try to use the existing motor noise from the robot is as a synchronization signal, but find it to be unreliable or distracting.

14.6.2 Visual Contact and Audience Appreciation

We also tested the effects of visual contact on audience appreciation. In this experiment, we filmed two pianists playing in two different improvisation settings each with the robot. We wanted to test how visual contact affects joint improvisation as judged by an audience. The physical setup was similar to the previous experiment, and the conditions were similar to two of those in the "Embodiment" manipulation, namely VISUAL and AUDITORY.

14.6.2.1 Experimental Setup

The pianists' sessions were videotaped, and from each session, a 30 second clip was extracted by choosing the 30 seconds after the first note that the robot or computer played. We posted these video clips onto a dedicated website, and asked an online audience to rate the clips on eleven scales. Each scale was a statement, such as "The robot played well", "the duo felt like a single unit", etc (see: Table 14.1 for all scales). The subjects were asked to rate the statement on a 7-point Likert scale between "Not at all" (1) and "Very much" (7). Subjects watched an introductory clip familiarizing them with the robot, and could play and stop the clip as many times as they liked.

For each pianist, the order of conditions was randomized, but the conditions were grouped for each pianist, to allow for comparison and compensation of each pianist's style.

In this preliminary study, we collected 30 responses, out of which 21 were valid, in the sense that subjects rates all performances of at least one pianist. The age of the respondents ranged between 25 and 41, and 58% identified as female.

14.6.2.2 Results

In order to compensate for each pianist's style, we evaluated the difference between conditions for each subject and each pianist. We then combined the results for both pianists across all subjects.

Table 14.1 shows the results of comparing the VISUAL condition to the AUDITORY condition. This comparison indicates the effect of visual contact between the pianist and the machine.

Table 14.1 Effects of visual contact on audience appreciation of a number of scales. T numbers indicate 1-sample T-Test with mean(x) =0 as the null hypothesis. * p < 0.05; ** p< 0.01; *** p<0.001.

Scale	VIS-AUD	
	$\bar{x} \pm \sigma$	T(38)
I enjoyed this performance	0.28 ± 1.28	1.36
The robot played well	0.92 ± 1.42	4.00 ***
The robot played like a human would	1.11 ± 1.25	5.37 ***
The robot was responsive to the human	0.54 ± 1.28	2.60 *
The human was responsive to the robot	1.08 ± 1.79	3.71 ***
The duo was well-coordinated and synchronized	1.00 ± 1.48	4.15 ***
The human seemed inspired by the robot	1.13 ± 1.90	3.67 ***
The robot seemed inspired by the human	0.67 ± 1.37	3.01 **
The two players felt connected to each other	0.97 ± 1.58	3.75 ***
The duo felt like a single unit	0.95 ± 1.63	3.58 ***
The duo communicated well	1.11 ± 1.74	3.85 ***

14.6.2.3 Effects of Visual Contact

We found a significant difference in audience appreciation of the improvisation session between the visual-contact and occluded conditions, on all scales but one (overall enjoyment). Specifically, we find that, even though the robot uses the same improvisation algorithm in all conditions, audiences felt that in the VISUAL condition the robot played better, more like a human, was more responsive, and seemed inspired by the human. In addition, we find that the human player, too, was rated as more responsive to the machine, and as more inspired by the robot. The overall rating of the duo as being well synchronized, coordinated, connected, coherent, and communicating was also significantly higher in the VISUAL condition.

These findings indicate that visual contact between human and robot contributes significantly to the audience's appreciation of robotic musicianship.

14.7 Live Performance

We have used the described robot and gesture-based improvisation system in a live performance before a public audience (as shown in Figure 14.5). The show occurred on April 17 2009 in Atlanta, GA, USA. The performance was part of an evening of computer music and was sold-out to an audience of approximately 160 attendants. It was structured around "Jordu", a Jazz standard by Duke Jordan. The first part was an adaptive and synchronized call-and-response, in which the pianist would prompt the robot with a number of renditions of the piece's opening phrase. The robot detected the correct phrase and, using preparatory gesture responded on beat. A shorter version of this interaction was repeated between each of the subsequent performance segments. The second phase used the introduction's last detected tempo to play a fixed-progression accompaniment to the human's improvisation. Then the robot started playing in opportunistic overlay improvisation taking tempo and chord cues from the human player while repeating an "opening-and-closing" breathing-like gesture, over which the rhythmic improvisation was structured. The next segment employed rhythmic phrase-matching improvisation, in which the robot adapted to the human's tempo, density, style, chord progression, and rhythmic phrases. Our gesture-based anticipatory approach enabled the robot to adapt without noticeable delay while maintaining an overall uninterrupted visual motion arc, and seem to be playing in interactive synchrony with the human player. An interesting result of this improvisation was a constant back-and-forth inspiration between the human and the robotic player. Since the robot's phrases were similar, but not identical to the human's phrases, the human picked up the variations, in return influencing the robot's next iteration of rhythms. Finally, a pre-programmed crescendo finale led to the end-chord, which was an anticipatory call-and-response, resulting in a perceived synchronous end of the performance. The overall performance lasted just under seven minutes. Video recordings of the performance [8, 9] were widely covered to acclaim by the press and viewed by an additional audience of approximately 40,000 online.

14.8 Conclusion and Future Work

We presented Shimon, an interactive improvisational robotic marimba player developed for research in Robotic Musicianship. We discussed the musically and visually expressive motor-control system, and a gesture-based improvisation system. The design of these systems stems from our belief, that musical performance is as much about visual choreography and visual communication, as it is about tonal music generation. We have implemented our system on a full human-robot Jazz performance, and performed live with a human pianist in front of a public audience. We are currently underway in an audience evaluation of gesture-based vs. algorithmic improvisation. Additionally, we are developing a novel predictive anticipatory system to allow the robot to use past interactions to generate preparatory gestures, based on our findings on anticipatory human-robot interaction [5, 7]. We also use Shimon to empirically study some of the core hypotheses of Robotic Musicianship. In this paper we evaluated the effect of embodiment on human-robot synchronization. In a preliminary study we found that visual contact accounts for some of the capability to synchronize to a fixed-rhythm interaction. However, we also found that humans can compensate for lack of visual contact and use rhythmic cues in the case where visual contact is not available. Visual contact is more valuable when the robot errs or changes the interaction tempo. It is also more valuable in slow tempos and delays, suggesting that using visual information in musical interaction is a relatively slow mechanism, or that the human's internal capability to beat-match is more accurate in faster tempos. In addition, our findings indicate that a visually occluded, but present, robot is distracting and does not aid in synchronization, and may even detract from it. In a study evaluating the effects of visual contact on audience appreciation, we find that visual contact in joint Jazz improvisation makes for a performance in which audiences rate the robot as playing better, more like a human, as more responsive, and as more inspired by the human. They also rate the duo as better synchronized, more coherent, communicating, and coordinated; and the human as more inspired and more responsive.

We plan to extend these preliminary studies to a wider audience, and in particular to also test them with subjects in a live audience. We are also currently adding a socially expressive head to Shimon. This will allow an additional channel of embodied and gesture-based communication, and adds a visual modality to the robot's perceptual system. We hope to extend the research described here by further testing the use of full-body robotic gestures and visual communication on synchronization and joint musical interaction.

References

1. Baginsky, N.A.: The three sirens: a self-learning robotic rock band (2004), http://www.the-three-sirens.info
2. Bainbridge, W., et al.: The effect of presence on human-robot interaction. In: Proceedings of the 17th IEEE International Symposium on Robot and Human Interactive Communication (RO-MAN) (2008)

3. Dannenberg, R.B., et al.: Mcblare: a robotic bagpipe player. In: Proceedings of the 2005 Conference on New Interfaces for Musical Expression (NIME), pp. 80–84 (2005)
4. Hoffman, G., Breazeal, C.: Robotic partners' bodies and minds: An embodied approach to fluid human-robot collaboration. In: Proceedings of the Fifth International Workshop on Cognitive Robotics, CogRob, AAAI 2006 (2006)
5. Hoffman, G., Breazeal, C.: Cost-based anticipatory action-selection for human-robot fluency. IEEE Transactions on Robotics and Automation 23(5), 952–961 (2007)
6. Hoffman, G., et al.: A hybrid control system for puppeteering a live robotic stage actor. In: Proceedings of the 17th IEEE International Symposium on Robot and Human Interactive Communication, RO-MAN (2008)
7. Hoffman, G., Breazeal, C.: Anticipatory perceptual simulation for human-robot joint practice: theory and application study. In: Proceedings of the 23rd AAAI Conference for Artificial Intelligence, AAAI (2008)
8. Hoffman, G.: Human-robot jazz improvisation (full performance) (2009a), `http://www.youtube.com/watch?v=qy02lwvGv3U`
9. Hoffman, G.: Human-robot jazz improvisation, highlights (2009b), `http://www.youtube.com/watch?v=jqcoDECGde8`
10. Kidd, C., Breazeal, C.: Effect of a robot on user perceptions. In: Proceedings of the IEEE/RSJ International Conference on Intelligent Robots and Systems, IROS (2004)
11. Lasseter, J.: Principles of traditional animation applied to 3d computer animation. Computer Graphics 21(4), 35–44 (1987)
12. Levenshtein, V.I.: Binary codes capable of correcting deletions, insertions and reversals. Soviet Physics Doklady 10, 707 (1966)
13. Meisner, S., Longwell, D.: Sanford Meisner on Acting, Vintage (1987)
14. Petersen, K., et al.: Toward enabling a natural interaction between human musicians and musical performance robots: Implementation of a real-time gestural interface. In: Proceedings of the 17th IEEE International Symposium on Robot and Human Interactive Communication, RO-MAN (2008)
15. Rowe, R.: Machine musicianship. MIT Press, Cambridge (2001)
16. Singer, E., et al.: Lemur GuitarBot: Midi robotic string instrument. In: Proceedings of the 2003 Conference on New Interfaces for Musical Expression (NIME), pp. 188–191 (2003)
17. Solis, J., et al.: The Waseda flutist robot no 4 refined IV: enhancing the sound clarity and the articulation between notes by improving the design of the lips and tonguing mechanisms. In: Proceedings of the IEEE/RSJ International Conference on Intelligent Robots and Systems (IROS), pp. 2041–2046 (2007)
18. Toyota: Trumpet robot (2004) `http://www.toyota.co.jp/en/special/robot`
19. Weinberg, G., Driscoll, S.: Toward robotic musicianship. Computer Music Journal 30(4), 28–45 (2006)
20. Weinberg, G., Driscoll, S.: The design of a perceptual and improvisational robotic marimba player. In: Proceedings of the 18th IEEE Symposium on Robot and Human Interactive Communication (RO-MAN), Jeju, Korea, pp. 769–774 (2007)

Chapter 15
Interactive Musical System for Multimodal Musician-Humanoid Interaction

Jorge Solis, Klaus Petersen, and Atsuo Takanishi

Abstract. The research on Humanoid Robots designed for playing musical instruments has a long tradition in the research field of robotics. During the past decades, several researches are developing anthropomorphic and automated machines able to create live musical performances for both understanding the human itself and for creating novel ways of musical expression. In particular, Humanoid Robots are being designed to roughly simulate the dexterity of human players and to display higher-level of perceptual capabilities to enhance the interaction with musical partners. In this chapter, the concept and implementation of an interactive musical system for multimodal musician-humanoid interaction is detailed.

15.1 Introduction

The research on musical robots has a long tradition since the golden era of automata. More recently, thanks to the technological advances on power computation, Musical Information Retrieval (MIR), Robot Technology (RT), etc. several researchers have been focusing on development of anthropomorphic robots [18, 21, 22, 24, 25] and interactive automated instruments [3, 4,13, 14, 17] capable of interacting with musical partners.

Jorge Solis
Faculty of Science and Engineering, Waseda University/Humanoid Robotics Institute,
Waseda University, 3-4-1 Ookubo, Shinjuku-ku, Tokyo, Japan
e-mail: solis@ieee.org

Klaus Petersen
Graduate School of Advanced Engineering and Science,
Waseda University, 3-4-1 Ookubo, Shinjuku-ku, Tokyo, Japan
e-mail: klaus@moegi.waseda.jp

Atsuo Takanishi
Faculty of Science and Engineering, Waseda University/Humanoid Robotics Institute,
Waseda University, 3-4-1 Ookubo, Shinjuku-ku, Tokyo, Japan
e-mail: contact@takanishi.mech.waseda.ac.jp

J. Solis and K. Ng (Eds.): Musical Robots and Interactive Multimodal Systems, STAR 74, pp. 253–268.
springerlink.com

In particular, automated instruments have been developed as an approach to introduce novel ways of musical expression. For this purpose, the term *machine musicianship* has been introduced as actuated mechanical sound generators controlled by computer musical models, perception and interaction algorithms [14]. More recently, the term *robotic musicianship* has been also introduced as the combination of a musical performance robot with perceptual and interaction capabilities [26].

On the other hand, Humanoid Robots capable of playing musical instruments have been developed as an approach to understand better the principles of human-robot interaction by emulating both the human dexterity required to perform an instrument and the intelligence to enable the interaction with musical partners at the same logical level of perception (this requires the robot to process both aural and visual information). The first attempt of developing an anthropomorphic musical robot was done by Waseda University in 1984 by the late Prof. Ichiro Kato. In particular, the WABOT-2 was capable of playing a concert organ. Then, in 1985, the WASUBOT built also by Waseda, could read a musical score and play a repertoire of 16 tunes on a keyboard instrument [5].

Even several humanoid musical performance robots have been developed up to now, very little research on Musician-Humanoid Interaction (MHI) has been done. Therefore; in this chapter, the preliminary efforts done towards implementation a novel musical interaction system is presented.

15.2 Musical-Based Interaction System

15.2.1 Design Concept

Conventionally, the Musical Performance Humanoid Robots (MP-HRs) are mainly equipped with sensors that allow them to acquire information about its environment. Based on the anthropomorphic design of humanoid robots, it is therefore important to emulate two of the human's most important perceptual organs: the eyes and the ears. For this purpose, the humanoid robot integrates in its head, vision sensors (i.e. CCD cameras) and aural sensors (i.e. microphones) attached to the sides for stereo-acoustic perception. In the case of a musical interaction, a major part of the typical performance (i.e. Jazz) is based on improvisation. In these parts musicians take turns in playing solos based on the harmonies and rhythmical structure of the piece. Upon finishing his solo section, one musician will give a visual signal, a motion of the body or his instrument, to designate the next soloist. Another situation of the musical interaction between musicians, is basically where the higher skilled musician has to adjust his/her own performance to the less skilled one. After both musicians get used to each other, they may musically interact.

Toward enabling the multimodal interaction between the musician and MP-HRs, the Musical-based Interaction System (MbIS) is introduced and described. The MbIS has been conceived for enabling the interaction between the MP-HRs (or/and musicians); as it is shown in Fig. 15.1. The proposed MbIS is composed by two levels of interaction that enables partners with different musical skill levels to interact with the MP-HR: intermediate-based and advanced-based level of interaction.

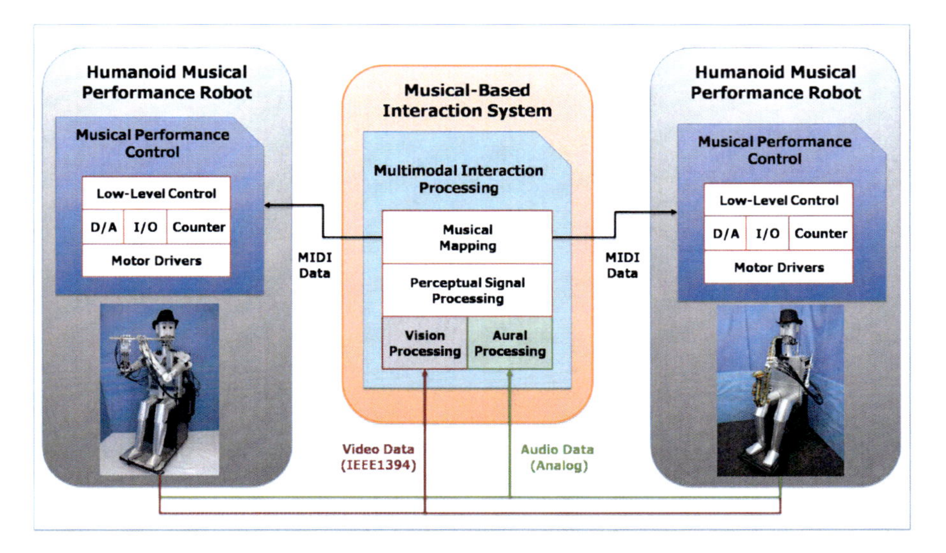

Fig. 15.1 The principle of the Musical-Based Interaction System for Musical Playing-Instrument Humanoid Robots. By implementing such a system, different kind of humanoid musical performance robots (and/or human musicians) may interact.

15.2.2 Musician-Humanoid Interaction

Until now, several researchers from different fields of Human-Robot Interaction [16], Musical Information Retrieval [7], etc. have been proposing algorithms for the development of interactive systems. Based on such research fields, different kinds of automated instruments and robotic players have been introduced (refer Chapters 9–14). However, when we talk at the level of humanoid robot, we are talking about not just analyzing the human performance, but also we are required to map those musical parameters into control parameters of the robot (*from perception to action*). This means that we are also required to take into account the physical constraints of the robot. Due to the complexity of doing this task, it is important to consider the inherent properties of the human players so that MP-HRs can interact at the same level of perception as musicians do. From this long-term research approach, it is more feasible to understand (from a scientific point of view) how humans can actually interact in musical terms. Of course, new ways of musical expressions can be discovered that otherwise would not be conceived by means of traditional methods.

Based on the above considerations, the system implementation of the MbIS is composed by different levels of interaction which makes it usable for musicians with different levels of skill [19]. In order to define the musician partner of each of the proposed levels of interaction, it is possible to relate the way the musicians are classified based on the level of expertise: Beginner (no prior experience with the musical instrument), Intermediate (a fair amount of experience with the basics of the instrument), Advanced (proficient traditional musicians) and Master (mastery of the instrument-playing technique). As a result, the MbIS has been proposed to include two-levels of interaction (Figure 15.2): intermediate-based and advanced-based levels of interaction. The intermediate-based level of interaction is designed to provide easy-to-learn controllers which have a strong resemblance to established musical environment (i.e. studio equipment). On the other hand, the advanced-based level of interaction allows for free control of the performance parameters. However, it also requires previous experience in robot-human interaction (i.e. certain initialization procedures to calibrate the software are required). Moreover, it is important to keep in mind that the humanoid musical performance robot is a complex system. It bears several limitations (as human players may have while playing instruments), which on the one hand are related to its shape and on the other hand are implied by its technical complexity. For this reason, the musician partner is also required to gain a substantial knowledge about the MP-HR. As a result, the musician partner can progressively go from the intermediate-based interaction to the advanced one.

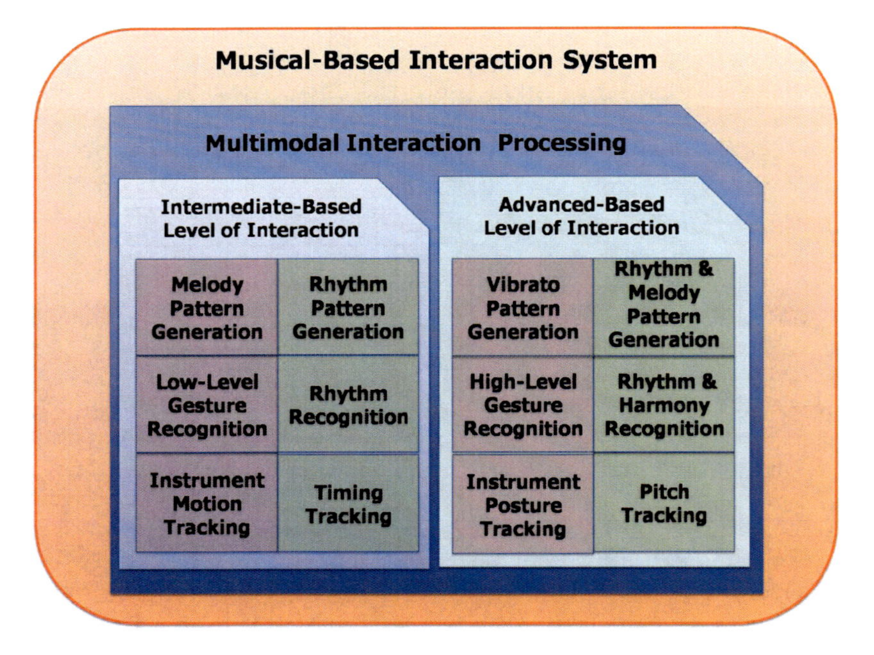

Fig. 15.2 In order to implement the MbIS for enabling musician-humanoid interaction, the proposed system is composed by two different levels of interaction.

The intermediate-based level of interaction is designed to process both visual and aural channels. Regarding the visual channel processing, the MP-HR should be able of processing visual cues from the musical partner. In order to facilitate its interaction with intermediate level players, a common technology found in music studios has been proposed. In particular, virtual buttons and faders have been designed as tools for interaction so that the musician can learn how to intuitively use them (no previous knowledge is required). In order to implement such virtual buttons and faders, it is required to detect the movements of the user by image processing. In particular, the instrument tracking method has been implemented to recognize areas of motion for activating the proposed controls. A low-level gesture is then used to trigger those controls to activate different kinds of melody patterns. On the other hand, the aural processing of the MP-HR analyzes the timing of a tone sequence. From this analysis, the recognized tempo is matched with a library of timing patterns that are previously stored on the robot's memory. Therefore, the recognition tempo method determines the best matching pattern and passes this information on to the mapping module, in order to generate an output performance by the robot.

The advanced-based level of interaction requires more experience in working interactively with the robot, but also allows for more advanced interaction controls during the musical performance. Regarding the visual processing, the MP-HR is capable of tracking the instrument posture (if the musician is holding a wind instrument, he/she can adjust two musical parameters at the same time, one by moving the instrument sideways and the other one by bending the instrument closer or further away from the robot). This high-level gesture is recognized and then it is used to modulate a particular musical parameter (i.e. amplitude of the vibrato, etc.) of the robot's performance. On the other hand, the aural processing of the MP-HR, the musician is allowed to select a harmony pattern for playing with the robot. For this, the robot analyzes the tonal content of the sequence played by the musician partner. Then, by means of the harmony recognition module, the MP-HR activates the correspondent rhythm and melody pattern.

15.3 Implementation Details

15.3.1 Vision Perception

15.3.1.1 Instrument Motion Tracking

The instrument motion tracking detects the movements of the wind instrument hold by the musician partner. If we have a continuous stream of video images, for every frame a difference image with the previous frame can be computed. In particular, a threshold can be defined to determine the changes from one frame to the next. This preliminary output can then be filtered by computing the running average over several of these images [27]. Due to the simplicity on its implementation, a robust tracking performance can be achieved while any musician without knowledge of vision processing can intuitively use it (due to its simplicity, no calibration procedure is required).

For this purpose, the proposed motion tracking algorithm is inspired by an interface extension called Eyetoy [23]. It enables players to control games by movements of their body in front of a small camera connected to the gaming console. In the case of the MbIS, it is required to extract the information about the musician partner's movement in a deliberate environment [9]. A related method called delta framing is employed in video compression [10]. Thus, if we have a continuous stream of video images, for every frame we calculate a difference image with the previous frame by using Eq. (15.1). By applying a threshold into the resulting image, a b/w bitmap of the parts in the video image that have changed from one frame to the next can be easily computed as Eq. (15.2). Finally, the result is filtered by running average over several of these images, as it is shown in Eq. (15.3).

$$p_r = \left| p_p - p_c \right| \tag{15.1}$$

p_r: absolute difference for the resulting pixel
p_p: pixel at the same position in the previous image
p_c: same pixel in the current image

$$p_r = \begin{cases} 0 & \text{if } p_c \leq t_r \\ 255 & \text{if } x \geq t_r \end{cases} \tag{15.2}$$

p_r: threshold pixel value
t_r : threshold level
p_c: same pixel in the current image

$$p_r = \alpha * p_p + (1 - \alpha) * p_c \tag{15.3}$$

p_r: average for the resulting pixel
p_p: pixel at the same position in the previous difference image
p_c: same pixel in the current image
α: Averaging factor

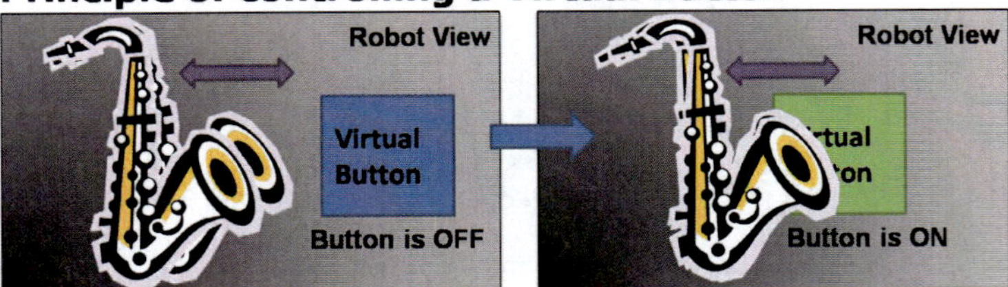

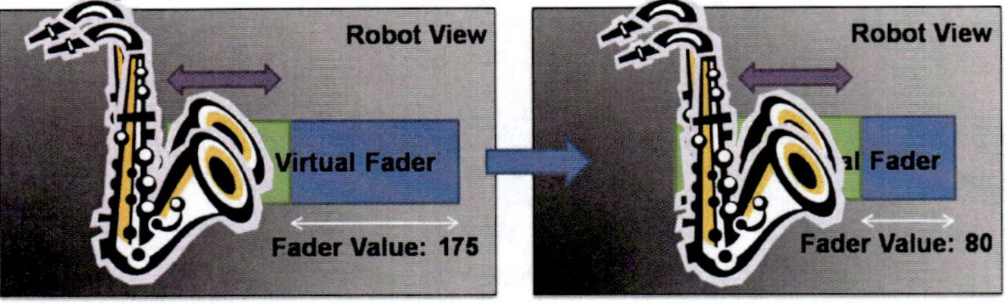

Fig. 15.3 Implementation of virtual buttons and faders resemble music studio controllers that can be manipulated by tracking the instrument motion.

By tracking the motion of the instrument, it is possible to define sensitive regions within the robot's gaze [9]. In music production, composers use switches and faders to control their electronic musical instruments, so it is possible to model such controls within the image space (Figure 15.3). The first controller is a simple virtual push-button, in functionality similar to a drum pad of an Akai MPC drum machine [1]. Such push button can be positioned anywhere in the video image. If a push button is triggered, a previously defined MIDI signal is sent to the MP-HR. The second controller is a virtual fader that can be used to continuously set a controller value. In this case, the position of a fader can be changed for example by a motion of the hand. For each change of the fader position, a MIDI controller message is sent to the robot. The fader slowly resets itself to a default position after it has been manipulated like a mechanical fader. This prevents a fader from remaining in an erroneous position that might have resulted from background noise. A fader can be deliberately positioned in the image and orientated in any angle to allow the user to easily adjust it to his control requirements and physical constraints.

15.3.1.2 Instrument Posture Tracking

The instrument posture tracking determines the orientation of the wind instrument hold by the musician partner. For this purpose, the color histogram matching [15] and particle tracking [2] were implemented. The combination of the two methods is an established way to follow an object with a certain color profile. The system is initialized manually by defining the starting positions of the player's hands (thus, further calibration process is not required). For the computation of the 3D data, the algorithm makes use of a stereo image mapping technique. Instead of calculating the complete depths map of the scene, four patches are found by the particle tracker. This approach saves resources due to the limited number of points being calculated. After the x-y-axis coordinates of the image patches are determined, the distance of a patch from the camera is computed. To achieve this, the difference between the x-position from the left camera image) and the x-position from the right camera image is computed. The larger the difference, the closer the patch is located to the camera.

In order to model the shape of the instrument (i.e. flute, saxophone, etc.) to be tracked, a simple line can be considered. The hands of the player are located on two spots along that line. The average of the position of the hands is recorded as the center position of the instrument. Similarly, the orientation can be computed by considering a line drawn from the center of one hand to the center of the other. The inclination of the line is the orientation of the instrument. There is no ambiguity about the position, as normally a player would not hold the instrument upside down. From the 2D coordinates of the four hand particles, the relative position, inclination and rotational angle of the instrument are computed. To compute the depth values of both hands, the z-transformation can be used, as it is shown in Eq. (15.4) and (15.5). Where, Δx_p is the distance between the x-coordinate of the patch in the left camera image (x_{pl}) and the right camera image (x_{pr}). Accordingly, z denominates the z-coordinate of the patch. The constant α is defined to adjust the value of Δz for further calculations. Inclination and rotational angle are obtained by transforming the Cartesian coordinates resulting from the object tracking into a cylindrical system.

$$\Delta x_p = \left| x_{pl} - x_{pr} \right|, \tag{15.4}$$

$$z = \frac{1}{\Delta x_p} * \alpha \tag{15.5}$$

Although new object coordinates are adapted only from the particle with the highest likelihood (the particle filter method works recursively), the information about the other particles is not lost (Figure 15.4). A particle with an initially lower than maximum likelihood is not discarded, but it can still propagate to gain more likelihood later. However, research on particle filters has shown that in case all particles are kept for the whole tracking run, all but one particle tend to be degraded to probabilities close to zero. There are several ways to counteract this behavior [2]. One of those approaches is the re-sampling method. After each new predict-update cycle, particles with a probability lower than a certain threshold are exchanged for newly initialized particles. This threshold, as well as the optimum number of particles to be used, can be manually determined.

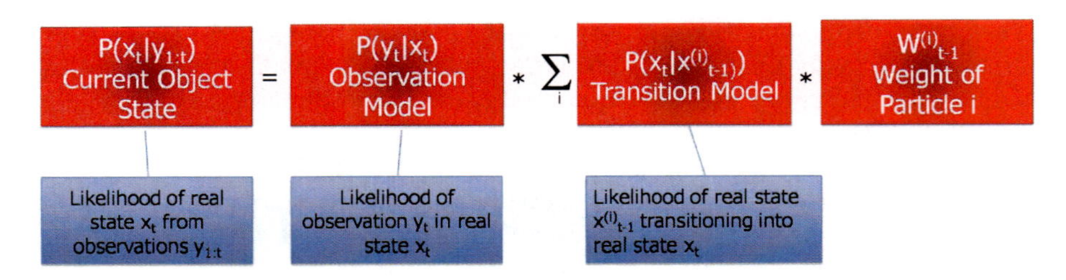

Fig. 15.4 Principle of the particle filter tracking implemented on the MbIS for tracking the orientation of a wind instrument hold by the musician partner.

15.3.2 *Aural Perception*

15.3.2.1 Timing Tracking

The timing tracking extracts the rhythmic information from the sound produced by the performance of the musician partner. The proposed algorithm basically determines the best matching candidate from a library of timing patterns that are saved as previous knowledge in the humanoid musical robot. For this purpose, in the sound waveform, separate notes are represented as distinguishable amplitude peaks. Such peaks are removed by using on Eq. (15.6). The duration of one tone impulse naturally is longer than a certain minimum time. In order to prevent very short noise peaks from falsely triggering, the sound wave is filtered by a running average calculation.

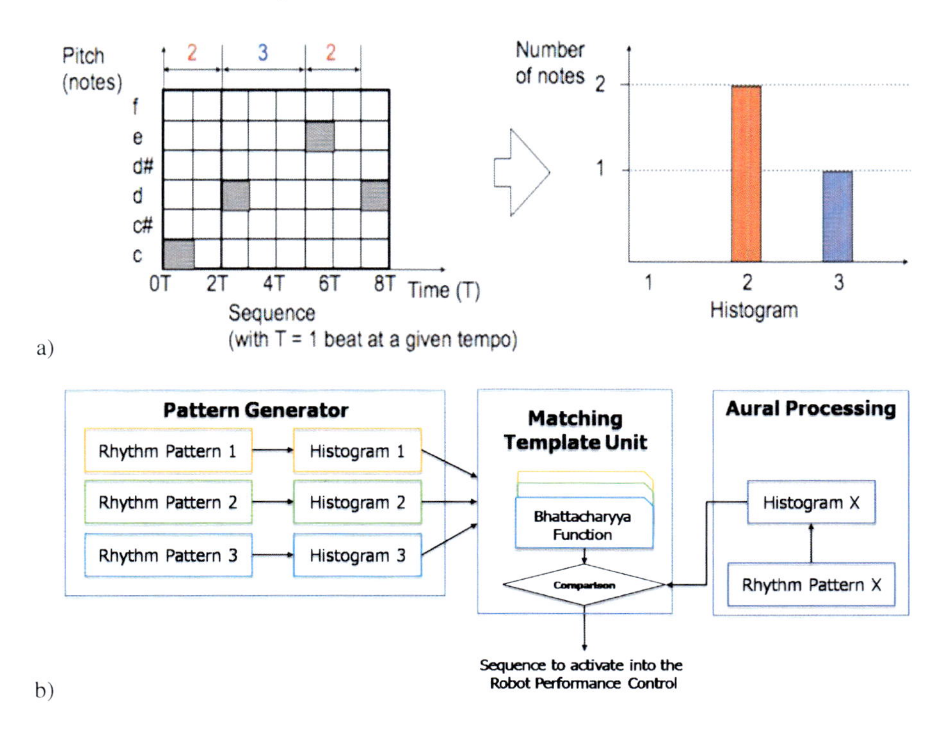

Fig. 15.5 Timing tracking recognition and mapping: a) Example of the histogram constructed for the timing tracking algorithm; b) Rhythm patter recognition.

The rhythm patterns have a certain length. To identify the most recently played pattern, it is not required all the previous sound inputs. Instead, a window is used so that it always contains only the most up-to-date part of the recorded music information. This window continuously slides forward as new data is acquired. The size of the window is the length of the longest rhythm pattern in the robot's pattern library. Regardless of which pattern is currently played by the musician partner, it will always completely fit inside the window. Each positive edge of the threshold sound wave in the time window represents a rhythmic pulse. To characterize the timing of this sequence of pulses as a whole, the time differences between adjacent pulses is computed [8]. By using this information, a histogram can be constructed with one bin representing one certain time difference (both axis of the histogram a normalized). This histogram is then compared to the histograms of the timing patterns in the library of the robot (Figure 15.5b). The similarity between two histograms is determined using the Bhattacharyya coefficient [6]. Such coefficient is computed by using Eq. (15.7); where p^i is the histogram of one library pattern, q resembling the sampled rhythm pattern and m express the histogram size. The best matching library pattern is identified by analyzing at its Bhattacharyya difference (Figure 15.5b). To prevent patterns from being falsely detected, a threshold has been defined. If the result of the pattern comparisons falls below this threshold, the robot does not recognize the input as a known rhythm. The result of the rhythmic analysis is therefore the best match from the rhythm pattern library.

$$a_t = \begin{cases} 0 & \text{if } i_t \leq m \\ i_t & \text{if } i_t \geq t_r \end{cases} \tag{15.6}$$

a_t: sound wave threshold value
m: threshold level
i_t: input sound wave

$$\rho\left[p^i, q\right] = \sum_{u=1}^{m} \sqrt{p_u^i q_u} \tag{15.7}$$

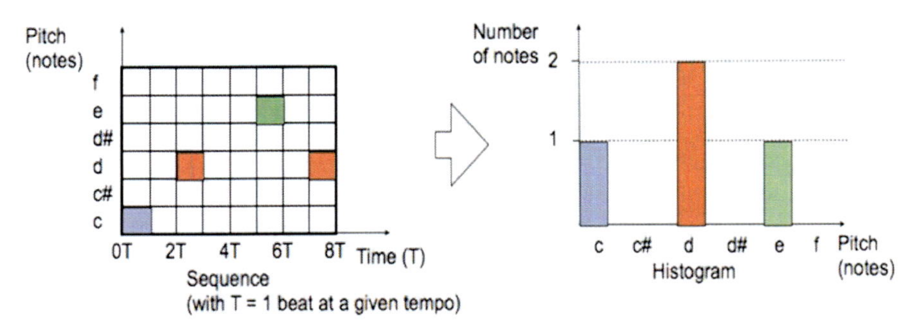

Fig. 15.6 Example of the histogram constructed for the pitch tracking algorithm.

15.3.2.2 Pitch Tracking

The pitch tracking extracts the harmonic components of the sound produced by the performance of the musician partner. In this case, the pitch information is recovered from the input data stream by applying a discrete short-time Fast Fourier Transformation (SFFT). For each sampled sound data (1024 sampled points), a Hann windowing function has been applied to smoothen the spectral leakage as it is shown in Eq. (15.8). Similar to the timing analysis, a running average was applied to adjacent frequency spectra with a threshold operation to reduce noise. If the threshold amplitude is retained by one or more peaks of the spectrum for long enough (not to be suppressed by the low-pass filter), the peak with the highest amplitude is identified as fundamental frequency. A recently registered pitch is approximated by the twelve-tone system note with the closest frequency. The value of this note is queued into the sequence window. When looking for harmonic information, the previous past data is considered only for the number of notes contained in the longest library pattern. The note information in the sequence window is gathered by generating a histogram from the pitch values (Figure 15.6). Again we match this histogram to the library histogram in order to find the best match. Information regarding which pattern was recognized is then forwarded to the mapping module.

$$w_i = a_i * 0.5 * \left(1 - \cos\left(\frac{2\pi i}{N-1}\right)\right) \tag{15.8}$$

w_i: resulting amplitude for sample i
a_i: input amplitude indexed with i
N: number of samples in the window

15.3.3 Musical Mapping

The mapping module translates the perceptual information (recognized through the interaction with the musician partner) into robot's actions (in musical terms). This output should make musical sense in a way that the musician can express himself/herself as freely as possible, while at the same time considering the physical limitations of the MP-HR. Due to the design complexity of the robot, we may observe different limitations such as: air volume of the lung, playing speed, modulation of other performance parameters (i.e. vibrato frequency), etc. The humanoid musical performance robot is a complex mechanical construction, built in order to emulate the human way of playing the flute. It thus naturally bears similar limitations to a human. It is important that when controlling or interacting with the humanoid, the musician does not drive the robot into these limitations, risking physical damage of the machine and leading to unnatural performance behavior. When the MP-HR processes the data coming from the musician partner through the visual channel, we may possibly map this data directly onto a musical performance parameter. This relationship can be formulated as Eq. (15.9). This equation contains the constant k representing a scaling factor to resize the sensor value $I(t)$ to an appropriate output value A. For example, given a virtual fader (implemented for the intermediate-based level interaction system), the maximum and minimum value of this control are predefined. Using that information, it is possible to condition k accordingly, so that the maximum input from the fader will never exceed the acceptable range for the performance parameter. However, some limitations of the robot are not time-constant. The capability of the robot to create an air-beam in order to play the wind playing-instrument, it will depend on the air volume left in the lung. Taking this into account we add time-dependence to k; as it is shown in Eq. (15.10). The parameter $T_{Breathing}$ indicates the duration that there will be air remaining in the lung, and thus a tone can be produced. The Eq. (15.11) expresses that the intended output of the humanoid musical performance robot needs to be conditioned with the breathing status of the lung. If the lung becomes empty, the equation constant $k(t)$ is set to zero.

$$A(t) = k * I(t) \tag{15.9}$$

$$A(t) = k(t) * I(t) \tag{15.10}$$

$$k(t) = \begin{cases} k & \text{if } t < T_{Breathing} \\ 0 & \text{if } t \geq T_{Breathing} \end{cases} \tag{15.11}$$

15.4 Preliminary Interaction Experiments with Musicians

In this section, the experiments carried out are focused on verifying the capabilities of the Advanced Level of the MbIS to enable musicians to interact with the WF-4RIV.

At first, the vision module has been tested. In this experiment, a musician has been asked to move the instrument in front of the robot camera and the instrument's orientation has been recorded in real-time meanwhile the vibrato amplitude of the robot sound is changed. The experimental results are shown in Figure 15.7. As a first approach, in order to keep the analysis as simple as possible only one tone is played. Basically, the average volume of the sound output becomes higher with less vibrato effect as the amount of air streaming through the glottis mechanism of the robot, which controls the vibrato; it is at these times higher and produces louder volume. As we may observe in the graph, the vibrato oscillates over this average value and changes its amplitude according to the orientation value calculated by the particle tracking algorithm (a related video can be accessed at [11]).

Finally, the aural module has been tested. In this experiment, a musician has been asked to choose between sequence A and sequence B of the My Favorite Things composed by John Coltrane. A graph of the results of this advanced interaction level aural recognition experiment is shown in Fig. 15.8. The robot generates an answer for each of the melodic questions posed to it by the musician, this time reproducing rhythm and melody. The pitch plot in Fig. 15.8b shows that the robot successfully imitates the phrases that are played by its partner. The flutist robot uses the patterns that are saved in its library. As the robot plays with the sequenced timing there are slight differences between the timing of one pattern of the same type played by the robot and one played the human musician. The answer from the robot follows 0.5s after the musician has played a pattern. This is the duration of silence that is necessary for the robot to detect if the player has finished one phrase and starts searching for matching patterns its library. The histogram analysis of the aural recognition system tolerates such inaccuracies. The system always picks the library pattern as an answer to the input from the musician that fits closest. So, as long as the timing does not differ as much as to make the histogram of one pattern more similar to an incorrect library pattern the recognition is done correctly (a related video can be accessed at [12]).

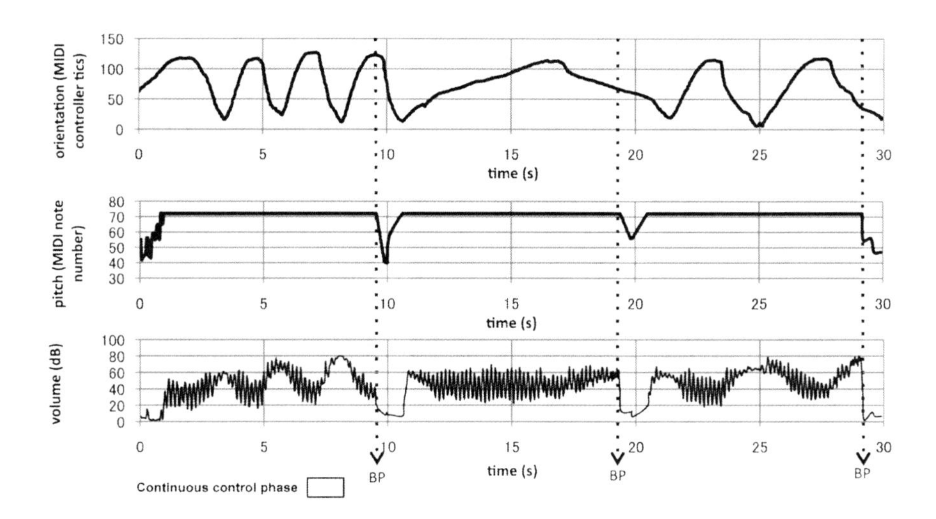

Fig. 15.7 Experimental results while interacting with the WF-4RIV meanwhile the musician's instrument orientation adjusts the vibrato amplitude of the robot's sound output (BP: Breathing Point).

The experimental results show the characteristics of the advanced level interaction system. They detail the functionality of the advanced level interaction mode, confirming that the system allows the user to control the musical performance of the robot in the intended way. What we saw from the advanced level interaction system graph as well is that there are certain systematical restrictions that need to be considered when using the system. The foremost limitation here is the breathing-in, breathing-out rhythm of the lung. This behavior can as usual be observed in the volume plot of the robot output (Figure 15.8d). During the time the robot is breathing out, a tone is generated, that can be manipulated by the user utilizing the controls provided by the interaction system. However, when the robot is breathing in, naturally, no tone is produced, which means that the user has to take a forced break in his control scheme. To be prepared for these interruptions in the flutist robot's play, the musician has to adapt his musical material in a way that is similar to creating musical material for a human (as a human does).

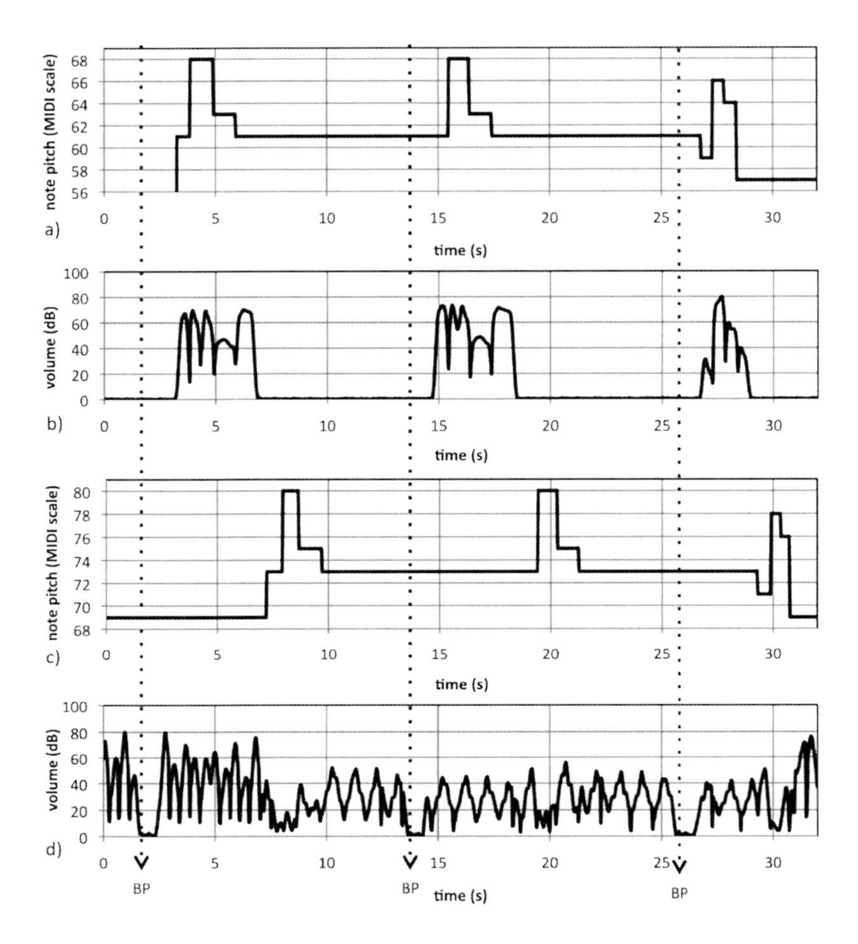

Fig. 15.8 Recorded input and output of the aural processing of the advanced level interaction system: a) is the pitch analysis plot of the musician's phrases, b) the amplitude plot of these patterns, c) the pitch analysis plot of the flute robot's response and d) the amplitude plot of that response.

15.5 Conclusions and Future Work

The musician-humanoid interaction research poses different kind of challenging issues from the point of view of human-robot interaction, aural/vision processing, perception, etc. In this chapter, the recent research efforts to implement a Musical-based Interaction System for Humanoid Musical Performance Robots were introduced. In particular, such a proposed system includes perceptual models to process incoming information (visual and aural) and mapping models to activate different behaviors on the robot. More recently, the authors have been focused on implementing sensor fusion methods and advanced mapping strategists into the MbIS to enable the flutist robot to interact more naturally with musicians.

From this research approach, we expect that the human musical interaction can be studied in detail from an engineering point of view by performing longer experiments under controlled conditions. On the other hand, the implementation of the MbIS into humanoid robots will certainly contribute for the better understanding of mechanism of human musical performance as well as for developing higher-skilled robots that can naturally interact with humans at the same level of perception.

Acknowledgments. A part of this research was done at the Humanoid Robotics Institute (HRI), Waseda University and at the Center for Advanced Biomedical Sciences (TWINs). This research is supported (in part) by a Gifu-in-Aid for the WABOT-HOUSE Project by Gifu Prefecture. This work is also supported (in part) by Global COE Program "Global Robot Academia" from the Ministry of Education, Culture, Sports, Science and Technology of Japan.

References

1. Akai: MPC, Reference Manual (2009), http://www.akaipro.com
2. Arulampalam, M.S., Maskell, S., Gordon, N., Clapp, T.: A tutorial on particle filters for online nonlinear / non-GaussianBayesian tracking. IEEE Transactions on Signal Processing, 174–188 (2002)
3. Hayashi, E.: Development of an automatic piano that produce appropriate: touch for the accurate expression of a soft tone. In: The Proc. of the IEEE/RSJ International Conference on Intelligent Robots and Systems - Workshop: Musical Performance Robots and Its Applications, pp. 7–12 (2006)
4. Kapur, A., Lazier, A., Davidson, P., Wilson, R.S., Cook, P.R.: The Electronic Sitar Controller. In: International Conference on New Interfaces for Musical Expression (2004)
5. Kato, I., Ohteru, S., Shirai, K., Matsushima, T., Narita, S., Sugano, S., Kobayashi, T., Fuji-sawa, E.: The robot musician WABOT-2 (Waseda Robot-2). Robotics 3(2), 143–155 (1987)
6. Nummiaro, K., Koller-Meier, E., Gool, L.V.: An adaptive color based particle filter. Journal of Image and Vision Computing 21, 99–110 (2003)
7. Orio, N.: Musical Retrieval: A tutorial and review. Foundations and Trends® in Information Retrieval 1(1), 1–90 (2003)
8. Petersen, K., Solis, J., Takanishi, A.: Development of a Aural Real-Time Rhythmical and Harmonic Tracking to Enable the Musical Interaction with the Waseda Flutist Robot. In: Proceedings of the IEEE/RSJ International Conference on Intelligent Robots and Systems, pp. 2303–2308 (2009)
9. Petersen, K., Solis, J., Takanishi, A.: Development of a Real-Time Instrument Tracking System for Enabling the Musical Interaction with the WF-4RIV. In: Proceedings of the IEEE/RSJ International Conference on Intelligent Robots and Systems, pp. 3654–3659 (2008)
10. Richardson, I.: H. 264 and mpeg-4 video compression: Video coding for next-generation multimedia, p. 320. Wiley, Chichester (2003)
11. Solis, J., Takanishi, A., et al.: MbIS's visual interaction video: http://www.takanishi.mech.waseda.ac.jp/top/research/music/flute/wf_4riv/index.htm

12. Solis, J., Takanishi, A., et al.: MbIS's aural interaction video: http://www.takanishi.mech.waseda.ac.jp/top/research/music/flute/wf_4rv/interaction_index.htm
13. Dannenberg, R.B., Brown, B., Zeglin, G., Lupish, R.: McBlare: A Robotic Bagpipe Player. In: Proc. of the International Conference on New Interfaces for Musical Expression, pp. 80–84 (2005)
14. Rowe, R.: Machine Musicianship. MIT Press, Cambridge (2004)
15. Saxe, D., Foulds, R.: Toward Robust Skin Identification in Video Images. In: 2nd International Conference on Automatic Face and Gesture Recognition, p. 379 (1996)
16. Breazeal, C., Takanishi, A., Kobayashi, T.: Social robots that interact with people. In: Siciliano, B., Khatib, O. (eds.) Springer Handbook of Robotics, pp. 1349–1422 (2008)
17. Singer, E.: LEMUR GuitarBot: MIDI robotic string instrument. In: Proc. of the International Conference on New Interfaces for Musical Expression, vol. 2003, pp. 188–191 (2003, 2007)
18. Shibuya, K.: Toward developing a violin playing robot: bowing by anthropomorphic robot arm and sound analysis. In: Proc. of the 16th Int. Conference on Robot & Human Interactive Communication, pp. 763–768 (2007)
19. Solis, J.: Robotic Control Systems for Learning and Teaching Human Skills, Ph.D. Dissertation, Perceptual Robotics Laboratory, Scuola Superiore Sant'Anna, Pisa, Italy, p. 226 (2004)
20. Solis, J., Marcheschi, S., Frisoli, A., Avizzano, C.A., Bergamasco, M.: Reactive Robots System: an active human/robot interaction for transferring skill from robot to unskilled persons. International Advanced Robotics Journal 21(3), 267–291 (2007)
21. Solis, J., Taniguchi, K., Ninomiya, T., Takanishi, A.: Understanding the Mechanisms of the Human Motor Control by Imitating Flute Playing with the Waseda Flutist Robot WF-4RIV. Mechanism and Machine Theory 44(3), 527–540 (2008)
22. Solis, J., Petersen, K., Yamamoto, T., Takeuchi, M., Ishikawa, S., Takanishi, A., Hashimoto, K.: Design of New Mouth and Hand Mechanisms of the Anthropomorphic Saxophon-ist Robot and Implementation of an Air Pressure Feed-Forward Control with Dead-Time Compensation. In: Proc. of the International Conference on Robotics and Automation, pp. 42–47 (2010)
23. Sony Computer Entertainment, Playstation eyeto (2008), http://www.eyetoy.com
24. Toyota Motor Corporation, Development of Partner Robots that Can Support an Aging Society: Creating an Era when Robots Provide Nursing Care and Perform Domestic Duties, http://www.toyota.co.jp/en/csr/report/08/stakeholder/02.html
25. Toyota Motor Corporation, Overview of the partner robots, http://www.toyota.co.jp/en/special/robot/
26. Weinberg, G., Driscoll, S.: Toward Robotic Musicianship. Computer Music Journal 30, 28–45 (2006)
27. Wren, C., Azarbayejani, A., Darrell, T., Pentland, A.: Pfinder: Real-Time Tracking of the Human Body. IEEE Transactions on Pattern Analysis and Machine Intelligence, 780–785 (1997)

Author Index

Springer Tracts in Advanced Robotics

Edited by B. Siciliano, O. Khatib and F. Groen

Further volumes of this series can be found on our homepage: springer.com

Vol. 74: Solis, J.; Ng, K. (Eds.)
Musical Robots and Interactive
Multimodal Systems
268 p. 2011 [978-3-642-22290-0]

Vol. 73: XXX

Vol. 72: Mullane, J.; Vo, B.-N.; Adams, M.;
Vo, B.-T.
Random Finite Sets for Robot Mapping
and SLAM
146 p. 2011 [978-3-642-21389-2]

Vol. 71: XXX

Vol. 70: Pradalier, C.; Siegwart, R.; Hirzinger, G.
(Eds.)
Robotics Research
752 p. 2011 [978-3-642-19456-6]

Vol. 69: Rocon, E.; Pons, J.L.
Exoskeletons in Rehabilitation Robotics
138 p. 2010 [978-3-642-17658-6]

Vol. 68: Hsu, D.; Isler, V.; Latombe, J.-C.;
Ming C. Lin (Eds.)
Algorithmic Foundations of Robotics IX
424 p. 2010 [978-3-642-17451-3]

Vol. 67: Schütz, D.; Wahl, F.M. (Eds.)
Robotic Systems for Handling
and Assembly
460 p. 2010 [978-3-642-16784-3]

Vol. 66: Kaneko, M.; Nakamura, Y. (Eds.)
Robotics Research
450 p. 2010 [978-3-642-14742-5]

Vol. 65: Ribas, D.; Ridao, P.; Neira, J.
Underwater SLAM for Structured
Environments Using an
Imaging Sonar
142 p. 2010 [978-3-642-14039-6]

Vol. 64: Vasquez Govea, A.D.
Incremental Learning for Motion Prediction
of Pedestrians and Vehicles
153 p. 2010 [978-3-642-13641-2]

Vol. 63: Vanderborght, B.;
Dynamic Stabilisation of the
Biped Lucy Powered by Actuators
with Controllable Stiffness
281 p. 2010 [978-3-642-13416-6]

Vol. 62: Howard, A.; Iagnemma, K.;
Kelly, A. (Eds.):
Field and Service Robotics
511 p. 2010 [978-3-642-13407-4]

Vol. 61: Mozos, Ó.M.
Semantic Labeling of Places with
Mobile Robots
134 p. 2010 [978-3-642-11209-6]

Vol. 60: Zhu, W.-H.
Virtual Decomposition Control –
Toward Hyper Degrees of
Freedom Robots
443 p. 2010 [978-3-642-10723-8]

Vol. 59: Otake, M.
Electroactive Polymer Gel Robots –
Modelling and Control of
Artificial Muscles
238 p. 2010 [978-3-540-23955-0]

Vol. 58: Kröger, T.
On-Line Trajectory Generation in Robotic
Systems – Basic Concepts for Instantaneous
Reactions to Unforeseen (Sensor) Events
230 p. 2010 [978-3-642-05174-6]

Vol. 57: Chirikjian, G.S.; Choset, H.;
Morales, M., Murphey, T. (Eds.)
Algorithmic Foundations
of Robotics VIII – Selected Contributions
of the Eighth International Workshop on the
Algorithmic Foundations of Robotics
680 p. 2010 [978-3-642-00311-0]

Vol. 56: Buehler, M.; Iagnemma, K.;
Singh S. (Eds.)
The DARPA Urban Challenge – Autonomous
Vehicles in City Traffic
625 p. 2009 [978-3-642-03990-4]

Vol. 55: Stachniss, C.
Robotic Mapping and Exploration
196 p. 2009 [978-3-642-01096-5]

Vol. 54: Khatib, O.; Kumar, V.;
Pappas, G.J. (Eds.)
Experimental Robotics:
The Eleventh International Symposium
579 p. 2009 [978-3-642-00195-6]

Vol. 53: Duindam, V.; Stramigioli, S.
Modeling and Control for Efficient Bipedal
Walking Robots
211 p. 2009 [978-3-540-89917-4]

Vol. 52: Nüchter, A.
3D Robotic Mapping
201 p. 2009 [978-3-540-89883-2]

Vol. 51: Song, D.
Sharing a Vision
186 p. 2009 [978-3-540-88064-6]

Vol. 50: Alterovitz, R.; Goldberg, K.
Motion Planning in Medicine: Optimization
and Simulation Algorithms for
Image-Guided Procedures
153 p. 2008 [978-3-540-69257-7]

Vol. 49: Ott, C.
Cartesian Impedance Control of Redundant
and Flexible-Joint Robots
190 p. 2008 [978-3-540-69253-9]

Vol. 48: Wolter, D.
Spatial Representation and
Reasoning for Robot
Mapping
185 p. 2008 [978-3-540-69011-5]

Vol. 47: Akella, S.; Amato, N.;
Huang, W.; Mishra, B.; (Eds.)
Algorithmic Foundation of Robotics VII
524 p. 2008 [978-3-540-68404-6]

Vol. 46: Bessière, P.; Laugier, C.;
Siegwart R. (Eds.)
Probabilistic Reasoning and Decision
Making in Sensory-Motor Systems
375 p. 2008 [978-3-540-79006-8]

Vol. 45: Bicchi, A.; Buss, M.;
Ernst, M.O.; Peer A. (Eds.)
The Sense of Touch and Its Rendering
281 p. 2008 [978-3-540-79034-1]

Vol. 44: Bruyninckx, H.; Přeučil, L.;
Kulich, M. (Eds.)
European Robotics Symposium 2008
356 p. 2008 [978-3-540-78315-2]

Vol. 43: Lamon, P.
3D-Position Tracking and Control
for All-Terrain Robots
105 p. 2008 [978-3-540-78286-5]

Vol. 42: Laugier, C.; Siegwart, R. (Eds.)
Field and Service Robotics
597 p. 2008 [978-3-540-75403-9]

Vol. 41: Milford, M.J.
Robot Navigation from Nature
194 p. 2008 [978-3-540-77519-5]

Vol. 40: Birglen, L.; Laliberté, T.; Gosselin, C.
Underactuated Robotic Hands
241 p. 2008 [978-3-540-77458-7]

Vol. 39: Khatib, O.; Kumar, V.; Rus, D. (Eds.)
Experimental Robotics
563 p. 2008 [978-3-540-77456-3]

Vol. 38: Jefferies, M.E.; Yeap, W.-K. (Eds.)
Robotics and Cognitive Approaches to
Spatial Mapping
328 p. 2008 [978-3-540-75386-5]

Vol. 37: Ollero, A.; Maza, I. (Eds.)
Multiple Heterogeneous Unmanned Aerial
Vehicles
233 p. 2007 [978-3-540-73957-9]

Vol. 36: Buehler, M.; Iagnemma, K.;
Singh, S. (Eds.)
The 2005 DARPA Grand Challenge – The Great
Robot Race
520 p. 2007 [978-3-540-73428-4]

Vol. 35: Laugier, C.; Chatila, R. (Eds.)
Autonomous Navigation in Dynamic
Environments
169 p. 2007 [978-3-540-73421-5]

Vol. 34: Wisse, M.; van der Linde, R.Q.
Delft Pneumatic Bipeds
136 p. 2007 [978-3-540-72807-8]

Vol. 33: Kong, X.; Gosselin, C.
Type Synthesis of Parallel
Mechanisms
272 p. 2007 [978-3-540-71989-2]